M. Hori H. Suga J. Baan E.L. Yellin (Eds.)

Cardiac Mechanics and Function in the Normal and Diseased Heart

With 155 Figures

Springer-Verlag Tokyo Berlin Heidelberg New York
London Paris Hong Kong

Dr. MASATSUGU HORI
The First Department of Medicine
Osaka University School of Medicine
Fukushima-ku, Osaka, 553 Japan

Dr. HIROYUKI SUGA
Department of Cardiovascular Dynamics
National Cardiovascular Center Research Institute
Suita, Osaka, 565 Japan

Prof. JAN BAAN
Laboratory of Clinical Physiology
Department of Cardiology
Leiden University Hospital
Leiden, The Netherlands

Prof. EDWARD L. YELLIN
Department of Cardiothoracic Surgery and
Department of Physiology and Biophysics
Albert Einstein College of Medicine
Yeshiva University
The Bronx, NY 10461, USA

ISBN-13: 978-4-431-68020-8 e-ISBN-13: 978-4-431-67957-8
DOI: 10.1007/978-4-431-67957-8

Library of Congress Cataloging-in-Puplication Data
Cardiac mechanics and function in the normal and diseased heart/M. Hori... [et al.], (eds.).
1. Heart – Muscle – Physiology. 2. Heart – Muscle – Patho-
physiology. I. Hori, M. (Masatsugu), 1945 – .[DNLM: 1. Heart – physiology. 2. Heart – physiopathology. WG 202 C266] QP113.2.C366 1989 612.1'71 – dc20 DNLM/DLC for Library of Congress

Softcover reprint of the hardcover 1st edition 1989

Typesetting: Koford Prints (Pte) Ltd., Singapore

Foreword

Cardiovascular dynamics is a field in which modelling and systems analysis have formed an extremely important discipline. For example, understanding of even such a fundamental function of the circulation as the relationship between central venous pressure and cardiac output has required evolution of a pertinent model based on years of exhaustive experimental investigations by Starling, Starr, and Guyton. Hemodynamic analyses of pulsatile pressures and flows in the arteries and veins have been a continuing challenge taken up by champions of fluid dynamics such as Frank, Wetterer, Taylor, and Wormersley, just to mention a few names, and some kind of model was always proposed as a conceptual framework. An even greater challenge to cardiovascular dynamicists was how to analyze the intermittent coupling of the ventricle and the arterial or venous vasculature through the valve. The availability of numerical solutions by computer and the recently evolved ventricular model with a time-varying elastance and a pressure-dependent internal resistance opened the way to analysis of this coupling. The ever-increasing speed of computers has also facilitated trips between the frequency and the time domain, even *on-line* for some experimental studies. This book contains many analyses dedicated to the interactions between the heart and the vasculature, providing the reader with findings at the cutting edge of current research in this field.

Systems physiology pertains not only to the analysis of interaction at the organ level; it should be applied to a system at any level of interest, including cells, cellular organelles, and biological molecules. Whenever there are subcomponents of a system, there is interplay which contributes to the determination of the behavior of the system as a whole. Thus, a cardiac muscle cell is as complex a system as the ventricle itself, or a singular branch of artery or part of its wall.

The recent worldwide advocacy of molecular and cellular physiology and pathophysiology by some funding agencies is outrageous. It may mislead young biomedical researchers into the blind faith that all the secrets of life and disease lie only in subcellular entities, and one-way explorations into the molecular world promise results of redeeming scientific value. True, the up-to-date knowledge gained by high technology is astounding and exciting. But microscopic details of any matters whether living or nonliving, have always fascinated scientists. A true *physiologist* has always had to recognize the indispensability of a systems approach in the attempt to make sense out of quickly accumulating piecemeal knowledge. This is true even in the simplest physicochemical world; we all know that thorough knowledge of hydrogen and oxygen atoms does not allow us to predict the nature of the water molecule. Therefore, despite (or, rather, because of) the current fascination with molecular and cellular physiology, a systems approach is all the more a desirable discipline for investigators in any branch of the sciences. Again, this book provides the physiologist reader with cardiovascular systems analyses at multiple levels. It illustrates how a systems approach can be applied to many different corners of the circulatory system's physiology.

Finally, the Cardiovascular System Dynamic Society consists of a very special breed of circulatory physiologists, biomedical engineers, cardiologists, and angiologists. They are all well equipped with rigorous measurement techniques and mathematical tools to deal with cardiovascular dynamics. They also see nature from a systems point of view and love to chew on such concepts as *matching* and *optimal control*. Inevitably, this book strongly reflects these features of the membership. I hope that the readers like this menu of "System Dynamics" because, after all, one pursues something if and only if he truly likes it.

Baltimore, 1989

Kiichi Sagawa
Professor
Department of Biomedical Engineering
The Johns Hopkins University
Medical School

It is with great sadness that the editors report the death of Dr. Kiichi Sagawa in August of 1989. His dedication and expertise will be sorely missed.

List of Contributors

The address can be found at the page numbers indicated

Table of Contents

Chapter A.
Cellular Aspect of Cardiac Mechanics

Masatsugu Hori

Since William Harvey's discovery of the pumping function of the heart, a number of investigators have searched for the answers to these questions: How is the contraction of the heart controlled? What happens in the normal and diseased hearts? Although the definite answers to these questions are not yet available, recent progress in cardiology, either in basic or clinical research, have provided us with a rapidly growing wealth of knowledge to help our understanding of mechanics of the living heart.

During the past two decades, biomedical science has revealed the important role of intracellular Ca^{2+} ($[Ca^{2+}]_i$) in the mechanical contraction of myocardial fibers. The contraction-relaxation cycle of a muscle is regulated by $[Ca^{2+}]_i$; the increase in $[Ca^{2+}]_i$ initiates contraction by removing the inhibitory effect of troponin C on the interaction of actin and myosin. However, the previous measurements of $[Ca^{2+}]_i$ have been performed in superfused muscle or enzymatically dissociated cells whose responses cannot be assumed to resemble those of the native tissue. E. Marban applied ^{19}F-NMR to the isolated, perfused heart loaded with the fluorinated Ca^{2+} indicator 5F-BAPTA to study the periodical changes of $[Ca^{2+}]_i$ during a cardiac cycle. He also described the relationship between coronary flow and Ca^{2+} transients to explain the "Gregg's phenomenon." In contractile dysfunction in ischemic hearts, Ca^{2+}-related abnormalities may play a causative role. Kusuoka et al. studied the relationship between $[Ca^{2+}]_i$ and the maximal Ca^{2+}-activated force in stunned isolated hearts. They revealed that contractile dysfunction in stunned myocardium is due to a decline in the maximal Ca^{2+}-activated force and a shift in Ca^{2+} sensitivity.

Although Ca^{2+} is a key mediator of contraction of the striated muscle, contraction machinery is necessary to generate a force. In both skeletal and heart muscles, the unit of contractile component, i.e., the sarcomere, is composed of an array of overlapping thick and thin filaments. Using a newly developed apparatus, Saeki et al. observed the isotonic segment length changes of isolated right ventricular papillary muscle in rats in response to stepped tension reduction. They analyzed the tension-velocity and length-tension relationships, which reflect the kinetics of the cross-bridges between actin and myosin filaments. The quantitative characteristics of the length-tension relationship has formed the basis of the sliding filament mechanism of contraction in striated muscle. This sliding theory, however, is criticized by T. Iwazumi in several aspects. He performed four entirely different experiments which suggest that the cross-bridge theory appears to be extremely inadequate to explain the obtained results. Thus, the mechanism of force generation during contraction is still controversial.

J.W.Kruger studied the sarcomere shortening and relengthening in a canine left ventricular myocyte and trabeculae by laser diffraction technique, analyzing the velocity-length relationship at various loads. He focused his interest on the late relengthening in relation to diastolic filling mechanism and concluded that the internal restoring force in the single cell can cause a negative pressure of approx. 10 mmHg in the ventricle. This theory that the restoring force may be largely attributed to the intracellular component may conflict with van Hoeven and Factor's idea that a contribution of the connective tissue, especially collagen matrix, to the elastic recoil is considerably large. The latter investigators postulated the role of intercellular and myocyte-to-capillary struts which provide short bridges to be important in diastolic alignment and ventricular filling both in normal and diseased hearts, e.g., ischemic and hypertrophied hearts.

Lastly, it is to be noted that cardiac proteins may be altered in the various diseased states; hence, mechanical function could also be affected. Yazaki et al. demonstrated redistribution of cardiac myosin isozymes in the overloaded humen heart and showed that the transition of α cardiac myosin heavy chain (HCα) to β type (HCβ) in the overloaded heart is induced by the expression of HCβ-gene.

In this first section, new insights of cellular aspects of myocardial function in normal and diseased state are discussed.

Ca^{2+} Transients in Perfused Hearts: Fundamental Properties of the Chemical Signals Underlying Ventricular Mechanics

EDUARDO MARBAN[1]

Summary. An increase in the intracellular free calcium concentration, $[Ca^{2+}]_i$, initiates contractile activation in the heart. Until recently, the changes in $[Ca^{2+}]_i$ during the cardiac cycle (the "Ca^{2+} transients") had eluded measurement. Here I describe the use of gated nuclear magnetic resonance (NMR) spectroscopy to achieve direct quantitation of the cyclical changes in $[Ca^{2+}]_i$ that underlie mechanics in perfused ferret hearts loaded with the fluorinated Ca^{2+} indicator 5F-BAPTA. $[Ca^{2+}]_i$ increased from approximately 200 n*M* in diastole to 750 n*M* or higher in early systole. In this preparation, the effects of changing coronary arterial flow on $[Ca^{2+}]_i$ can be investigated simultaneously with measurements of high-energy phosphate concentrations by ^{31}P-NMR. When hypoperfusion is induced such that the pressure generated by contraction falls without metabolic evidence of ischemia ("Gregg's phenomenon"), a decrease in the amplitude of Ca^{2+} transients underlies the observed fall in pressure. This down-regulation of Ca^{2+} transients constitutes the cellular basis for Gregg's phenomenon, and serves to decrease energy demand during low-flow ischemia.

Introduction

The application of various techniques to measure $[Ca^{2+}]_i$ in isolated heart muscle and in single cardiac cells has led to considerable advances in our understanding of excitation-contraction coupling and other Ca^{2+}-mediated processes (see [1] for review). During each cardiac cycle, $[Ca^{2+}]_i$ rises quickly (within 100 ms) from about 100 n*M* to values an order of magnitude higher, then decays back to baseline. The increase in $[Ca^{2+}]_i$ initiates contraction by removing the inhibitory effect of troponin C and tropomyosin on the interaction of actin and myosin. Nevertheless, major conceptual gaps remain, particularly regarding the regulation of $[Ca^{2+}]_i$ in hearts perfused via the normal arterial circulation. Almost all measurements of $[Ca^{2+}]_i$ have been performed in superfused muscle or enzymatically dissociated cells whose responses cannot be assumed to resemble those of the native tissue. In particular, the effects of alterations in coronary perfusion can only be crudely mimicked in isolated muscle or cells. Nuclear magnetic resonance (NMR) spectroscopy can detect signals from fluorinated Ca^{2+} indicators [2, 3] and has already yielded useful measurements of time-averaged $[Ca^{2+}]_i$ in perfused hearts [4, 5], but the lack of time resolution has restricted further progress.

[1] Division of Cardiology, Department of Medicine, The Johns Hopkins University School of Medicine, Baltimore, MD 21205, USA

Here I will summarize results obtained in collaboration with Masafumi Kitakaze, Martin Pike and V.P. Chacko [6] demonstrating the use of gated ^{19}F-NMR to obtain time-resolved measurements of $[Ca^{2+}]_i$ during the cardiac cycle in perfused ferret hearts. While still in its infancy, this technique has already yielded unique insights into Ca^{2+} regulation and energy metabolism during normal and decreased coronary flow.

Strategy for Obtaining Time-Resolved Measurements of $[Ca^{2+}]_i$

Perfused ferret hearts were placed in an NMR spectrometer (8.46 Tesla) for the simultaneous measurement of NMR spectra and isovolumic left ventricular pressure [4]. The perfusate contained (in m*M*): 108 NaCl, 5 KCl, 1 $MgCl_2$, 5 HEPES (pH adjusted to 7.4 with NaOH), 20 Na acetate, and 10 glucose, equilibrated with 100% O_2. Hearts were paced at 0.9–1.6 Hz and maintained at a temperature of 30°C. Loading with the Ca^{2+} indicator 5F-BAPTA, the 5,5'-difluoro derivative of 1, 2-bis (*o*-aminophenoxy) ethane-N, N, N', N'-tetraacetic acid (5F-BAPTA), was achieved by addition of the cell-permeant acetoxymethyl ester form [2] to the perfusate for 30-60 min [6].

The NMR methods have been described fully [4, 6] and need not be repeated here. I will only note that the excitatory radiofrequency pulses and the subsequent NMR data acquisition were gated according to a programmable delay from the time of the pacing stimulus, using two pacemakers. The first, which set the overall cycle length, supplied a synchronization signal to the spectrometer and to a second pacemaker which stimulated the heart with no delay. For gated NMR data acquisition, a delay relative to the pacemaker signal was included in the programmed pulse sequence so that the radiofrequency pulse could be imposed at any time during the cardiac cycle. One pulse was applied during each cycle, so that the total interpulse delay approximated 1 s (the exact value depending on the pacing rate). At each point in the cardiac cycle, 100–1200 consecutive gated scans were averaged to achieve an acceptable signal-to-noise ratio. A family of ^{19}F (or ^{31}P) spectra was obtained throughout the cardiac cycle by sampling sequentially at various delay settings. $[Ca^{2+}]_i$ can be calculated according to the equation $[Ca^{2+}] = K_d \cdot [B]/[F]$, where [B] and [F] represent the concentrations of the indicator bound to Ca and free in the cytoplasm, respectively; these in turn are proportional to the areas under readily-identifiable peaks in the ^{19}F spectra (see next section and [2]). We used the K_d of 285 n*M* previously measured at 30°C in EGTA-buffered solutions [4] to calibrate our signals.

Oscillations in $[Ca^{2+}]_i$, But Not in High-energy Phosphates, During the Cardiac Cycle

Fig. 1 illustrates the findings of a typical experiment. The top row shows isovolumic pressure recorded during the acquisition of gated spectra at the various times during the cardiac cycle designated by the letters **a–d**. Under the pressure record are ^{19}F (*left column*) and ^{31}P (*right column*) NMR spectra obtained at the times indicated on the

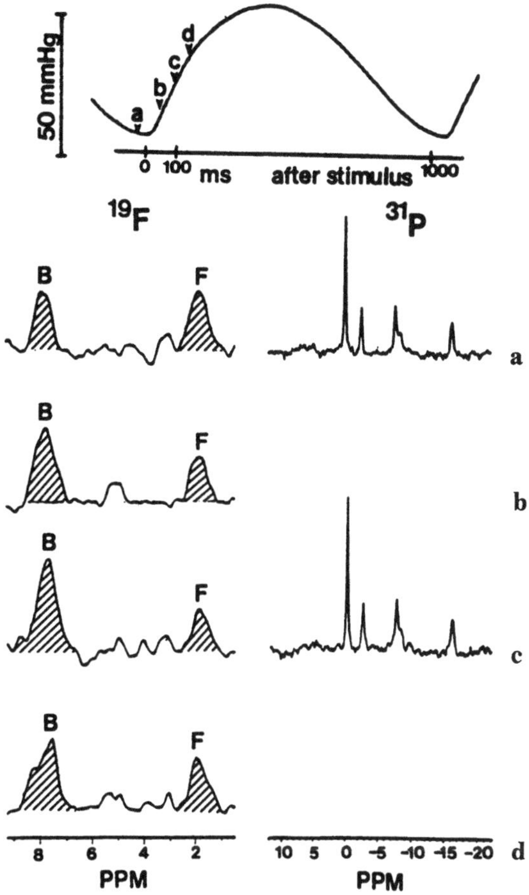

Fig. 1.a-d. Changes in gated NMR spectra during the cardiac cycle. *Top panel:* Isovolumic left ventricular pressure in a ferret heart paced at 1.06 Hz in 8 m*M* $[Ca]_o$. NMR spectra were acquired at the four times indicated on the pressure record: **a** 10 ms prior to stimulation; **b** 75 ms, **c** 100 ms, and **d** 125 ms after stimulation. Below, the left-hand column shows gated ^{19}F spectra recorded at times **a** through **d**, as indicated. The bound (B) and free (F) peaks of 5F-BAPTA exhibit distinct chemical shifts at ~ 8 and 2 ppm, respectively, downfield from a standard of 1 m*M* 6-*F*-tryptophan at 0 ppm (not shown here). The right-hand column shows gated ^{31}P spectra acquired at times **a** and **d** in the same heart. The major peaks correspond to phosphocreatine (PCr, 0 ppm) and ATP (the three peaks upfield from phosphocreatine); inorganic phosphate (P_i) is hardly visible at 4-5 ppm under these conditions

pressure record: **a**, 10 ms before the stimulus, and, after the stimulus, **b** (75 ms), **c** (100 ms), and **d** (125 ms). The ^{19}F spectra show two peaks [2-5] that have been highlighted by shading, one at ~2 ppm corresponding to free 5F-BAPTA (F) and the other at ~8 ppm corresponding to 5F-BAPTA bound to Ca (B). $[Ca^{2+}]_i$ is proportional to the ratio of the areas under the two peaks ([B]/[F]). At time **a**, just before the pacing stimulus, the bound peak is smaller than the free peak ([B]/[F]) = 0.82), corresponding to a diastolic $[Ca^{2+}]_i$ of 233 n*M*. Just 75 ms after the stimulus (**b**), however, the spectrum exhibits dramatic changes: the bound peak now exceeds the free ([B]/[F] = 2.2), yielding a $[Ca^{2+}]_i$ of more than 600 n*M*. The greatest value for [B]/[F] (2.5) during the cardiac cycle was found 100 ms after the pacing stimulus (**c**), yielding an estimated peak systolic $[Ca^{2+}]_i$ of about 700 n*M*. By 125 ms after the stimulus (**d**), $[Ca^{2+}]_i$ had already declined to less than 500 n*M*. The changes were quite reproducible: another diastolic spectrum obtained afterwards (not shown) was comparable to that in **a**.

Interestingly, high-energy phosphate concentrations showed no such oscillation. When we kept the gating parameters unchanged and acquired the ^{31}P spectra shown in the right-hand column of Fig. 1, no difference was apparent between **a** and **c**; both display the normal pattern of phosphorus NMR spectra in well-oxygenated hearts [7], with prominent phosphocreatine and ATP peaks. The finding of cyclical changes in one set of nuclei but not another in the same heart indicates that the changes in the ^{19}F spectra do not arise from artifacts related to contraction. Indeed, we expect $[Ca^{2+}]_i$ to vary by roughly an order of magnitude during the cardiac cycle [1]; on the other hand, cyclical changes in phosphorus metabolites have been sought but not found in ferret hearts [8], and even in rat hearts at high work loads the observed fluctuations amount at most to ± 20% [9–11].

Quantitative Aspects of Ca^{2+} Transients

From ^{19}F spectra acquired at several times during the cardiac cycle, we mapped out the change in $[Ca^{2+}]_i$ during each contraction. Fig. 2 shows a family of ^{19}F NMR spectra stacked one upon another according to the time delay from the pacing stimulus (*receding axis*). In the first spectrum, obtained just before stimulation, the free peak at 2 ppm slightly surpasses the bound peak (8 ppm). After the stimulus at

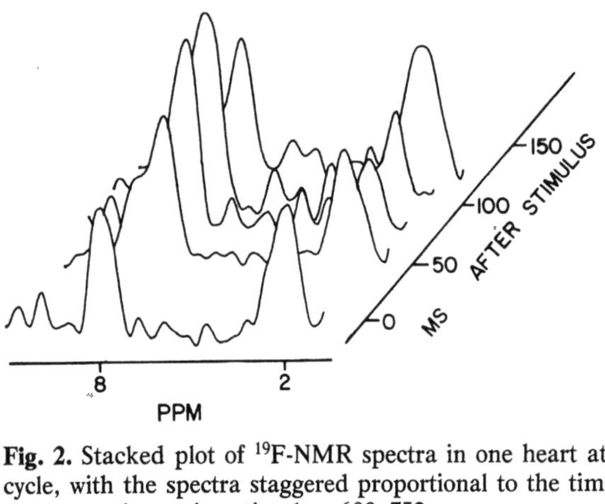

Fig. 2. Stacked plot of ^{19}F-NMR spectra in one heart at different times during the cardiac cycle, with the spectra staggered proportional to the time at which they were acquired with respect to the pacing stimulus. 600–752 scans were accumulated for each spectrum. Preparation MF76; pacing rate 1.00 Hz, 8 m*M* $[Ca]_o$. Reproduced from [6], by permission of the American Heart Association

time 0, the bound peak increases rapidly and reaches a maximum within 100 ms, while the free peak becomes quite small. Thereafter, the signals return towards baseline. Such data form the basis for the Ca^{2+} transients plotted in Figs. 3 and 4. Fig. 3a shows the Ca^{2+} transient from a heart perfused with 2 m*M* $[Ca]_o$ along with the corresponding record of pressure developed by the left ventricle. $[Ca^{2+}]_i$ rises sharply from just under 200 n*M* in diastole to a maximum of about 1 μM. The Ca^{2+} transient in this heart peaks at 75 ms, well before the corresponding peak in contractile pressure. A similar delay in maximal twitch force relative to the peak of the Ca^{2+}transient has been observed repeatedly with the Ca^{2+}-activated photoprotein aequorin in papillary muscles (e.g., [12]). The absolute values for diastolic and systolic $[Ca^{2+}]_i$ are also remarkably similar with the two indicators [13], although the rate of decay of the transients is definitely slower with 5F-BAPTA (see [6] for a discussion of this apparent discrepancy).

Because of Ca buffering by intracellular 5F-BAPTA, twitch pressure after loading with the indicator is quite low in 2 m*M* $[Ca]_o$ (Fig. 3a). To approximate more closely the pressures achieved during physiologic excitation-contraction coupling, we often raised $[Ca]_o$ to 8 m*M* [4], which in this example augmented the developed pressure to 68 mmHg (Fig. 3b, *top row*). In association with the increase in pressure, systolic $[Ca^{2+}]_i$ increased dramatically, now peaking at approximately 2.4 μM. In contrast, end-diastolic $[Ca^{2+}]_i$ remained virtually unchanged, and the overall shape of the Ca^{2+} transients was similar at both $[Ca]_o$. For all 24 hearts studied in 8 m*M* $[Ca]_o$,

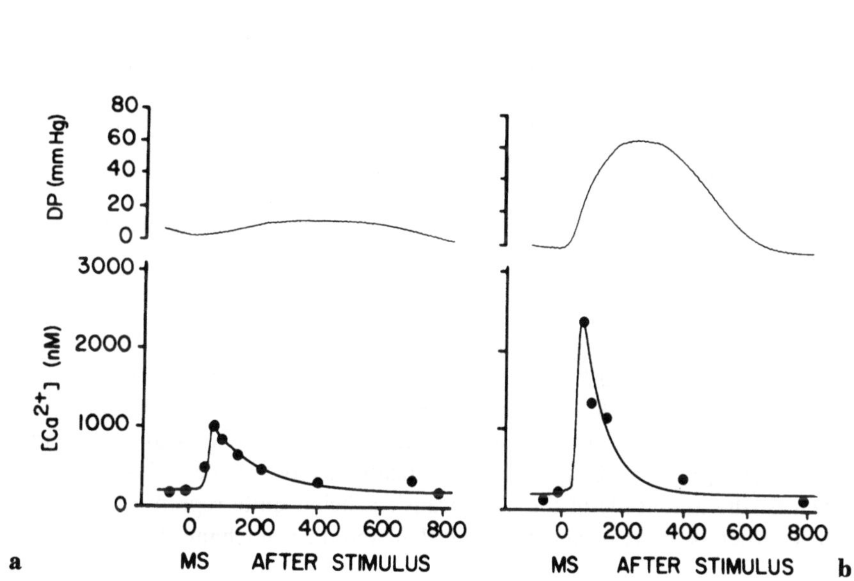

Fig. 3.a, b. Effects of increasing extracellular Ca concentration on developed pressure (*upper row, DP*) and Ca^{2+} transients (*lower row*). **a** 2 m*M* $[Ca]_o$; each point was calculated from a gated spectrum obtained over 183–1061 scans. The Ca^{2+} transients in this and subsequent figures were fit (*solid line*) by eye to the sum of cosine and rising exponential functions, which predominate during the rising phase, and a declining single exponential which prevails after the peak of the transient. **b** 8 m*M* $[Ca]_o$, with each point based on a spectrum acquired over 600–1059 scans. Preparation MF62 paced at 1.18 Hz. Reproduced from [6], by permission of the American Heart Association

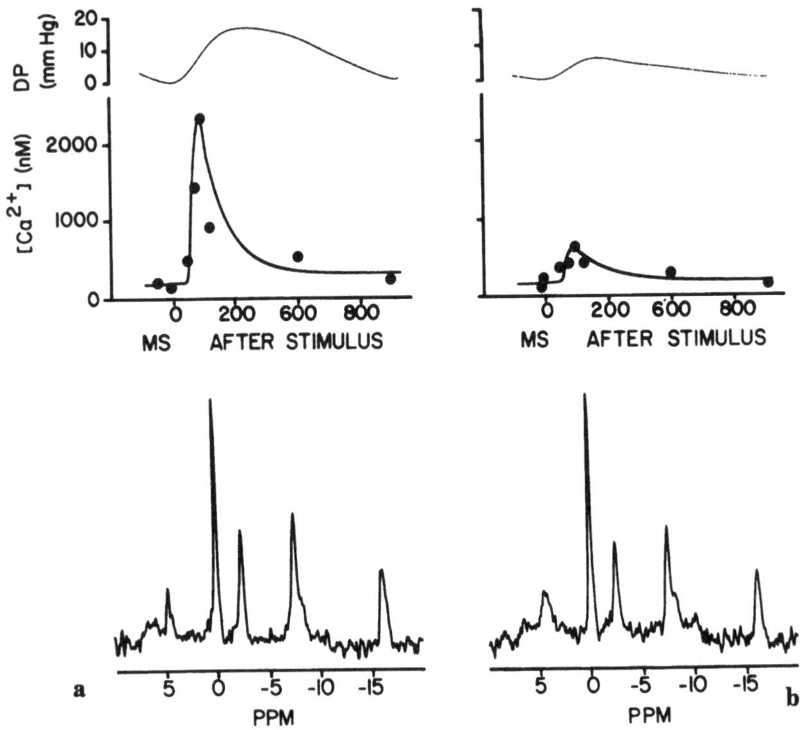

Fig. 4a, b. Developed pressure (*top row, DP*), Ca^{2+} transients (*middle row*) and ^{31}P-NMR spectra (*bottom row*) during normal and decreased coronary perfusion. **a** Records obtained with coronary flow rate set at 18 ml/min. **b** Data during the steady state effect of decreasing the flow rate to 10 ml/min. The spectra used to calculate the Ca^{2+} transients in this experiment each consisted of 263–594 scans. For the ^{31}P spectra, the peaks are attenuated by saturation by the following factors: PCr, 0.69; ATP, 0.90; P_i, 0.73. Preparation MF64 paced at 1.35 Hz in 8 m*M* $[Ca]_o$. Reproduced from [6], by permission of the American Heart Association

the Ca^{2+} transient spanned from a low of 210 ± 20 n*M* (mean ± SEM) in diastole to peak of 1.68 ± 0.23 μ*M* in systole. The values for systole, being further from the K_d than those for diastole, are more susceptible to misestimation due to noise: small differences in the area under the free peak will give rise to large apparent differences in the [B]/[F] ratio. Despite this qualification, most of the variability was among hearts, the amplitude of the Ca^{2+} transient being quite reproducible in any given heart when $[Ca^{2+}]_o$, pacing rate, and flow were kept unchanged.

Changes in Perfusion and their Effects on Ca^{2+} Transients

Our experimental preparation is uniquely well-suited to address questions related to coronary flow and to energy metabolism, since we can manipulate perfusion while measuring $[Ca^{2+}]_i$, high-energy phosphate concentrations, and intracellular pH (from ^{31}P-NMR spectra [14]). We have exploited these capabilities of our system to investigate the mechanism of the fall in myocardial force generation during reduced coronary perfusion. As coronary flow is decreased, contractile pressure can fall by 50% or more with little or no change in high-energy phosphate concentrations [15]. Unlike the contractile failure of hypoxia or severe ischemia which is at least partially due to the intracellular accumulation of inorganic phosphate [7], the decrease in contractile force during a moderate reduction of arterial flow occurs without appreciable inorganic phosphate buildup [15]. This behavior of the heart in response to a moderate decrease in coronary flow has been termed "Gregg's phenomenon" [16] and may share a common basis with the reversible pump failure ("hibernating myocardium") reported in some patients with coronary artery disease [17].

The experiment illustrated in Fig. 4 sheds some light on the cellular basis of this phenomenon. Panel **a** shows ventricular pressure (*top row*) and the Ca^{2+} transient (*middle row*) during perfusion at a coronary flow rate of 18 ml/min (at which level coronary perfusion pressure equalled 80 mmHg). The corresponding ^{31}P-NMR spectrum (a, *bottom row*) has a normal appearance [7], with an intracellular pH equal to 7.07. Flow was then decreased until developed pressure reached a new steady state at roughly 40% of control (**b**, *top row*; new flow rate = 10 ml/min). Under these conditions, the Ca^{2+} transient (**b**, *middle row*) was markedly decreased in amplitude, but the ^{31}P-NMR spectrum (*bottom row*) showed only minor changes. Intracellular pH did fall to 6.97 in **b**, but this mild degree of acidosis can itself account for only a minor fraction of the observed decrease in pressure [15]. By comparison to experiments such as that in Fig. 3 in which pressure was altered by changing $[Ca]_o$, the decline in the amplitude of the Ca^{2+} transient during low flow is seen to be quite sufficient to explain the decrease in force generation. The finding of a decrease in systolic $[Ca^{2+}]_i$ during low-flow ischemia with little observable alteration of energy metabolism has now been confirmed in four other experiments.

Limitations and Strengths of This Approach for Measuring $[Ca^{2+}]_i$

Gated acquisition has allowed us to overcome partially the lack of time resolution that restricted previous estimates of $[Ca^{2+}]_i$ in perfused hearts using NMR. Consideration of the various factors which determine the time resolution leads to a conservative estimate of 10–20 ms [6], which is ≤2% of our usual cycle length. Because of the relatively low concentration of ^{19}F in the heart, several minutes of signal averaging (≥ 100 pulses) are still required for each gated spectrum to exhibit a satisfactory signal-to-noise ratio. This restricts the measurements of Ca^{2+} transients to steady-state interventions: the Ca^{2+} transients must not be changing throughout the time in which any given family of spectra is being collected. Thus, it has been possible for us to measure transients during normal and moderately decreased coronary perfusion, since both are stable situations, but it is unlikely that we will be able to use the technique in its present state of development to follow beat-to-beat changes in Ca^{2+} transients, as expected during progressive stages of digitalis toxicity or during total ischemia.

Signals related to $[Ca^{2+}]_i$ in perfused hearts can also be measured from the fluorescence produced by Ca^{2+} indicators chemically similar to 5F-BAPTA [18,19]. The strengths and limitations of this approach are, in many respects, complementary to those of NMR spectroscopy [4]. The use of the fluorescent indicator indo-1 in particular offers good time resolution even without signal averaging, at intracellular concentrations that produce little Ca buffering [19]. These favorable properties make it possible for indo-1 to track moment-to-moment changes in Ca^{2+} transients. Nevertheless, the technique is subject to several limitations that its users are careful to acknowledge. Fluorescence arises not only from the reaction of the indicator with Ca^{2+}, but also from complexes with heavy metals and from endogenous compounds and Ca^{2+}-insensitive forms. Movement artifacts can be minimized but not entirely eliminated. Finally, only localized subepicardial or subendocardial regions of the heart are readily accessible. The NMR approach, while requiring signal averaging and careful consideration of the consequences of Ca buffering, offers the unique opportunity to sample a number of nuclei in rapid alternation. This has enabled us not only to measure signals arising from the fluorinated Ca^{2+} indicator but also from endogenous high-energy phosphates, all in the same heart. No other approach, even in isolated muscle or single cells, offers the exceptional opportunity to measure $[Ca^{2+}]_i$ and the reactants of energy metabolism simultaneously.

The observation that Ca^{2+} transients are decreased during low-flow ischemia suggests a fundamental mechanism whereby energy demand can be reduced in ischemia, at the expense of a substantial decrease in myocardial force generation. Since $[Ca^{2+}]_i$ rises much less during each beat, the energy required for cycling intracellular Ca^{2+} and for activation of the contractile proteins (which can account for more than 80% of myocardial oxygen consumption [20]) will be substantially diminished. The elucidation of this mechanism provides a vivid preview of the potential usefulness of this technique for investigating the regulation of the strength of contraction of the heart during normal and altered coronary perfusion.

References

1. Blinks JR, Wier WG, Hess P, Prendergast FG (1984) Measurement of Ca^{2+} concentrations in living cells. Prog Biophys Mol Biol 40: 1-114
2. Smith GA, Hesketh RT, Metcalfe JC, Feeney J, Morris PG (1983) Intracellular calcium measurements by ^{19}F NMR of fluorine-labelled chelators. Proc Natl Acad Sci USA 80: 7178-7182
3. Metcalfe JC, Hesketh TR, Smith GA (1985) Free cytosolic Ca^{2+} measurements with fluorine labelled indicators using ^{19}F NMR. Cell Calcium 6: 183-195
4. Marban E, Kitakaze M, Kusuoka H, Porterfield JK, Yue DT, Chacko VP (1987) Intracellular free calcium concentration measured with ^{19}F NMR spectroscopy in intact ferret hearts. Proc Natl Acad Sci USA 84: 6005-6009
5. Steenbergen C, Murphy E, Levy L, London RE (1987) Elevation in cytosolic free calcium concentration early in myocardial ischemia in perfused rat heart. Circ Res 60: 700-707
6. Marban E, Kitakaze M, Chacko VP, Pike MM (1988) Ca^{2+} transients in perfused ferret hearts revealed by ^{19}F NMR spectroscopy. Circ Res
7. Kusuoka H, Weisfeldt ML, Jacobus WE, Zweier J, Marban E (1986) Mechanism of early contractile failure during hypoxia in intact ferret heart: evidence for modulation of maximal Ca^{2+}-activated force by inorganic phosphate. Circ Res 59: 270-282

8. Morris GM, Allen DG, Orchard CL (1984) High-time-resolution ^{31}P NMR studies of the perfused ferret heart. Advances in Myocardiology 5: 27–38
9. Fossel ET, Morgan HE, Ingwall JS (1980) Measurement of changes in high-energy phosphates in the cardiac cycle by using gated ^{31}P nuclear magnetic resonance. Proc Natl Acad Sci USA 77: 3654–3658
10. Wikman-Coffelt J, Sievers R, Coffelt RJ, Parmley WW (1983) The cardiac cycle: regulation and energy oscillations. Am J Physiol 245, H354
11. Kusuoka H, Inoue M, Tsuneoka Y, Watari H, Hori M, Abe H (1985) Augmented energy consumption during early systole as a mechanism of cyclical changes in high-energy phosphates in myocardium assessed by phosphorus nuclear magnetic resonance. Jpn Circ J 49: 1099–1107
12. Yue DT (1987) Intracellular [Ca^{2+}] related to rate of force development in twitch contraction of heart. Am J Physiol 252: H760–H770
13. Wier WG, Yue DT (1986) Intracellular [Ca^{2+}] transients underlying the short-term force-interval relationship in ferret ventricular myocardium. J Physiol (Lond) 376: 507–530
14. Flaherty JT, Weisfeldt ML, Bulkley BH, Gardner TJ, Gott VL, Jacobus WE (1982) Mechanisms of ischemic myocardial cell damage assessed by 31-phosphorus nuclear magnetic resonance. Circulation 65: 561–571
15. Jacobus WE, Pores IH, Lucas SK, Kallman CH, Weisfeldt ML, Flaherty JT (1982) The role of intracellular pH in the control of normal and ischemic myocardial contractility. In: Nuccitelli R, Deamer DW (eds) Intracellular pH: its measurement, regulation and utilization in cellular function, Alan R Liss, New York, pp 537-565
16. Feigl E (1983) Coronary physiology. Physiol Rev 63: 1-205
17. Braunwald E, Rutherford JD (1986) Reversible ischemic left ventricular dysfunction: evidence for the "hibernating myocardium". J Am Coll Cardiol 8: 1467–1470
18. Lattanzio FA, Pressman BC (1986) Alterations in intracellular calcium activity and contractility of isolated perfused rabbit hearts by ionophore and adrenergic agents. Biochem Biophys Res Commun 139: 816-821
19. Lee HC, Smith N, Mohabir R, Clusin WT (1987) Cytosolic calcium transients from the beating mammalian heart. Proc Natl Acad Sci USA 84: 7793–7797
20. Suga H, Hisano R, Goto Y, Yamada O, Igarashi Y (1983) Effect of positive inotropic agents on the relation between oxygen consumption and systolic pressure-volume area in canine left ventricle. Circ Res 53: 306–318

Mechanics of the Sarcomere

Tatsuo Iwazumi[1]

Summary. Studies on the mechanics of sarcomeres at the level of single myofibrils or single myocytes revealed properties not predicted by the cross-bridge theory of muscle contraction. This paper summarizes four major aspects:

1. Passive compression of the sarcomere showed two inflections in the force length curve; one was at the length where opposing thin filaments meet at the center of the sarcomere, and the other occurred where the Z-discs butt against the thick filaments, but the former inflection showed an unexpected direction.
2. When all sarcomeres in the preparation were uniform, at constant levels of free calcium concentration, the active forces increased with increasing sarcomere length up to 3.6μm, at the point where the overlap between the thick and thin filaments nearly ends.
3. At length perturbation frequencies higher than 50 Hz, protein structures in the sarcomere showed increasing stiffness with increasing frequency. However, the stiffness arising from the active force had no frequency dependence in the same frequency range.
4. Single sarcomeres are expected to produce force fluctuations, arising from cross-bridge cycling, whose rms value is on the order of 1% of the mean force. Fluctuation measurements at a sensitivty at least 100 times greater than the expected level revealed nothing.

Passive Compression Force vs Sarcomere Length Relations

All experiments were performed using the system developed by the author [1]. Transducer specifications (Table 1) have been improved by about a factor of two over those published previously.

The preparations used in the four sections differed substantially, and the ones used for this section were either enzymatically dispersed single cardiac myocytes or manually dissected very thin trabeculae from rats [2, 3], typically 20 μm wide. The segment length between force and length control transducers was about 10 μm so that the preparations could be compressed without buckling. Mounting the preparation securely onto the transducers with a segment length of only 10 μm using silicone adhesive required further technical development and the method has been described [4]. The entire segment was observed throughout the experiment using a TV camera and was recorded on videotape; the image of the segment was analyzed off-line to

[1] Department of Medical Physiology, University of Calgary, Calgary, Alberta, T2N 4N1 Canada

Table 1. Transducer specifications

Length controller	
Resolution (without servo)	0.1 Å/[Hz]$^{1/2}$ over 0–2 kHz bandwidth
Resolution (with servo)	10 Å rms over 0–50 kHz bandwidth
Step response (with servo)	10 μs
Frequency Response	2 kHz without servo, 50 kHz with servo
Stiffness (with servo)	10 mg (w)/μm
Force transducer	
Resolution (without servo)	0.05 ng (w)/[Hz]$^{1/2}$ over 0–2 kHz bandwidth
Resolution (with servo)	1 ng (w)/[Hz]$^{1/2}$ over 50 kHz bandwidth
Frequency response	2 kHz without servo, 50 kHz with servo
Maximum Force (with servo)	5 mg (w)
Stiffness (with servo)	10 mg (w)/μm

measure sarcomere length. The distance between the two transducers could be measured in real time but was not used in data analysis because of slight elongation of the glued joints (joint stiffness was typically 5 mg (w)/μm).

Since the stroke distance of the length controller was limited to ± 3 μm, the length changes were made manually (in this experiment only) by slowly turning the micrometer which supported the length controller, while measuring the passive force. In active force measurements sarcomere length was set at 10% increments by scaling the sarcomere image on a TV monitor.

Fig. 1 shows the passive force vs sarcomere length relation for a single myocyte and Fig. 2 shows such a relation for a very thin trabecula. Initially preparations were held at resting length then repeatedly compressed and stretched 3–5 times with a period of about 5 seconds. The stress relaxation and irreversible elongation of the cellular structure occurred when the myocyte was stretched more than 2.6 μm in sarcomere length; a looping of the curve was observed. In the trabecula, hardly any looping was seen after applying an excessive stretch force as shown in Fig. 2. Apparently, the epithelium of the trabecula is so strong that it completely prevents further extension. Note that in the myocyte the resting sarcomere length is slightly longer than it is in the trabecula and the stretch limit where the passive force suddenly increased is also longer. This is probably due to the epithelium which exerts a slight compression on the internal structure and adds substantial stiffness.

The most interesting observation in Fig. 1 is that the curve breaks at a resting length of about 2 μm, and the slopes of the curve above (stretch) and below (compression) are clearly different. By applying small length perturbations and by measuring the resulting force changes, the discontinuity was confirmed independently; the stiffness below 2 μm was about one half of the stiffness above that point. It has been suggested [5] that the force vs sarcomere length curve has a break at 2 μm because of mechanical interference which tends to prevent double overlap of thin filaments at the center of the sarcomere. If this explanation were true, we would expect that passive compression would produce in a curve which has a break at 2 μm and bends downward, the stiffness below 2 μm being greater than that above 2 μm. There are alternative explanations. It can be argued that an elastic structure may become taut

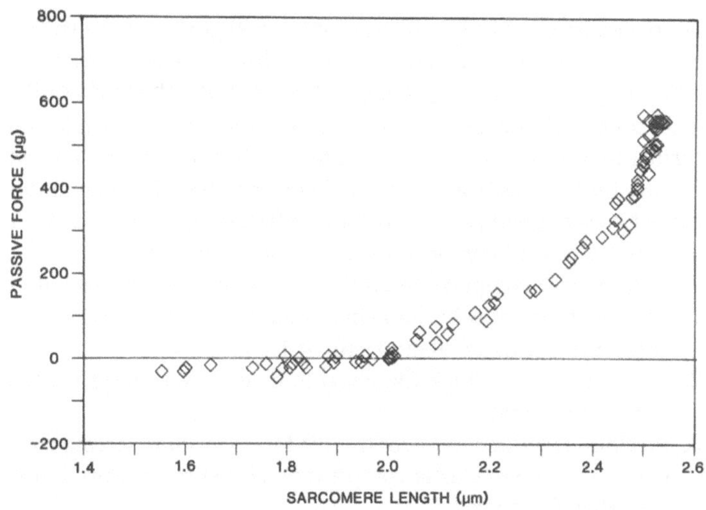

Fig. 1. Passive force vs sarcomere length relation of a single rat myocyte about 20 × 10 μm in cross section

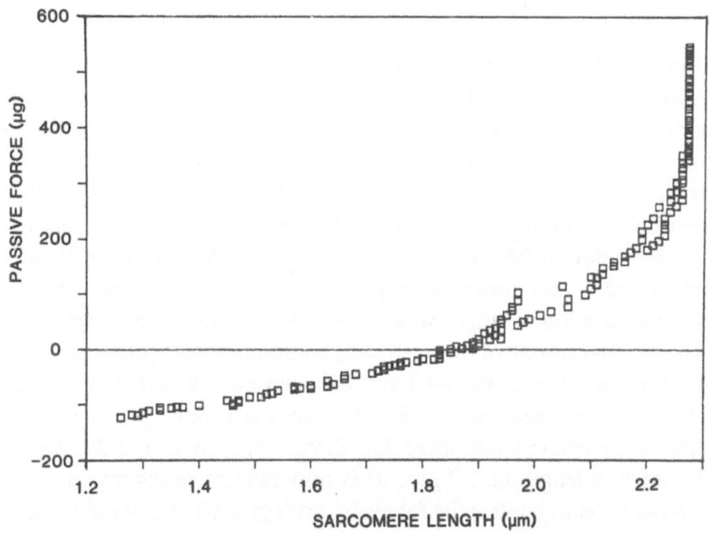

Fig. 2. Passive force vs sarcomere length relation of a very thin rat trabecula. The diameter of the preparation was approximately 22 μm including the epithelium

when stretched but become completely limp when compressed; in this case, the stiffness seen below 2 μm would be entirely due to mechanical interference between opposing thin filaments. This concept would not explain the precipitous force reduction seen below 2 μm in the tetanized frog single fiber. The magnitude of the stress that resists thin filament to double overlap is about 40 g(w)/cm^2 even at 1.6 μm in cardiac myocytes. This value is too small to account for the observed force diminution, since active stress produced by the myocyte at pCa 5.5 at the same sarcomere length was found to be about 500 g(w)/cm^2. In trabeculae, resistance to compression is borne mostly by the epithelium which, however, accounts for only 100–200 g(w)/cm^2 at 1.6 μm depending upon the thickness of the epithelium.

At about 1.6 μm the myocyte buckled while the trabecula could be compressed further to almost 1.2 μm. Apparently, the buckling of the myocyte is caused by the Z-discs butting against the thick filaments, but the magnitude of the resistance cannot be very great since an obvious inflection downward at 1.6 μm has never been seen in the passive compression curves of trabeculae.

From passive compression studies of sarcomeres we conclude that:

1. The resistance of thin filaments to double overlap is very small, the stress being only on the order of 40 g(w)/cm^2 at 1.6 μm.
2. The resistance of thick filaments to double overlap (or breaking through the Z-disc) is also small.
3. Precipitous reduction of contractile forces in short sarcomere lengths (less than 2 μm) is not due to elastic compression of a stiff structure but due to reduction of the active force itself.

Active Force vs Sarcomere Length Relations

Single or double myofibrils were mechanically dissected from bullfrog atrial cells using super-fine tungsten needles mounted on remotely controlled micromanipulators of special design. The myofibril was mounted between the length control and force transducers at a segment length of 20–50 μm. Activations were accomplished by passing solutions of pCa 6.0, 5.5, and 5.0 through a chamber into which two transducers and a myofibril were immersed. The chamber volume was about 50 μl and the solution flow was precisely controlled by a servo mechanism. To construct active force vs sarcomere length curves the myofibril length was initially adjusted so that the sarcomere length was 2.2 μm in relaxing solution (pCa 9.0). A contraction was initiated by one of the activating solutions and maintained for 10–20 s. Relaxing solution was then introduced and the sarcomere length was changed to a new value while in the relaxed state. This activation-relaxation cycle was repeated for each pCa value at sarcomere lengths of 2.2, 2.55, 2.9, 3.25, 3.6, 3.25, 2.9, 2.55, returning to the starting length of 2.2 μm. This protocol was necessary to account for continual deterioration of contractile forces by averaging two values of force at each sarcomere length (except at 3.6 μm). Activating solutions with pCa values less than 5.0 always destroyed myofibrils by inducing excessive sarcomere distortions and rapid force deterioration which prevented completion of a protocol cycle. In fact, very few myofibrils survived even at pCa 5.0. When myofibrils were in good condition, the active forces rose to the final level in 1–2 s but as they deteriorated the rise time became progressively longer. Poor quality myofibrils with visibly distorted and inhomogeneous sarcomeres (caused by damage during microdissection or excessive activations) were very slow to develop force (requiring 30 s or more), developed less force, and deteriorated much more quickly than high-quality myofibrils with clear and uniform sarcomere patterns. In this respect, the difficulty of maintaining the quality of myofibrils containing only 10–20 sarcomeres activated at pCa 5.0

illustrates the impossibility of maintaining the quality of conventional skinned fiber preparations containing millions of sarcomeres activated by low pCa solutions.

Fig. 3 shows the force-length relation of a high-quality myofibril [6]. A 100% relative force corresponds to about 5 kg(w)/cm², a very high value compared to 1–2 kg(w)/cm² usually found in ordinary skinned fibers. In other words, in large preparations, sarcomeres do not work at their full potential either due to damage inflicted on some of them or due to incomplete activation. Since each myofibril produced a maximum force of about 50 μg(w), the number of myofibrils in a very small preparation can be found readily by measuring the maximum force. A conspicuous property of the curves in Fig. 3 is the monotonic increase of force with increasing sarcomere length, i.e., the less the overlap length, the greater the force. This directly contradicts the cross-bridge theory [7, 8]. Experiments have shown that active forces beyond 3.6 μm fall precipitously, but such extended sarcomere lengths cannot be included in normal protocols since the myofibril, once overstretched, develops sarcomere distortions and loss of A-band contrast, indicating the filament misregistration which lead to non-uniform contractions afterwards when activated at shorter sarcomere lengths.

In summary, reliable active force vs sarcomere length relations are extremely difficult to obtain at pCa values lower than 5.5 due to the difficulty of maintaining a very high degree of sarcomere homogeneity. If the structural and functional integrity of all sarcomeres in the preparation is maintained, the maximum contractile force should reach 5 kg/cm²; submaximal force should increase with increasing sarcomere length (up to the length where the overlap between the thick and thin filaments disappears) and then fall precipitously beyond that length.

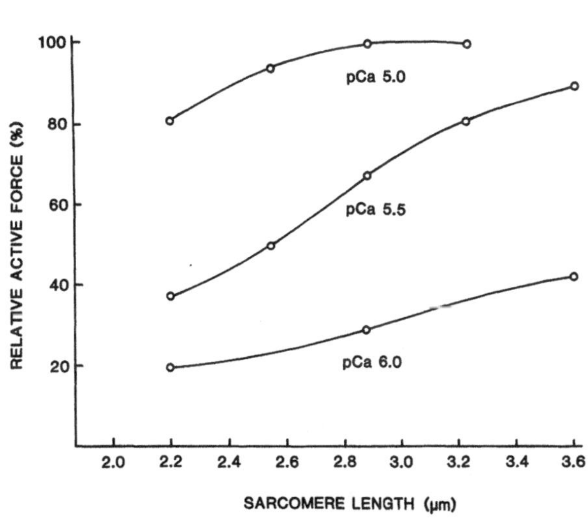

Fig. 3. Active force vs sarcomere length relation of an atrial myofibril from bullfrog. 100% relative force corresponds to about 5 kg/cm²

Stiffness Frequency Spectrum of the Sarcomere

Stiffness is the tangent at a point on a force-length curve and has been an important topic in the investigation of molecular events in the sarcomere. One would expect sarcomeres to stiffen when cross-bridges form mechanical linkages between thick and thin filaments; the stiffness will be maximal in rigor, and intermediate in active contractions because fewer cross-bridges will be bound to the thin filament. Therefore, the stiffness should be proportional to the number of linked cross-bridges. The elastic properties of proteins and most other solids exhibit frequency dependence, becoming stiffer at higher frequencies.

Since the length control and force transducers described previously are extremely sensitive and have a large bandwidth, they are ideally suited for elastic transfer function measurements when small vibrations are applied at one end of an elastic body and the resulting changes in force detected at the other. The amplitude of vibrations in one half unit sarcomere should be much less than one period of cross-bridge spacing (14 nm); for example, a segment length 10 sarcomeres long requires a vibration amplitude of less than 100 nm peak to peak. There are three methods of constructing a transfer function: step, sinusoidal, and white noise. Here, synthesized white noise consisting of 400 sinusoids of equally spaced frequencies with random phases was used. By synchronizing the clock frequency of the noise generator and that of the spectrum analyzer, it was possible to compute a complete frequency spectrum of a transfer function in just one period of the lowest frequency of the spectrum. In contrast, Gaussian white noise requires many runs for spectrum averaging, a definite shortcoming when dealing with short-lived phenomena.

Manually microdissected myofibrils from rat psoas muscle were used in this experiment. Since the vibration transmission through water will interfere with the measurement, the myofibril was stretched to 2.64 μm sarcomere length to raise the passive stiffness in the relaxed state so that the transmission through the myofibril is much greater than that through water. Transfer functions were measured in the relaxed state at pCa 9.0, in the contracted state at pCa 5.5 and in the rigor state. Fig. 4 shows three stiffness spectra in these states. Phase angles were eliminated in the figure because they were essentially constant from 40–500 Hz. A noticeable dip at 10 Hz in the stiffness spectrum during contraction (b) arises from the active force producing process interacting with applied vibrations. When the spectrum during relaxation is subtracted from that during contraction, one finds that the difference spectrum (i.e., the incremental stiffness spectrum) is frequency independent above 50 Hz. On the other hand, the spectrum during rigor is substantially different. If isometric contraction is brought about by partial cross-linking between the thick and thin filament, the shape of the contraction spectrum should be intermediate between the relaxed and rigor spectra. Frequency independence of the incremental stiffness spectrum clearly indicates that the incremental stiffness arises not from cross-linking of the protein structure but from a process such as electrostatic interaction which does not involve any mechanical structure.

In summary, the frequency spectra of incremental force during contraction provide important information for understanding contractile processes. Frequency independence of the incremental stiffness at frequencies above 50 Hz renders strong evidence that the stiffness increase during contraction is not due to mechanical cross-linking between the thick and thin filament.

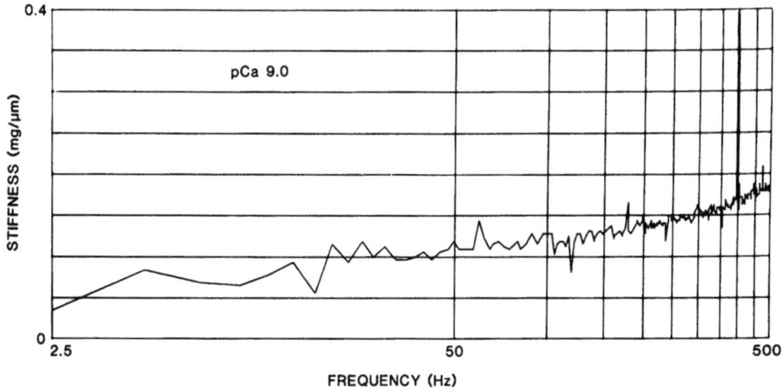

a

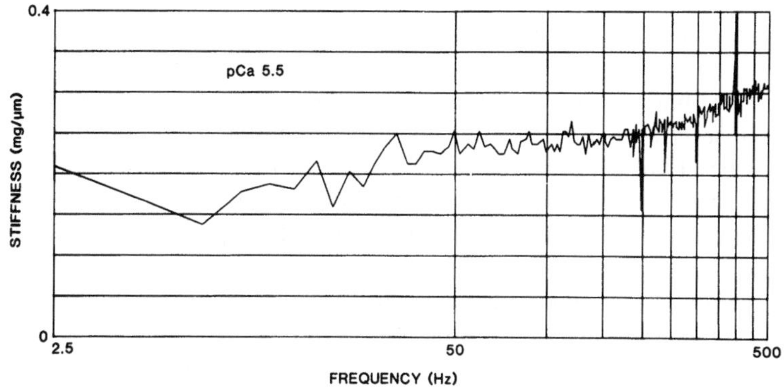

b

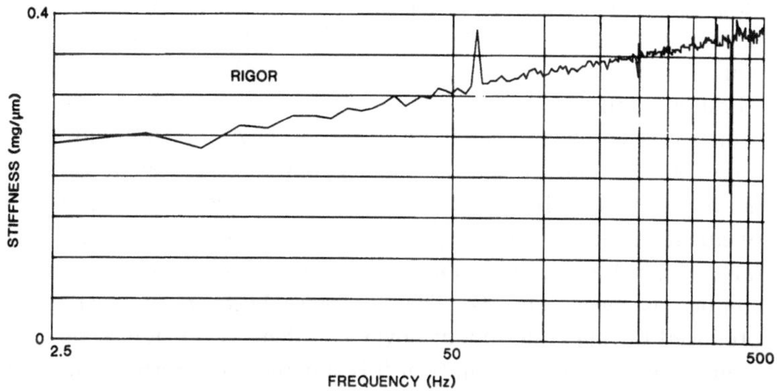

c

Fig. 4.a Stiffness spectrum of a relaxed myofibril from rat psoas muscle. **b** Stiffness spectrum of the same myofibril in contraction at pCa 5.5. **c** Stiffness spectrum of the same myofibril in rigor

Force Fluctuations in Sarcomere

The essence of cross-bridge theory is the cyclic attachment and detachment of cross-bridges at binding sites on the thin filament in conjunction with the enzymatic hydrolysis of ATP. Therefore, the total force acting on a thin filament is the sum of individual impulsive cross-bridge forces. It has been generally assumed that each cross-bridge acts independently of the others. This statistical independence leads to a conclusion that the total sarcomere force will fluctuate randomly with time about a mean value. By choosing a model of cross-bridge kinetics from the many proposed [7, 9-13] one can calculate a root-mean-square (rms) value for force fluctuations [14]. Regardless of the model chosen, the rms value increases with the square-root of the number of cross-bridges in a half-sarcomere. The number varies with the extent of overlap; at full overlap it is about 50,000. The estimated rms value varies depending upon the assumptions used — it is on the order of 1% of the mean force. Since the actually measured maximal force of single myofibrils was about 50 μg (w), the rms value is about 500 ng(w). This fluctuation rms value is a total power, i.e., a value obtained by integrating a spectral curve over entire frequencies. The shape of the spectral density depends upon the adenosine triphosphatase (ATPase) kinetics involved but it can be approximated by the first-order process (20 dB/decade slope in the spectrum) with a corner frequency corresponding to the ATPase rate. Assuming a rate of 10/s (Fig. 4b), a fluctuation spectrum can be predicted as shown in Fig. 5.

Manually microdissected myofibrils from *Rana pipiens*, bullfrogs, turtles, rabbits, and rats were used in this experiment [15]. The results were the same in all animals and are represented by those from a rat psoas containing two single myofibrils. As in the elastic transfer function measurements, the myofibril was mounted between the length control and force transducers except that the length control transducer was used as another force transducer (since these transducers are of identical design and their functions can be alternated). This method greatly reduces environmental vibrations because the transducers sense myofibril force in-phase while building vibrations and acoustic vibrations are sensed out-of-phase; by adding two force signals it is possible to eliminate almost all environmental vibrations. Without this noise cancellation scheme the force signal spectrum contains a prominent broad noise peak from 10–40 Hz even if the complete apparatus is mounted on a pneumatic isolation table. Fig. 6 shows typical force fluctuation spectra during rest (*broke curve*) and contraction (*solid curve*). Each spectrum was averaged over 16 continuous measurements. The difference between the two spectra is the noise produced during contraction; obviously, there is no significant difference between them. The level of uncertainty of this statement is approximately 0.1 ng(w)/[Hz]$^{1/2}$. Comparing this result with Fig. 5, we may safely conclude that the predicted fluctuation level is at least 100 times (40 dB) too large. The rise of noise power at frequencies below 50 Hz in Fig. 6 comes from fluctuations in the refractive index of the air in the illuminator housing. The possibility that force fluctuations could not be detected because all cross-bridges were locked up with the thin filaments during highly stable isometric contractions was dismissed on the basis of the following: (a) elastic transfer function measurements done on the same myofibril to demonstrate frequency-independent incremental stiffness as described in the previous section, and (b) slow stretch and release contractions which prohibit cross-bridges to hang on the same binding sites. In the latter, averaged fluctuation spectra were obtained by slowly stretching and releasing the myofibril 16 times during rest and during contraction by 5% of the slack length (about 50 nm per half sarcomere) which covers a distance of about three cross-bridge periods. The result was identical to Fig. 6.

In summary, the force fluctuation study demonstrated that the level of the fluctuations, if they exist, is at least 100 times smaller than the level predicted by cross-bridge theory. In order to accommodate this finding, the theory must be modified to include the assumption that cross-bridges move in extraordinary synchronism and complementarity among themselves so that when one cross-bridge makes contact with the thin filament another one becomes detached, resulting in almost no ripples in the sum of their forces. Whether or not such extreme cooperativity in molecular motion is possible is open to discussion.

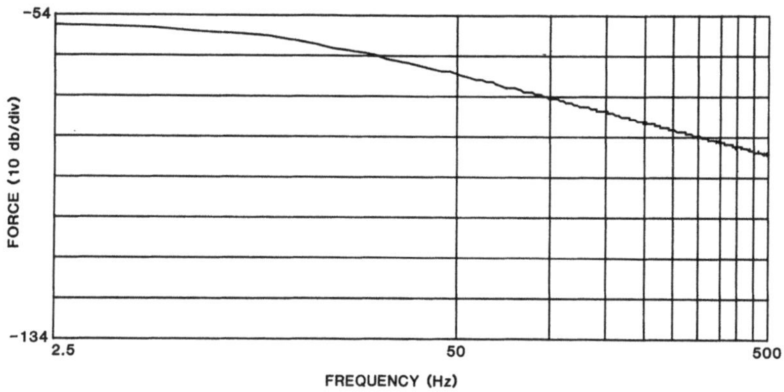

Fig. 5. Active force fluctuation spectrum expected from cross-bridge cycling during contraction. 0 dB = 0.1 mg (w), −100 dB = 1 ng (w). *div*, division

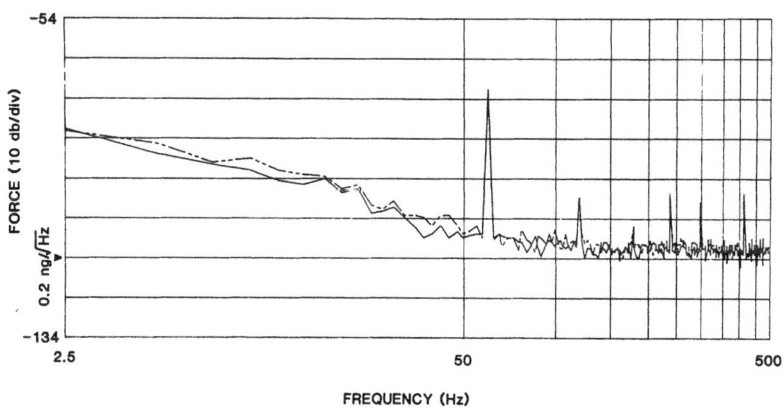

Fig. 6. Actually measured force fluctuations during rest (*broken curve*) and during contraction (*solid curve*). The difference (square root of square difference) of these curves gives the incremental fluctuations due to contractions. *div*, division

Conclusion

Results of four entirely different experiments designed to test the predictions of the cross-bridge theory are presented in this article. Many different interpretations of the results are possible, but the cross-bridge theory appears to be extremely inadequate to explain any of the results presented above within the framework of reasonable assumptions. I believe that it is time to reevaluate critically the cross-bridge theory and many other theories of muscle contraction. After all, scientific theories are made to be tested, not to be admired as dogma.

References

1. Iwazumi T (1987) High-speed and ultra-sensitive instrumentation for myofibril mechanics measurements. Am J Physiol 252: C253-C262
2. Iwazumi t, ter Keurs HEDJ (1987) Mechanical properties of single cardiac myocytes and ultra small trabeculae of rat. Am Heart Assoc Annual Meeting
3. ter Keurs HEDJ, Iwazumi T (1988) Restoring forces in rat cardiac myocytes and ultra-small trabeculae. Biophys J 167a
4. Iwazumi T (1987) A new method to attach single cardiac myocytes to transducers. Biophys J 51: 4791a
5. Gordon AM, Huxley AF, Julian FJ (1966) The variation in isometric tension with sarcomere length in vertebrate muscle fibres. J Physiol (Lond) 184: 170–192
6. Iwazumi T (1988) The mechanism of ascending limb of the length-tension relation in heart muscle. In: ter Keurs HEDJ, Noble MIM (eds) Starling's law of the heart: revised. Martinus Nijhoff, pp 18–27
7. Huxley AF (1957) Muscle structure and theories of contraction. Prog Biophys Biophys Chem 7: 255-318
8. Huxley AF, Simmons RM (1971) Proposed mechanism of force generation in striated muscle. Nature 233: 533-538
9. Lymn RW, Taylor EW (1971) Mechanism of adenosine triphosphate hydrolysis by actomyosin. Biochemistry 10: 4617-4624
10. Hill TL (1974) Theoretical formalism for the sliding filament model of contraction of striated muscle. Part 1. Prog Biophys Mol Biol 28: 267–340
11. Hill TL (1975) ibid. Part 2. Prog Biophys Mol Biol 29: 105-159
12. Eisenberg A, Greene LE (1980) The relation of muscle biochemistry to muscle physiology. Annu Rev Physiol 42: 293-309
13. Eisenberg A, Hill TL (1985) Muscle contraction and free energy transduction in biological systems. Science 227: 999-1006
14. Borejdo J (1980) Tension fluctuations in contracting myofibrils and their interpretation. Biophys J 29: 49–64
15. Iwazumi T (1988) Myofibril tension fluctuations and molecular mechanism of contraction. In: Sugi H, Pollack GH (eds) Molecular mechanism of muscle contraction. Plenum, pp 595-608

Rapid Relengthening in Isolated Cardiac Cells and the Origin of Diastolic Recoil

John W. Krueger[1]

Summary. Late relengthening in both intact muscle and in the isolated cardiac muscle cell is characterised by two distinct phases that represent separate intracellular compartments or steps in the relaxation process. The rapidity of the first component facilitates the rapid relaxation that is required to promote early filling of the heart. Conversely, early filling will be slowed by any factor that increases loading or sarcomere length. Late diastolic properties in normal myocardium represent an independent mechanism which probably reflects extrusion of calcium ion from the cell rather than its reuptake by the intracellular compartments. The magnitude of just one internal restoring force in the single cell can be estimated to be on the order of 0.2–0.4 mN/mm^2. This value probably represents a *lower* limit to the true intracellular restoring force, and it is not known whether the cell lengthens in a truly elastic fashion. Application of a simple model of the ventricle indicates that, even at this minimal level, the cellular restoring force would account for a negative pressure of ~10mmHg. Thus the actual magnitude of the negative pressures developed during early filling in the myocardium is largely consistent with the component which originates by lengthening of the isolated cell. The true restoring force that arises in the cell may be much larger, and it is worth considering the possibility that extracellular structures counteract the cellular restoring force. The quantitative parallel between cellular and myocadial restoring forces does not exclude the extracellular matrix as a restoring force between layers of the heart but, for the moment, the predictions based on a role for the elastic distortion of the latter are not easy to reconcile with the mechanics of the isolated muscle.

Introduction

Contemporary research in cardiac function has rekindled fresh interest in the mechanisms by which the heart generates negative pressures when it relaxes at very small volumes [1]. The common explanation is that shortening (or perhaps contraction) of the muscle fibers distorts elements of the myocardium *elastically*. The potential energy stored by contraction is then reimparted to the sarcomeres to relengthen the muscle fibers via "diastolic recoil". Yet the precise origin of diastolic recoil has not been resolved, nor is it in fact clear that the recoil is *truly* elastic. In view of the structural complexity of the heart, one central and unsettled issue is the physical

[1] Division of Cardiology, The Albert Einstein College of Medicine, 1300 Morris Park Avenue, Bronx, NY 10461, USA

basis by which intra- and extracellular structures influence the shortening of the cardiac cell, and *vice versa*. Fortunately, sufficient advances have occurred in describing the mechanics of the myocardium, the behavior of the sarcomere in the isolated muscle, and the properties of the single cell, to assess the origin of restoring forces in the heart. Naturally, a related question is how the properties of its isolated components contribute to the mechanics of the whole heart.

Characterization of Relengthening in Isolated Myocytes

Relengthening of the unattached cell provides an insight into relaxation that is not complicated by force generation, heterogeneity or extraneous compliance. Several common features of relaxation that were first apparent in the isolated cell are illustrated in Fig. 1. The simplest and yet still the most striking behavior of the unattached cell is that it relengthens vigorously after shortening [2]. Attachment for force measurements in single cells has remained a difficult task, and so characterizations of cell contraction almost universally rely on the measurement of the extent of shortening and the maximum speed of sarcomere motions. Yet sarcomere motions reflect several factors including the internal load and the myoplasmic calcium dynamics, and so a precise interpretation for the basis of relengthening can be somewhat illusive. More specific information is obtainable by examining also the manner in which the velocity of the sarcomere's motions depend upon its length (Fig. 1b). It has been observed routinely that the maximum velocity of relengthening is proportional to the maximum velocity of shortening, yet both parameters peak at different lengths. Any *supposed* elastic element that is distorted by shortening would give rise to a force which should (a) slow shortening, (b) increase the speed of relengthening, and (c) increase the ratio (b/a) of relengthening to shortening velocities at any length. The dependence of this "velocity" ratio upon length is also shown as an inset in Fig. 1b. The ratio increases to a maximum at short lengths, as might be expected from the influence of a length-dependent intracellular load that resists shortening of the sarcomere and that promotes relengthening.

Changes in contractility identify some useful distinctions between these measures of relengthening dynamics that prevent the attribution of a cellular restoring force to a simple length-dependent elastic recoil. While the peak velocity of sarcomere shortening is enhanced by contractility, the peak velocity of relengthening often reaches a plateau to remain constant during a significant portion of relengthening, and the slope of the respective velocity-length trajectory can appear to be largely independent of length (Fig. 1c) [3]. Consequently the peak of the velocity ratio is increased and is shifted to *shorter* lengths while the ratio is actually reduced at any length (Fig. 1d) [3]. Any purely elastic restoring force should be the same at any length and, therefore, shortening should decelerate more rapidly at lower levels of contractility or shorter sarcomere lengths. By a similar rationale, relengthening should decelerate more slowly at shorter lengths. The simple shift in the velocity-length trajectory in both shortening and relengthening (Fig. 1c) challenges the assumption that a relengthening in the unattached cell can be treated as a length-dependent elastic discharge of potential energy.

Separate and distinct phases of relengthening. The velocity-length trajectory shows that the slowing of shortening and the acceleration and deceleration of lengthening appear to be largely dependent on myofilament sliding rather than an internal load that arises from sarcomere length. Whenever the velocity-length trajectory is linearly related to length, its *slope* provides an insight about relaxation in terms of a rate constant for relengthening (Fig. 1b). (Intriguingly, this rate constant appears

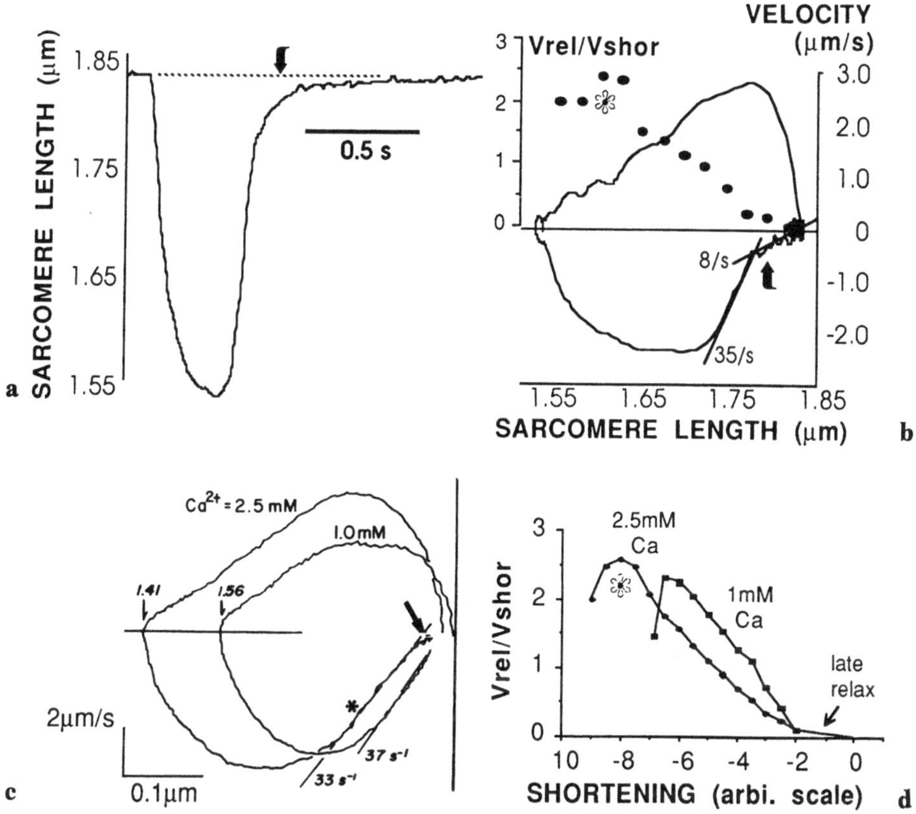

Fig. 1a-d. Dynamics of relengthening in single, unattached ventriculocytes. **a** and **b** Sarcomere shortening and relengthening in a canine left ventricular endocardial myocyte ($[CaCl_2] = 2.0$ m*M*, stimulus rate 0.5 Hz). **b** The velocity-length trajectories of cell in **a**. The velocity of shortening is upward. The slopes of the last, linear phases of the relengthening trajectory can be related to a rate constant, since the general solution to the equation $\delta(SL)/\delta t = m(SL) + b$ results in an exponential, $SL(t) = e^{(mt+b)}$. Separate components of late relengthening (*arrows*) can be approximated by two such rate constants. The superimposed points represent the instantaneous ratio of the velocities of relengthening/shortening which peak *flower symbol* at that length where acceleration of relengthening stops. **c** Influence of contractility on the velocity trajectory of a guinea pig ventriculocyte (adapted from [3, Fig. 2] by permission of the publisher). **d** Calcium increases the peak of the relengthening velocity ratio *flower symbol*, but shifts it to shorter lengths. Cells were electrically stimulated to contract (34°C) and sarcomere length was measured 500 to 1000 times per second using a precision edge detector in **b** [21] or by FM detection in **a** [22]

similar to that predicted from the time course of isometric tension decline in the intact heart.) Thus the deceleration in the velocity trajectories provides a convenient demonstration that relaxation consists of at least two separate and potentially independent processes. For example, an increase in the frequency of stimulation can speed relengthening but decrease the isolated cell's initial length [2, 3]. Consequently, quite separate intracellular mechanisms govern early and late filling of the heart, and only the last will influence its end diastolic properties. The transition to the slow, last phase of relaxation does not appear to be related to membrane repolarization, and it has been attributed to extrusion of cell calcium by the sodium-calcium exchanger [4]. Occasionally, cells will shorten rapidly but relengthen very slowly, and in these circumstances late relengthening is characterised by only one rate constant for deceleration. Overreliance on just one parameter of either sarcomere motion or of relaxation itself may lead to an incomplete understanding for the full relaxation process.

Relation of the cell to relengthening in intact muscle. A separation in two phases of isotonic relaxation also occurs in intact muscle when the sarcomere length exceeds 1.9–2.0 μm and so this aspect of relengthening can not be attributed to overlapping of the thin filaments or to a restoring force that is unique to muscle tissue [5]. Afterload minimizes the difference between the last components of relengthening by slowing the first and speeding the second [5]. The acceleration of the earlier component would enable more rapid relaxation in the myocardium with a lower afterload. Conversely, any factor that increases afterload might impair the early diastolic filling.

The peak velocity of isotonic relengthening is increased by extracellular [Ca^{2+}] in both isolated muscle and single cells but, unlike the cell, the slope of the muscle's velocity-relengthening trajectory is steepened dramatically [6]. One possibility is that the effect of intracellular calcium dynamics on relaxation is sensitive to the presence of cross-bridges. Also consistent with the influence of cross-bridge is the fact that the rate constant of relengthening is slowed by afterload [5]. To the extent that the deceleration of relengthening in muscle or cells reflects the myofilaments' sensitivity to calcium, the steeper slope of the isotonic velocity-relenthening trajectory need not represent a rate limiting step imposed by the intracellular reuptake of calcium. With one exception, as described below, there appears to be little *intrinsic* difference between relengthening in the isolated muscle and the unattached cell.

Cellular Restoring Forces and the Isolated Muscle Preparation

As the cell isolation methods improve, the reported rest length of the sarcomere has been found to increase to that length (SL $\geq$ 1.93μm) that characterizes the relaxed cardiac myofibril [7]. Precise determination of the true rest length of the sarcomere in intact muscle is limited by the difficulties of measuring "zero" tension. The inactive sarcomere buckles at lengths that are 5%–10% shorter than its rest length in both the isolated cell [2] and in intact muscle [8]. Moreover, the rest length of the sarcomere in isolated cells which are selected for stable contractile behavior is 1.98 μm for rat ventriculocytes [9]. This value is in excellent agreement with the mean sarcomere length of 2.01 μm that is observed in the left ventricular free wall of the open chest rat [10].

These striking *quantitative* parallels between sarcomere length in unattached cells, isolated muscle, and cardiac tissue indicate that the sarcomere is not prestressed to any significant degree in the *normal*, unloaded, diastolic ventricular myocardium. The existence of resident internal stresses within unattached cardiac tissue might have been inferred from the adverse influence of size of isolated fragments upon sarcomere shortening [11]. This latter interpretation, however, is predicated on several untested assumptions: (1) that elastic interactions exist between cells in tissue; (2) that they are of sufficient magnitude to account for the differences in shortening; and (3) that the size of the isolated fragment does not alter the access of calcium to the myofibril when activation is used to test the effect of size.

Most physiologic observations have reemphasized that striations shorten and relengthen uniformly in both intact muscle and in isolated cells. However, it has been found that striations buckle at slack length in exceptionally thin, isolated papillary muscle preparations [12]. The fact that the cells buckle is one indication that the force which relengthens the cardiac muscle arises within the cell rather than the extracellular matrix.

Differences in the dynamics of sarcomere relengthening in muscle and cells also suggest that extracellular elements do not promote — and perhaps actually resist — relengthening of the cell. The maximum extent and speed of sarcomere shortening in the isolated cell is similar to those levels that occur in intact muscle [2, 3]. However, in the normal isolated cell the peak velocities of relengthening and shortening are always about equal [2, 3, 9]. The proportionality between the velocities of unloaded shortening and relengthening in the isolated cell is in striking contrast to the freely shortening intact muscle, where the velocity of relengthening is appreciably less than its maximum speed of shortening [3]. (A recent comparison of cellular properties to intact hearts also comments about the slowing of relengthening in the myocardium [9].) Based upon circumstantial evidence, it has been suggested that relengthening in isolated muscle is due to the presence of extracellular connections which are elastically distorted by shortening [13]. If true, extracellular elements should always serve to speed relengthening when muscle shortening increases [13]. Yet, as shown in Fig. 2, when external forces are reduced to very low levels, relengthening slows dramatically in muscle even though its extent of shortening is increased [3, 6, 14]. The slowed relengthening of the total preparation may, in part, represent delayed shortening in viscoelastic structures stretched by contraction at its damaged ends, but the buckling of the sarcomere [12] and the selective slowing in the dynamics of *its* relengthening [3] supports the view that the force that restores the shortened cardiac muscle to its rest length originates in the cell.

Estimate of an internal load in the isolated, intact cell. The restoring force in mechanically skinned cardiac cells has been measured to be 4 ± 2% of maximum contractile tension [7]. Application to the cell with intact membranes is tempered by the fact that the contractile lattice is expanded in skinned cells where (1) wavy myofibrils do not occur unless the inactive sarcomere is shorter than 1.6 μm, and (2) relengthening is exceptionally slow [7]. The origin and the magnitude of the internal load that resists shortening and promotes relengthening in the *intact* cell is not

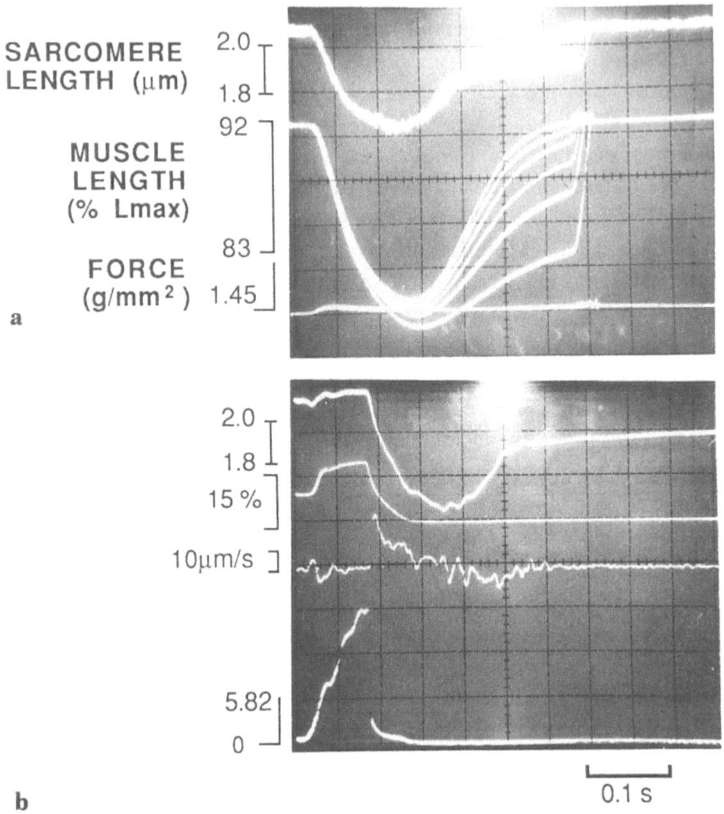

Fig. 2a, b. Differences between internal and external relengthening in isolated cardiac muscle. **a** The early relengthening of the sarcomeres is largely dissociated from that of the muscle; application of a force ≤ 0.7 mN/mm^2 (1% of the peak isometric tension at Lmax, maximum length) rapidly abolishes the difference in late relaxation. **b** Sarcomeres return spontaneously to rest length after extreme shortening, yet the overall length of the preparation remains shortened. Sarcomere length was measured by light diffraction in isolated right ventricular trabeculae (rat. 30°C, 24/min, [$CaCl_2$] = 1.9 m*M*)

known. One internal force that exists *between* sarcomeres in the cardiac cell can be estimated from the nonuniform, asynchronized motions which occur when the cell contracts unphysiologically. The *nonsymmetrical* distribution of sarcomere lengths in the asynchronous contraction (Fig. 3a) indicates that shortening lengthens inactive sarcomeres that are connected in series. Since the unattached cell is influenced only weakly, if at all, by the glass substrate of the observation chamber, any lengthening by inactive elastic elements reflects the force produced by the shortening sarcomeres.

In general, the relation between shortening and lengthening about the mode in the nonsymmetrical distribution of sarcomere length is about 1:0.5 (Fig. 3). Since the overall extensibility of the cardiac cell is generally known [15], the force which must be generated by elements that shorten by 0.5 μm can be estimated from the lengthened sarcomeres to be on the order of 0.4 mN/mm^2 (refer to Legend, Fig. 4). This force which resists sarcomere shortening in the isolated cell appears to be of the same order as the 1%-2% level of force that markedly accelerates relengthening of the inactivated, shortened sarcomere in intact muscle [8].

Importantly, any load that extends some elements is equal to the *net* force produced by the shortening sarcomere, and so it will represent only *one* component of a restoring force in the intact cell (Fig. 3c). Resting sarcomere length does not change when the cell's membranes are removed [7], so that the resistance to shortening by hydrostatic pressures is probably negligible. Since the peak extent of sarcomere shortening is similar in the synchronous and asynchronous contractions, the actual load computed from asynchronous stretch is only a fraction of the true restoring

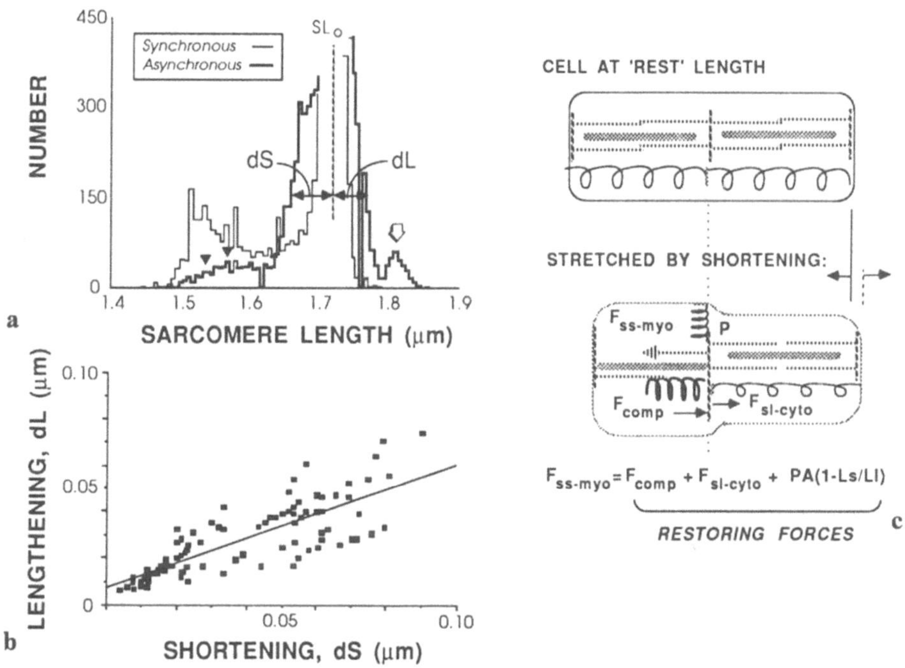

Fig. 3a-c. Demonstration of an internal load in an intact cell that resists shortening and would contribute to sarcomere relengthening. **a** The nonuniform distribution of sarcomere lengths in asynchronously shortening, unattached cell. The cell's striation pattern was visualized with a photodiode array, and the localized sarcomere length was determined 500/s by FM detection in a 2µm wide "window" that was displaced automatically along the length of the cell. The distribution of sarcomere lengths in the asynchronously shortening cell (*heavy lines*) is broader than in the synchronously contracting cell (*finer lines*). In the former case, the number of lengthened (*open arrow*) and shortened (*filled arrows*) elements appears equal (peak amplitude = 900 in 10,000 samples). **b** The nonsymmetrical distribution of local shortening and lengthening is compared at various amplitudes (e.g., *horizontal arrows* in **b**) in histograms from 7 cells. The mean of the slopes and the correlation coefficients of individual histograms were 0.622 ± 0.29(SD) and 0.946 ± 0.045, respectively y = 0.53x, r = 0.8. **c** An explanation for asynchronized lengthening in unattached cells. The active force generated by the myofilaments of shortening sarcomeres ($F_{ss\text{-}myo}$) at the Z-line is balanced by a myofibrillar compressive force (F_{comp}), a cytoskeletal force in the lengthened sarcomeres ($F_{sl\text{-}cyto}$), and a force due to the component of hydrostatic pressure transmitted by ends of the cell which have different cross sections force of the sarcomere. Thus, additional restoring forces may be recruited when sarcomere relengthening is synchronized, and they can be conservatively estimated to be on the order of 1-2 mN/mm^2. (An estimate based upon the data of others [7, Fig. 2] indicates that inactive myofibrils which have been previously shortened to 1.2μm can generate a sustained restoring force of 2-4 mN/mm^2.)

Relation between cellular and myocardial restoring forces. The relevance of a cellular restoring force to a negative filling pressures in the heart is further supported by application of a simple model of the myocardium [16] to estimate the corresponding wall stress. Fig. 4 depicts the relation between intraventricular pressure and volume that would occur at selected levels of constant wall stress in the endocardial layer. This simple application of the LaPlace relation to negative intraventricular pressures reveals an important point: i.e., by itself a decrease in intraventricular volume will make the intraventricular pressure more negative without necessarily requiring an increase in the myocardial restoring stress. The simple fact that ventricular filling pressure is reduced at end systolic volume *is not sufficient* to demonstrate that there has been an increase in the potential energy stored elastically in the wall of the heart!

Computations based on the negative pressure-volume relations obtained when ventricular filling is prevented by programmed mitral occlusion [1] show that volume dependence of intraventricular negative pressure requires an increase in the restoring stresses of the wall (Fig. 4a). However, the maximum negative filling pressure of -13 mmHg can be produced by a intramyocardial restoring force of only ~0.5 mN/mm^2, a value which is only 0.5% of the peak active tension that the isolated muscle can produce, and 5% of end systolic wall stress. If the rest length of the cardiac sarcomere is uniformly 1.9 μm in the unstressed heart, then the maximum peak negative filling pressure would occur at a sarcomere length of 1.4 μm in the endocardium and 1.8 μm in the epicardial layers, and the average sarcomere length will approach 1.6 μm at the minimal end systolic volume. Fig. 4 further compares (1) the wall stress that would account for the observed volume dependence of intraventricular pressures with (2) the minimal level of wall stress that which would originate from the internal shortening of the isolated cell, as detailed in the figure legend.

This analysis does not consider potential viscoelastic interactions and it admittedly oversimplifies the complex geometry and fiber orientation of the heart. The approach does, however, show that the magnitude of "diastolic recoil" is certainly consistent with a cellular origin of the myocardial restoring force. In fact, it may be worth considering the implications of these simple results, i.e., the possibility that the restoring forces in the cell may be *greater* than those which actually exist in the myocardium.

Evaluation of Extracellular Mechanisms for Cadiac Restoring Forces. In evaluating the cellular contribution to diastolic recoil, it is worth reconsidering the evidence for an alternate interpretation, i.e., that either (1) contraction or (2) shortening stores a potential energy in the myocardial connective tissue. Extracellular elements that transmit active tension would store the energy of contraction as a series elastic component in muscle. In the hearts of small animals, however, it is possible to decrease

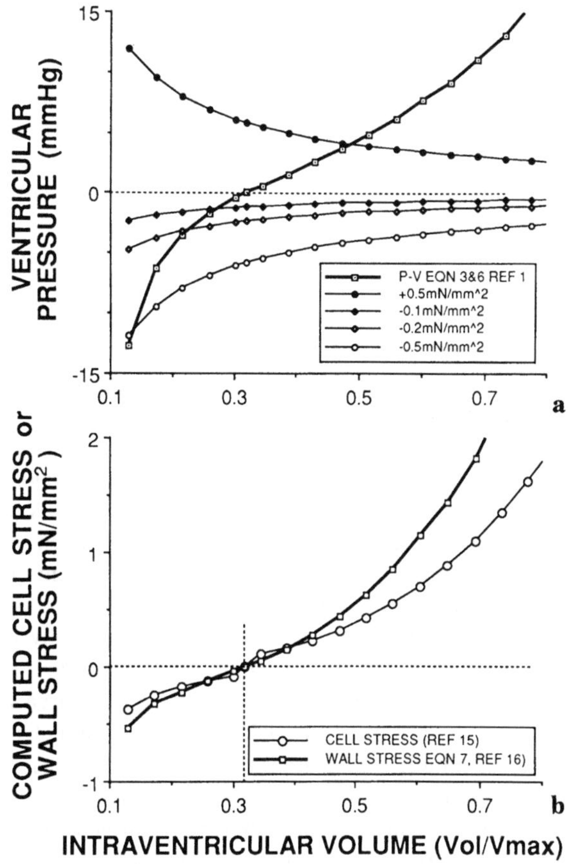

Fig. 4a, b. Comparison between ventricular and cellular restoring forces in the heart. **a** Relations between intraventricular pressure (P), volume (V), and mean wall stress (WS) computed by assuming that $WS = 1.36\ (g/cm^2.mmHg)\ P(mmHg)/[(1 + V_{max}/V)^{2/3} - 1]$ (from [16], Eq. (7)). The heavy lines denote the experimental (±) pressure-volume relations computed from equations developed by others [1]: $(+)P = -14\ln[(116 - V)/79]$; $(-)P = 5.1\ln[(V - 13)/24]$. ($V_o$ = 37 ml; V_{max} = 116 ml; V_{min} = 13 ml, myocardial volume = 113 ml (mean values from [1], Table 2). **b** Comparison between wall stress and cellular restoring stress. One restoring force attributable to the shortening cell (σ_n) was computed by a modification of the equation for cell stress from differential expressions of strain of Fish et al [15]: $\sigma_n = -0.829e^{10.07\ln(R[\delta SL + SLo]/Lo)}$ where R equals the mean slope of the distribution between lengthening and shortening sarcomeres shown in Fig. 3. This minimal restoring force does not consider the true restoring force within the sarcomere which is likely to be larger. Sarcomere length was assumed to be 1.9 μm in the unstressed volume of the heart (V_o).

volume with sufficient speed to measure accurately the elasticity of the myocardium: In this case the elasticity of the active myocardium very nearly equals that which is measured for the sarcomere [17]. The series elastic recoil is equivalent to 10 nm/half sarcomere (i.e., ≤1% muscle length) in an isolated cardiac muscle preparation [18] where developed tension can exceed 100 mN/mm^2 tension. In contrast, the elasticity attributable to radial strut-like connections between cells requires that they be stretched by 680 nm, or 17%, to produce lower levels of force [19]. The transmission of active force by extension of such compliant, radially oriented links between cells would be inconsistent with the greater tension produced at longer sarcomere lengths where the initial separation between cells decreases and, in active contraction, does not *increase* as required for greater force generation. An analogous problem exists in explaining the effect of velocity of shortening on force generation at any given sarcomere length when the separation between cells — and extension of lateral connections — does not change. The shape of the isometric myogram is governed by the most compliant elements in the isolated muscle preparation: in fact, the shape of truly isometric contraction in cardiac muscle is remarkably like that seen in the single skeletal muscle fiber. Last, contracture after the separation of the intercalated discs results in extreme internal shortening and the occurrence of large gaps between cells in the isometrically held cardiac muscle preparation (Krueger and Zhao, unpublished observations). Since pericellular elements are unable to prevent internal shortening, the mechanism by which they physiologically sustain active tension generation between cells is unclear. So far, there is no *direct* evidence that contraction alone can store potential energy by distortion of extracellar elements.

Another possibility is that extracellular elements that spiral helically around the heart cell can both resist stretch and cause the elastic recoil in heart muscle [13]. However, the isolated cell shortens isovolumically [2]. Thus, considering a cylindrically shaped cell of representative dimensions ($24 \times 100\mu m$), its circumference would increase by only 11.8% while its surface area would decrease by 10.6% when its length is shortened by 20%. While any one helix can either shorten or lengthen when cell length is decreased, the constraints imposed by its initial pitch mean that *it is unlikely to do both*. Moreover, sarcomeres no longer relengthen spontaneously after contraction although they resist strech normally and transmit contractile tension in cardiac muscle that has been freeze-dried [20]. Freezing disrupts the cell membranes, and so it is unlikely that a unique *extracellular* framework can simultaneously resist extension and relengthen the cardiac muscle.

It is not clear how distortion of extracellular connections would account for the inotropic increase in negative intraventricular pressure at any given end systolic volume [1]. In contrast, relengthening in cells provides a mechanism by which restoring forces may be varied by inotropic state and synchrony as explained earlier. The contribution of a cellular restoring force to diastolic recoil depends on the relation between the rest length of the cell and the surrounding tissue; it would be expected to play a role only at high levels of contractility or low end systolic volume. Accordingly, the cellular origin of restoring forces in the heart provides a simple explanation why enlarged hearts fill less efficiently. In contrast, the contribution to early filling by the storage of potential energy in the extracellular matrix (by contraction, by shortening, or by means of both events) ought to be independent of heart size. There is no *a priori* reason in the latter case to expect early filling to be compromised just because the unstressed volume of the heart is enlarged. An interesting test of the proposed role for distortion of the extracellular matrix would be to then ask whether chronically enlarged hearts can still developed negative pressures when peak ejection does not shorten the sarcomere to less than its rest length in the isolated cell?

Acknowledgments. Supported, in part, by HL 21325. I thank Hong Zhang, Adam Denton, and Gerard Siciliano for technical support, and the editors for their kind invitation to contribute to these proceedings.

References

1. Nikolic S, Yellin EL, Tamura K, Vetter H, Tamura T, Meisner JS, Frater RWM (1988) Passive properties of canine left ventricle: diastolic stiffness and restoring forces. Circ Res 62: 1210–1222
2. Krueger JW, Forletti D, Wittenberg B (1980) Uniform sarcomere shortening behavior in isolated cardiac muscle cells. J Gen Physiol 76: 587–607
3. Krueger JW, London B, Siciliano G (1988) Separability of relaxation indices in isolated ventriculocytes. In: Clark WA, Decker RS, Borg TK (eds) Biology of isolated adult cardiac myocytes, New York, Elsevier Science, pp 406–409.
4. Clusin WT (1981) The mechanical activity of chick embryonic myocardial cell aggregates. J Physiol (Lond) 320: 149–174
5. LeCarpentier Y, Martin J-L, Claes V, Chambaret J-P, Migus A, Antonetti A, Hatt P-Y (1985) Real-time kinetics of sarcomere relaxation by laser diffraction. Circ Res 56: 331–339
6. Strobeck JE, Bahler AS, Sonnenblick EH (1975) Isotonic relaxation in cardiac muscle. Am J Physiol 229: 646–651
7. Fabiato A, Fabiato F (1976) Dependence of calcium release, tension generation and restoring forces on sarcomere length in skinned and cardiac cells. Eur J Cardiol 4 (Suppl): 13–27
8. Ter Keurs HEDJ, Rijnsburger WH, van Heuningen R (1980) Restoring forces and relaxation in rat cardiac muscle. Eur Heart J 1 (Suppl A): 67–80
9. Harding SE, Viscovo G, Kirby M, Jones MS, Gurden J, Poole-Wilson PA (1988) Contractile response of isolated adult rat and rabbit cardiac myocytes to isoproterenol and calcium. J Mol Cell Cardiol 20: 635–647
10. Grimm AF, Lin H-L, Grimm BR (1983) The pattern of sarcomere lengths through the left ventricular free wall: differences between open and closed chest rats. Basic Res Cardiol 78: 560–570
11. Winegrad S, McClellan G, Robinson T, Lai NP (1976) Variable diastolic compliance and variable Ca sensitivity of the contractile system in cardiac muscle. Eur J Cardiol 4 (Suppl): 41–46
12. Gay WR, Jr, Johnson EA (1967) An anatomical evaluation of the myocardial length-tension diagram. Circ Res 21: 33–43
13. Robinson TF, Factor SM, Sonnenblick EH (1985) The heart as a suction pump. Sci Am 254(6): 84–91
14. Strauer BE (1973) Force-velocity relations of isotonic relaxation in mammalian heart muscle. Am J Physiol 224: 431–434
15. Fish D, Orenstein J, Bloom S (1984) Passive stiffness of isolated cardiac and skeletal myocytes in the hamster. Circ Res 54: 267–276
16. Suga H, Sagawa K (1979) Graphical estimation of ventricular wall force and stress from pressure-volume diagram. Am J Physiol 236: H787–H789
17. Schiereck P, Boom HBK (1978) Left ventricular active stiffness: dependency on tissue and inotropic state. Pflugers Arch 374: 135–143
18. Delay MJ, Vassallo DV, Iwazumi T, Pollack GH (1979) Fast response of cardiac muscle to quick length changes. In: Sugi H, Pollack GH (eds) Cross-bridge mechanisms in muscle contraction. Baltimore, University Park Press, pp 71–83
19. Robinson TF, Factor SM, Capasso JM, Wittenberg BA, Blumenfeld OO, Seifter S (1987) Morphology, composition and function of struts between cardiac myocytes of rat and hamster. Cell Tissue Res 249: 247–255

20. De Beer EL, Grundeman LF, Wilhelm AJ, Caljouw CJ, Klepper D, Schiereck P (1988) Caffeine supresses length dependency of Ca^{2+} sensitivity of skinned striated muscle. Am J Physiol (Cell Physiology 23) 254: C491–C497
21. London B, Krueger J (1986) Contraction in voltage-clamped internally perfused single heart cells. J Gen Physiol 88: 475-505
22. Myers J, Tirosh R, Jacobson RC, Pollack GH (1982) Phase locked-loop measurement of sarcomere length with high time resolution. IEEE Trans Biomed Eng 29: 463–466

Isotonic Segment Dynamics of Isolated Cardiac Muscle

Yasutake Saeki, Tetsuro Horikoshi, and Kouichi Shiozawa[1]

Summary. To determine dynamic properties of contracting cardiac muscle, we studied the isotonic segment length changes of isolated rat right ventricular papillary muscles in response to step tension reductions during Ba^{2+} contracture and the chemically skinned Ca^{2+} -activated state, using a newly developed segment control system. The segment length responses comprised four different phases, which were consistent with those properties of heart muscle in Ba^{2+} contracture previously characterized using step and sinusoidal length changes [4–6, 12]. The first phase is a rapid and minute shortening during the tension reduction. The second phase is a slow further shortening, and the third a slow lengthening. The fourth is an extremely slow and mild shortening. Increasing temperature slightly decreased the amplitude and markedly shortened the duration of segment length response in the second and third phases. Unlike that of the activated muscle, the segment length responses of the resting and the rigor muscle changed in a nearly stepwise fashion. From these results, we conclude that the transient segment length response following a rapid tension reduction is mostly determined by the kinetics of the attachment and detachment of cross-bridges between actin and myosin filaments. Lastly, we discuss the length- and activation-dependence of segment length response, particularly of the velocity of segment shortening, in terms of the cross-bridge kinetics.

Introduction

Dynamic characteristics of contracting muscle have been estimated from the transient response of activated muscle to change in length or tension. The analysis of transient response of isolated cardiac muscle is complicated by the fact that it normally twitches rhythmically, varying its properties periodically. Recently the tension responses following step and sinusoidal length perturbations and the length responses to step tension reductions have been analyzed by several different groups [1–3], including ours [4–7], under the condition in which the tension level has been fixed to a constant value. These perturbation studies indicated that the kinetics of the attachment and detachment of cross-bridges between actin and myosin filaments can be measured and defined under various contractile states, promising new insight into the mechanical and chemical kinetics of heart muscle contraction. However, a difficulty encountered in interpreting the results exclusively in terms of the cross-bridge kinetics is that at least one end of the preparation is damaged by the clamp

[1] Department of Physiology, Tsurumi University, School of Dental Medicine, 2–1–3 Tsurumi, Tsurumi-ku, Yokohama, 230 Japan

used to connect it to a mechanical apparatus. During tension development, the central portion of the muscle shortens by stretching the damaged end [8–11]. To circumvent this, several groups including ours have developed techniques to measure and control segment length [8, 10–12] or sarcomere length [9, 13, 14] in the central region of the preparation. Donald et al. [10] and Shibata et al. [12] inserted two tungsten pins transversely into the papillary muscle, and measured the distance between them as the segment length by the optical method. However, this method seems to require high skill to eliminate some errors associated with the transverse inhomogeneity, which will prevent the two pins from moving in parallel, while the pin insertion may itself produce new damage. On the other hand, the laser diffraction techniques for measuring sarcomere length [9, 13, 14], although not accompanied by the difficulties encountered in the segment length measurement, nevertheless have some disadvantages. This method requires an extremely thin preparation and monitors a very small central region. During contraction, especially under isotonic conditions, the region to be monitored may change significantly. Unless the preparation is homogeneous all over the regions monitored, this method, too, will produce some errors. We [11] have developed a new method which allows us to measure and control muscle length, segment length, and tension. Although our method is based on that of Julian et al. [8], the details of our technique are somewhat different. Briefly, two markers of thin black film were attached in parallel on the central surface of the muscle with silicone grease and the distance between them (i.e., segment length, about 30% of the whole muscle length) was measured by a Hamamatsu Photonics Width Analyzer (C1170) and TV Camera (C1000) coupled to a triocular microscope with a scanning frequently rate of 16.53 kHz and controlled with a servo-actuator.

For clear identification of cross-bridge behavior, therefore, using the newly developed apparatus we analyzed the isotonic segment length changes of isolated rat right ventricular papillary muscle (2.1–2.9 mm long and 250–390 μm in diameter). The response to step tension reductions at two different initial lengths SL_{max} and ML_{max} and at a temperature of 22°C was measured, paying close attention to the tension-velocity relationship and the length-tension relationship, during barium (Ba^{2+}) contracture and in the saponin-treated calcium (Ca^{2+})-activated state.

Isotonic Segment Dynamics in Ba^{2+} Contracture

To obtain a stable Ba^{2+} contracture with no spontaneous twitch, the twitching muscle was first incubated in Ca^{2+}-free Tyrode's solution, as has been previously reported [4]. In Fig. 1, the *arrow* on the *left* indicates the onset of this procedure. After twitch contraction had disappeared completely, an amount of $BaCl_2$ was added to the Ca^{2+}-free Tyrode's solution (*arrow* in *center*, Fig. 1) so that the concentration of Ba^{2+} would be 0.3 m*M*. The muscle then lapsed into a state of stable contracture. After the stable contracture was obtained, the servo control system was switched from muscle length isometric to isotonic mode (*arrow* on *right*, Fig. 1) at the muscle length of ML_{max}, the length at which the developed twitch tension was maximal. Then, the muscle was released stepwise from the steady tension (Tc) to new steady tensions of varying magnitude (Tr). As clearly seen in Fig. 1, the segment showed

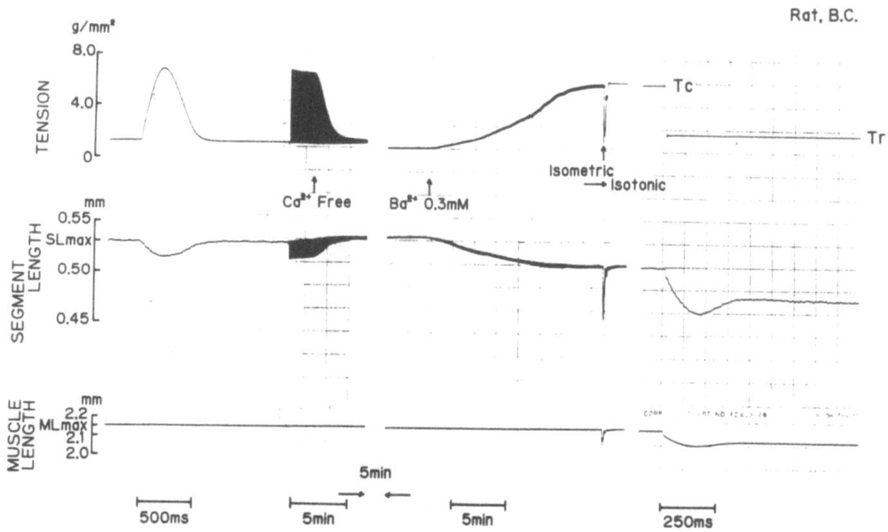

Fig. 1. Process to induce Ba^{2+} contracture (*B.C.*), and the segment length change (*middle trace*) and the muscle length change (*lower trace*) in response to a step tension reduction (*upper trace*) from the steady contracture tension (*Tc*) to a new steady tension (*Tr*)

marked internal shortening from the segment length of SL_{max} (the segment length at which the muscle length was ML_{max} in the resting state) during the tension development (5%–8% of segment length shortening at 0.3 m*M* Ba^{2+}) despite the fact that the muscle length was held constant at ML_{max}, as has been already reported [8–10]. The segment length response (*middle trace* of *right extreme*, Fig. 1) comprised the four different phases, as has been observed in the muscle length responses to step tension reduction [6] and can be expected from the tension response to a step change in length in the previous studies [1, 5]. The first phase is a prompt shortening during the tension reduction. The second phase is a slow further shortening, the peak of which is shorter than the new quasi-steady segment length at the new tension (Tr), and the third phase is a slow lengthening. The fourth phase is an extremely slow and mild shortening toward a new steady segment length under the new tension (Tr). The third phase was often followed by a damped oscillation in length, as shown in Fig. 2. These profiles and the time course of the segment length transient were identical to those of the muscle length transient (*bottom trace* of *right extreme*, Fig. 1). However, the length change relative to the initial length was more than twofold greater in the segment than in the muscle except for the first phase. These results are in agreement with the findings in the previous sinusoidal length perturbation experiment [12] that the frequency spectrum of dynamic stiffness is identical in its shape but different in amplitude between the central segment and the whole muscle; the results imply that the damaged ends introduce significant errors into the amplitude, but do not change the time course of the segment length transient, probably contributing to the segment length transient as a pure series elastic component. The percentage of the segment length shortening in the first phase was less than 1% of initial segment length and much smaller than that of the muscle length shortening in the

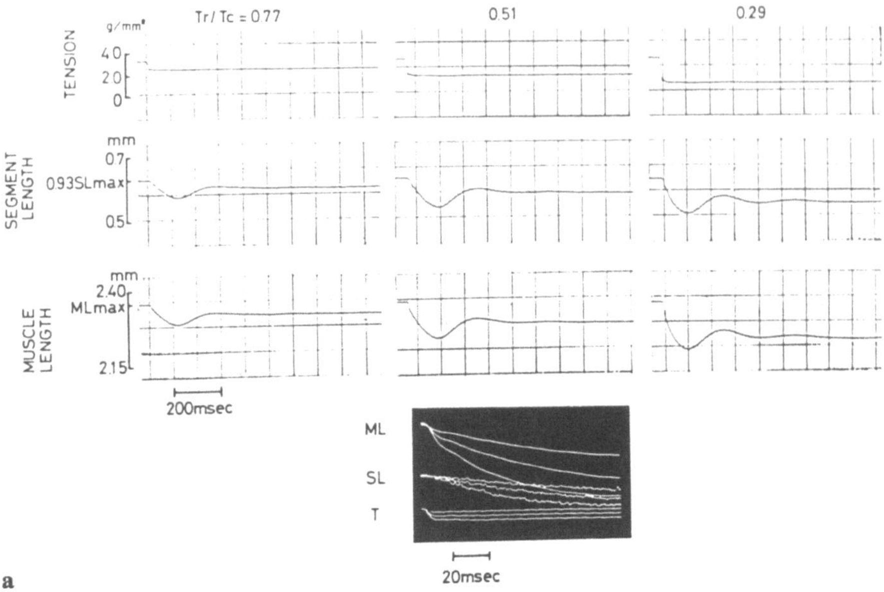

a

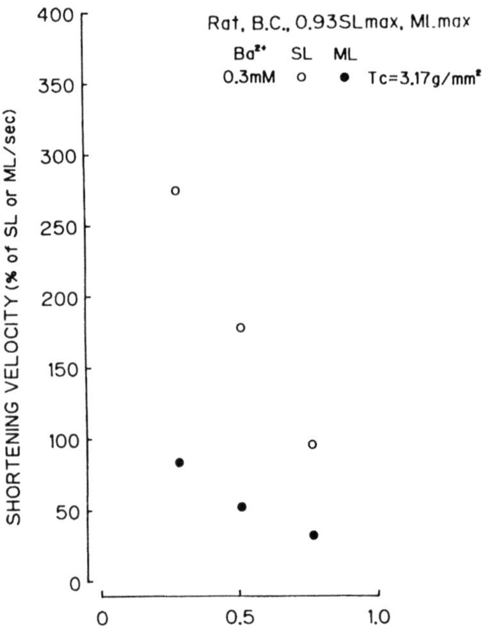

b

Fig. 2a. The segment length responses (*middle pen-recorder trace*) and the muscle length responses (*lower pen-recorder trace*) in response to three different amplitude of step tension reduction, Tr/Tc = 0.77, 0.51, and 0.29 (*upper pen-recorder trace*). The initial parts of these traces are shown in the bottom traces recorded by an oscilloscope with a faster time scale. **b** Relative tension-shortening velocity relationships measured from the segment (○) and muscle (●) length changes between 30 and 50 ms after the onset of tension reductions at the fixed initial length (segment length = 0.93 SLmax, muscle length = MLmax). *B.C.*, barium contacture; *Tc*, steady contracture tension; *Tr*, steady reduced tension first phase (about 3% of initial muscle length), as can be estimated from the actual segment traces in Figs. 2 and 4 which have virtually no clear first phase. This finding is quite consistent with the results derived from studies on sarcomere dynamics [9, 13]. Previous investigators who studied tetanized cat right ventricular papillary muscle [2] and EGTA-treated ventricular papillary muscle of the guinea pig [3] were aware of the presence of an oscillatory length response to similar step tension reduction, and recognized it to be an active characteristic rather than a passive property. However, they did not study the long-term length transient systematically. Instead, they paid attention primarily to short-term isotonic shortening over the initial 80 ms to obtain the tension-velocity relationship.

Increasing temperature decreased the amplitude and markedly shortened the duration of the length response in the second and third phase. Unlike that of the activated muscle, the length responses of the resting and of the rigor muscle changed in a nearly stepwise fashion.

Fig. 2a shows the segment length responses (*middle traces*) and the muscle length responses (*lower traces*) for three different amplitudes of step tension reduction, Tr/Tc = 0.77, 0.51, and 0.29 (*upper traces*) at the initial segment length of 0.93 SL_{max} (or ML_{max}). The amplitude of length transient was increased and the time course shortened slightly with increasing amplitude of tension reduction. Fig. 2b shows the tension-velocity relationship in the segment (○) and the muscle (●) determined from the records shown in Fig. 2a. The shortening velocities were measured from the length changes between 30 and 50 ms after the onset of tension reduction in the oscilloscope traces shown in the *bottom* of Fig. 2a, and plotted against the tension (Tr) relative to the initial tension (Tc), Tr/Tc. Note that the shortening velocity of the segment is more than double that of the muscle. This clearly indicates that the damaged ends introduce significant errors into the measurement of the isotonic length transient, explaining why the velocity of muscle shortening reported in previous studies [2, 3, 10] is quite low compared with the velocity of sarcomere shortening [9, 13, 14].

Fig. 3 represents an example of the effects of amplitude of tension reduction on the segment shortening velocity at two different initial segment lengths, SL_{max} (see Fig. 1) and 0.95 SL_{max} (or ML_{max}). As clearly seen in this figure, the velocity was lower at longer initial segment length, though the extrapolated value of unloaded shortening velocity appears to be the same (as discussed later). These features were commonly observed in each muscle studied. The difference in the velocity of shortening at two different initial segment lengths may partly arise from the presence of the length changes of the parallel element. At longer segment lengths where the tension is partly borne by the parallel element, as with shortening in the second phase, the tension will be transferred from the parallel element to the contractile element. The opposite seems to be true for the lengthening in the third phase. At shorter segment lengths, other factors including internal resistance and restoring force, factors well reviewed previously [15], appear to become important determinants of transient length response (i.e., the shortening velocity). The contribution of the deactivation by shortening [15] to the transient length response would not be significant, since Bodem and Sonnenblick [16] showed that both tetanization and lowered temperature (22°C) eliminated or substantially reduced deactivation in twitching rabbit papillary muscle. Forman et al. [2] in tetanized cat right ventricular papillary muscle

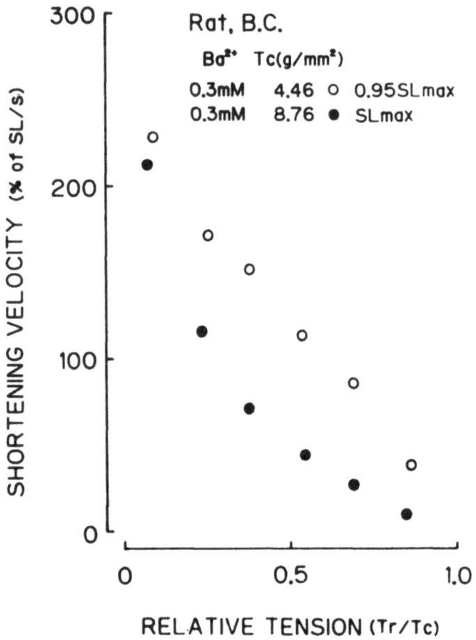

Fig. 3. Relative tension-shortening velocity relationships measured from the segment length changes between 30 and 50 ms after the onset of tension reductions at two different initial segment lengths, SLmax and 0.95 SLmax (or MLmax). *SL*, segment length; *Tc*, steady contracture tension; *Tr*, steady reduced tension

and Pollack and Krueger [9] in twitching rat papillary muscle reported that the extrapolated maximum velocities and isometric tensions diminished together in almost direct proportion to muscle shortening (over the muscle length range from 75%-90% of ML_{max} in the Forman et al. study, and over the sarcomere length range from 1.6-2.3 μm in the Pollack and Krueger study). According to Daniels et al. [14], however, in twitching rat right ventricular trabeculae the maximal velocity of sarcomere shortening was constant at sarcomere lengths between slack length (1.85 μm) and 2.3 μm where the developed tension changed from 60% to 100% of the maximal; the maximum velocity decreased with sarcomere length below 1.85 μm to 0 at 1.6 μm.

When the initial segment length was increased, the transient segment response decreased its amplitude and lengthened its time course. These results seem to indicate the length-dependent deceleration of cross-bridge kinetics, being contrary to those of the previous length perturbation experiments [1, 5, 12] that the kinetics of cross-bridge are independent of the initial length. However, one cannot exclude the possibility that nonlinear passive properties, which do not appear on the tension responses to the small length perturbations (less than 1% of muscle length) in the previous studies [1, 5, 12], come into play in the present isotonic experiment which involves large changes in length.

Isotonic Segment Dynamics in Chemically Skinned Muscle

To obtain a chemically skinned preparation, the muscle was first incubated in Ca^{2+}-free Tyrode's solution and then treated for one hour with saponin (250 μg/ml) dissolved in the relaxing ATP-salt solution. Prior to activation of muscle with Ca^{2+}-containing ATP-salt solution, the bath was perfused for about 10 min with the relaxing ATP-salt solution with no saponin. Other experimental protocols were exactly the same as in the experiment for the muscle in Ba^{2+} contracture.

Fig. 4a shows an example of effects of four different amplitudes of tension reduction on the segment length at the initial segment length of SL_{max} (see Fig. 1) in the chemically skinned muscle activated with pCa (pCa = -log [Ca^{2+}]) 5.2. The profiles of segment length response were quite similar to those of the muscle in Ba^{2+} contracture, indicating that these oscillatory segment length responses are not peculiar characteristics observed only in muscle activated with Ba^{2+}. This is quite consistent with our previous finding [7] that the transient tension responses for step length perturbations of Ba^{2+}-activated glycerinated cardiac muscle are similar to those of Ca^{2+}-activated glycerinated cardiac muscle.

The significant differences between the segment and muscle length response and the length dependence of the segment length transients, which were observed in the muscle in Ba^{2+} contracture and described in the previous section, were also observed in this chemically skinned preparation.

Shown in Fig. 4b are the effects of amplitude of tension reduction on the segment shortening velocity at three different activation levels, pCa 5.4, 5.2, and 4.4 (saturation concentration) at the initial segment length of SL_{max} in a single preparation from which the data in Fig. 4a were obtained. The velocity of shortening was significantly increased with increasing activation level from pCa 5.4 to 5.2, but tended to saturate with a further increase in activation level to pCa 4.4, though the developed tension still continued to increase, agreeing qualitatively with the results of Daniels et al. [14] that unloaded sarcomere shortening velocity reaches a maximum value with an amount of Ca^{2+} at which the developed tension has obtained only about 50% of its maximum value. These characteristic features were commonly observed in each muscle studied, though the data scattered rather widely from muscle to muscle. Maughan et al. [3] in EGTA-treated ventricular papillary muscle of the guinea pig and De Clerck et al. [17] in single rat ventricular cells activated by Ca^{2+} released by iontophoresis in an EGTA-buffered solution reported that both the shortening velocity and the developed tension depend on the level of Ca^{2+} activation over a similar range of Ca^{2+} concentration. However, their technique is fundamentally a compromise in that they measured the muscle length rather than the segment length, thus the relevance of the data can be questioned.

Shown in Fig. 4c are the effects of tension reduction on the amplitude of shortening from the initial segment length to the quasi-steady segment length at which the transient respose stopped (see *inset*). All the amplitude data were obtained from the segment length response in the muscle from which the data in Fig. 4b were obtained, and plotted against the tension reductions in the same way as in Fig. 4b. As clearly seen in this figure, the amplitude was increased with increasing activation level. This means that the higher the activation level, the greater the segment length shortening which is required to induce a given amount of relative tension reduction. This

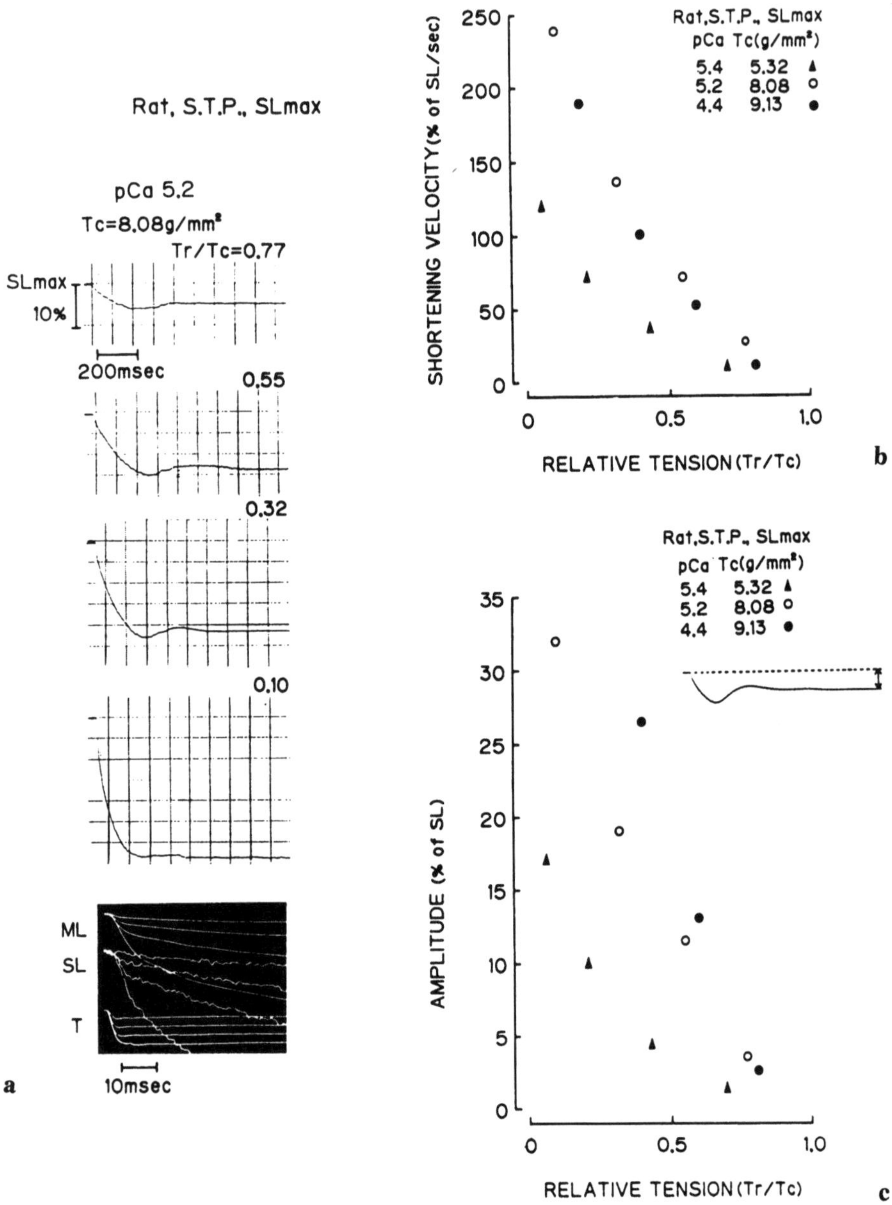

Fig. 4a. The segment length changes in response to four different amplitudes of step tension reduction, Tr/Tc = 0.77, 0.55, 0.32, and 0.10. The initial parts of these traces are shown in the bottom traces recorded by an oscilloscope with a faster time scale. **b** Effects of three different activation levels (pCa 5.4, 5.2, and 4.4) on the relative tension-shortening velocity relationship measured from the segment length changes between 30 and 50 ms after the onset of tension reductions at the initial segment length of SLmax in a single preparation from which the data shown in Fig. 4a were obtained. **c** Effects of three different activation levels (pCa 5.4, 5.2, and 4.4) on the amplitudes from the initial segment length to the length at which the transient segment length changes stopped. All data were obtained from the segment length responses in the muscle from which the data Fig. 4b were obtained. *STP*, saponim treated preparation; *SL*, segment length; *Tc*, steady contracture tension; *Tr*, steady reduced tension

finding is in agreement with the results of Maughan et al. [3] and seems to confirm the well-known activation-dependent length-tension relation [18].

Although the data are not shown, increasing activation level markedly increased the amplitude and lengthened the time course of the segment length response in the second and third phases, particularly at relatively large tension reductions, as has been reported in our previous study [19]. These results seem to suggest that the activation-dependent increase of shortening velocity in cardiac muscle is due to the co-operative behavior of more cross-bridges, as in the case of the shortening activation [20], which gives rise to positive feedback during shortening and thus lengthens the time course of the segment length transient (i.e., further shortening). In previous studies [3, 17], in contrast, the acceleration of cross-bridge reaction kinetics has been proposed. Since the time course of tension transient in response to small step length changes has been reported to be independent of activation level [1, 7], the rate of cross-bridge cycling appear to be insensitive to the changes in activation level under quasi-isometric conditions where the shortening activation is likely to be less prominent.

All the activation level dependences of the segment length response observed in chemically skinned preparations, described here, were also observed in the muscle in Ba^{2+} contracture.

References

1. Steiger GJ (1971) Stretch activation and myogenic oscillation of isolated contractile structures of heart muscle. Pflugers Arch 330: 347–361
2. Forman R, Ford LE, Sonnenblick EH (1972) Effect of muscle length on the force-velocity relationship of tetanized cardiac muscle. Circ Res 31: 195–206
3. Maughan DW, Low FS, Alpart NR (1978) Isometric force development, isotonic shortening, and elasticity measurements from Ca^{2+}-activated ventricular muscle of the guinea pig. J Gen Physiol 71: 431–451
4. Saeki Y, Sagawa K, Suga H (1978) Dynamic stiffness of cat heart muscle in Ba^{2+}-induced contracture. Circ Res 42: 324–333
5. Saeki Y, Sagawa K, Suga H (1980) Transient tension responses of heart muscle in Ba^{2+} contracture to step length changes. Am J Physiol 238: H340–H347
6. Saeki Y, Shibata T, Yanagisawa K (1984) Isotonic length transient of cat heart muscle in Ba^{2+}-induced contracture. Am J Physiol 247: H548–H555
7. Saeki Y, Kato C, Horikoshi T, Yanagisawa K (1984) Effects of Ba^{2+} on the mechanical properties of glycerinated heart muscle. Pflugers Arch 400: 235–240
8. Julian FJ, Sollins MR, Moss RL (1976) Absence of a plateau in length-tension relationship of rabbit papillary muscle when internal shortening is prevented. Nature 260: 340–342
9. Pollack GH, Kreuger JW (1976) Sarcomere dynamics in infarct cardiac muscle. Eur J Cardiol 4: 53–65
10. Donald TC, Reeves DNS, Reeves RC, Walker AA, Hefner LL (1980) Effect of damaged ends in papillary muscle preparations. Am J Physiol 238: H14–H23
11. Saeki Y, Totsuka T, Kato C, Yanagisawa K (1986) Length-tension relation of cat heart muscle studied by a segment-control method. Heart Vessels 2: 208–212
12. Shibata T, Hunter WC, Yang A, Sagawa K (1987) Dynamic stiffness measured in central segment of excised rabbit papillary muscles during barium contracture. Circ Res 60: 756–769
13. ter Keurs HEDJ (1983) Calcium and contractivity. In: Drake-Holland AJ, Noble MIM (eds) Cardiac metabolism. Wiley, pp 73–99

14. Daniels M, Noble MIM, ter Keurs HEDJ, Wohlfart B (1984) Velocity of sarcomere shortening in rat cardiac muscle: relationship to force, sarcomere length, calcium and time. J Physiol 355: 367-381
15. Alpert NR, Hamrell BB, Mulieri LA (1979) Heart muscle mechanics. Annu Rev Physiol 41: 521-537
16. Bodem R, Sonnenblick EH (1974) Deactivation of contraction by quick releases in the isolated papillary muscle of the cat. Effects of lever damping, caffeine and tetanization. Circ Res 34: 212-225
17. De Clerck NM, Claes VA, Brutsaert DL (1977) Force velocity relations of single cardiac muscle cells. J Gen Physiol 69: 221-241
18. Jewell BR (1977) A re-examination of the influence of muscle length on myocardial performance. Circ Res 40: 221-230
19. Saeki Y, Totsuka T, Kato C, Yanagisawa K (1987) The influence of activation level on the tension-velocity relationship in cardiac muscle segment. Automedica 9: 131
20. Bozler E (1975) Mechanical control of the time-course of contraction of the frog heart. J Gen Physiol 65: 329-344

Pathology of the Cardiac Collagen Matrix: Mechanical and Functional Effects

KAREN H. VAN HOEVEN and STEPHEN M. FACTOR[1]

Summary. The normal structure of the cardiac collagen matrix includes intercellular and myocyte-to-capillary struts which provide short bridges important in diastolic alignment and ventricular filling, collagen weaves which maintain structural integrity, coiled perimysial fibers which coordinate function in the beat-to-beat cardiac cycle, and the epimysium, surrounding the muscle to provide resistance to overstretch.

The pathologic changes in the collagen matrix observed with various cardiomyopathic processes have been discussed. Early changes in the struts and weaves contribute to the diastolic dysfunctions characteristic of these cardiomyopathies. When myocardial ischemia occurs in the absence of necrosis, early focal loss of connective tissue components is observed. Return of normal cardiac function following an ischemic insult is postulated to result after repair of the connective tissue matrix.

The Syrian hamster provides an interesting model to study connective tissue changes associated with a chronic cardiomyopathic process. Cardiac dysfunction out of proportion to the myocellular alterations is noted. Losses of connective tissue elements surrounding areas of necrosis are thought to contribute to the functional aberrations. Further studies to elucidate the pathogenetic mechanisms involved in connective tissue loss in this animal are necessary, and may be helpful in elucidating the 'final common pathway' of connective tissue injury observed in virtually every chronic cardiac pathologic process.

Introduction

Within the last fifteen years, there has been a veritable explosion of interest in the role of connective tissue in the normal function of the heart and its contribution to pathologic changes, both structural and physiologic. Rather than a passive skeletal framework, connective tissue acts in a dynamic interplay with myocytes in coordinating contractile activities, maintaining structural integrity, and in altering function following injury to the heart.

The following will review briefly the structural and functional roles of connective tissue in the normal heart. Then the focus will turn to several pathologic alterations in abnormal cardiac tissue.

[1] Department of Pathology, Albert Einstein College of Medicine, 1300 Morris Park Avenue, Bronx, NY 10461, USA

The Normal Cardiac Skeleton

The traditional model of the connective tissue skeleton involves an oversimplified compartmentalization of the myocytes: the entire muscle wrapped in an envelope of the epimysium, groups of myocytes packaged in the perimysium, and individual cells separated by the endomysium. Ultrastructural studies, using both scanning electron microscopy (SEM) and transmission electron microscopy (TEM) have revealed this schema to be much more complex.

Between adjacent myocytes, short bridges or 'struts' (Fig. 1) composed predominantly of collagen type I fibrils [1] are present which anchor into the basal laminae, just lateral and almost perpendicular to the Z bands [2, 3]. During diastole when ventricular volume increases and the myocellular cross section decreases [4], the intercellular struts appear taut. In systole, with myocellular shortening, these struts appear slack. Rather than simply separate cells, these struts appear to tether the cells, preventing intercellular disengagement and assuring proper diastolic alignment [3]. Furthermore, there is evidence to suggest that struts store systolic energy, transferring it to diastole to facilitate ventricular expansion, creating a negative pressure which acts as a suction pump to pull atrial blood into the ventricles [5]. These struts not only serve as intercellular bridges, but also interconnect myocytes and capillaries. While the insertion into the basement membrane of the capillary approaches a right angle, the struts insert into myocellular basement laminae at a steeper angle [2]. Unlike the intercellular struts, these myocyte-to-capillary struts appear slack during diastole and become taut during systole. Physiologically, these struts are thought to act in preventing capillary collapse during systole, thus allowing forward blood flow to proceed despite high pressures [3].

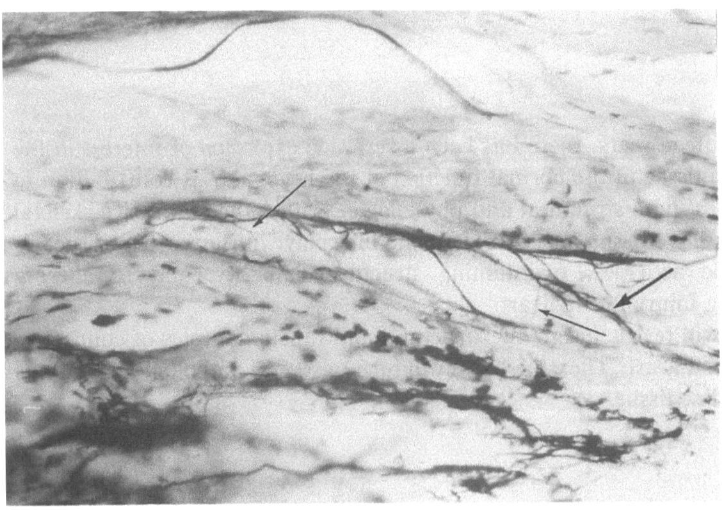

Fig. 1. A silver impregnated section of myocardium fixed under stretch reveals numerous intercellular struts crossing the interstitium to connect the lateral surfaces of two myocytes. Both relatively thick struts (*thick arrows*) and very fine struts (*thin arrows*) can be seen with this technique (Del Rio y Hortega stain, × 250)

Interconnecting portions of three or more myocytes are bundles of collagen fibers, termed weaves, which attach to the basement laminae of the myocytes by short struts [2] and to adjacent weaves. By chemically blocking the formation of the weave in young rats, the heart becomes deformed into a globular shape. Thus, the weave is thought to contribute to gross structural maintenance of the myocardium [3].

Ultrastructural and histological studies of perimysial fibers have revealed that they are arranged in a coiled fashion, most often parallel to the long axis of two or more myocytes (Fig. 2). They interconnect to each other, forming an anastomotic network which anchors at one end into the fibrous annulus near the cardiac base. At the apex, a knot-like web of fibers forms the second anchoring point. With increasing intraventricular pressures, the perimysial fibers progressively straighten, ultimately with disruption and fraying at very high pressure. Perimysial fibers are thought to play a role in maintaining structural coordination within the normal cardiac cycle, and preventing overstretch particularly during times of elevated end-diastolic pressures [6]. They may also be critical for energy storage and ventricular recoil after chamber emptying.

Lastly, collagen fibers of the epimysium surround a muscle. In the papillary muscle, these fibers are arranged in a criss-cross almost checkerboard fashion at slack length, but when passively stretched, they align themselves parallel to the long axis of the muscle, creating resistance to further stretch [7].

Interstitial connective tissue amounts to 1.1% [8] to 2% [9] of the normal left ventricular wall mass. Collagen content, measured by its hydroxyproline constituent, does not increase with aging through the sixth decade in human hearts without cardiac disease [10].

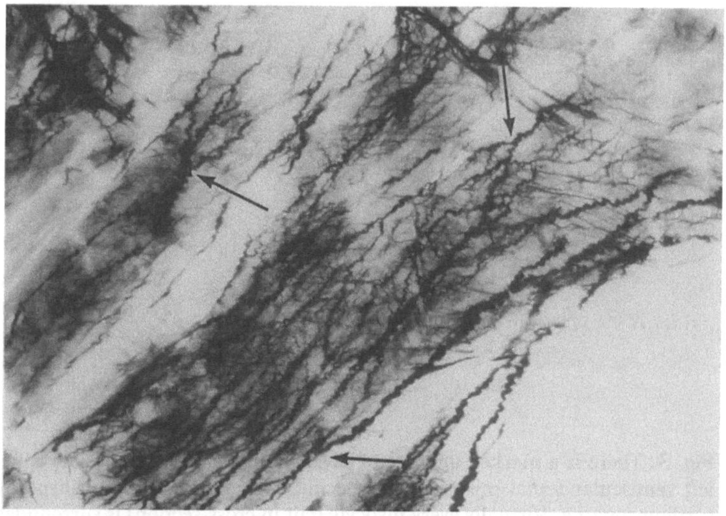

Fig. 2. A rat left ventricular papillary muscle reveals multiple longitudinally oriented coiled perimysial fibers running parallel to myocytes. Branching (*arrows*) can be seen throughout the tissue. The attachments of these coils to the myocytes and to other connective tissue elements cannot be appreciated at this magnification (Del Rio y Hortega stain, × 100)

Pathology of the Cardiac Collagen Matrix in Hypertrophy

It has long been recognized that cardiac tissue has only finite responses to chronic injury. Myocellular hypertrophy and fibrosis are the characteristic histologic hallmarks of chronic heart disease with resultant gross enlargement of the heart, often with the functional residual of clinical heart failure.

In assessing myocellular alterations, *muscle fiber orientation* is *maintained* in the hypertrophied pressure-overloaded left ventricle [11]. Furthermore, in patients with aortic stenosis and/or insufficiency, myocardial stiffness is not 'influenced by left ventricular muscle mass or muscle fiber size, but is increased in the presence of massive left ventricular interstitial fibrosis' [9].

Increases beyond normal amounts of myocardial fibrosis have been quantitatively confirmed in almost every cardiomyopathy, including congestive cardiomyopathy [9], hypertrophic cardiomypathy (Fig. 3), and cardiomyopathies associated with hypertension [8], aortic stenosis, and aortic insufficiency [9]. At the ultrastructural level, some hearts of hypertensive patients show an increase in the diameter of the collagen bundles which form the struts and weaves. Other hearts reveal replacement of the normal weave by sheets of collagen, with the functional result of increased stiffness in the ventricle [12].

Of particular interest as a structural and functional correlate are recent noninvasive assessments of early pathophysiologic left ventricular alterations in hypertension. Notably, before any evidence of systolic impairment in hypertensive

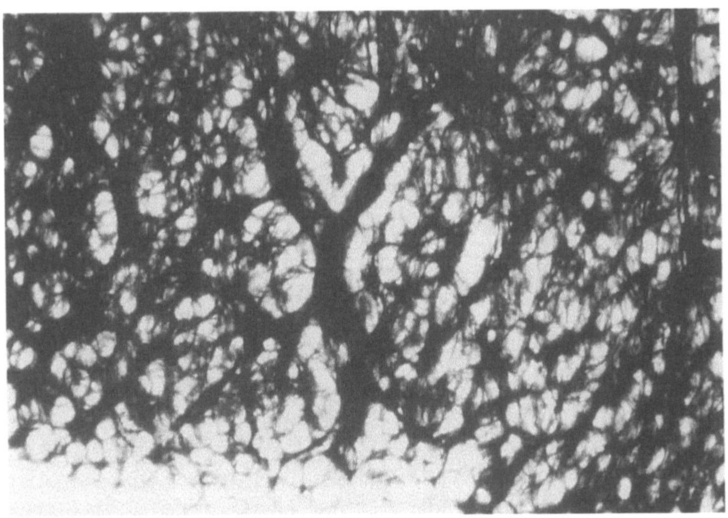

Fig. 3. There is a marked increase of silver-impregnated endomysial connective tissue in this left ventricular septal myomectomy specimen from a patient with hypertrophic, obstructive cardiomyopathy. Virtually every myocyte (cut in cross-section) is surrounded by silver positive connective tissue. This stainable material consists of increased interstitial fibrosis also demonstrable with trichrome stain, as well as matrix connective tissue. The pericellular fibrous tissue may contribute to the diastolic dysfunction characteristic of this condition, by causing a "stiff" poorly compliant myocardium (Del Rio y Hortega stain, × 100)

patients, left ventricular filling time is measurably decreased. An inverse correlation between left ventricular mass and peak left ventricular filling time has been noted [13]. As stated earlier, the primary functions of struts are thought to be the maintenance of *diastolic* alignment and active participation in left ventricular filling.

Despite the plethora of observations regarding the proliferation of collagen in various cardiomyopathies, the exact pathogenetic mechanisms for this fibrotic response are still unclear. Speculations abound; what is clear, however, is that increases in the collagen matrix are the 'final common pathway' and that a mechanism to explain its proliferation must be operative in virtually every cardiomyopathic process, whatever its initial provocation.

An Animal Model for Cardiomyopathy: The Syrian Hamster

The Syrian hamster provides a model of an inherited progressive myopathy which is characterized by multifocal cardiac myocytolysis and myocellular necrosis surrounded by histologically normal myocardium in animals older than one month of age. Later, larger, more confluent areas with associated fibrosis and calcification are noted. Congestive heart failure and death occur within one year. The pathogenesis of the cardiac lesions and failure has not been elucidated to date. What has been observed, however, is a pathophysiologic dysfunction disproportionate to the observed morphologic alterations [14, 15].

The connective tissue of the heart of the cardiomyopathic Syrian hamster has been investigated recently [15]. In all animals greater than 90 days of age, a marked loss of connective tissue elements is noted in a circumferential distribution surrounding foci of necrosis, creating a halo effect when the tissue is stained exclusively for connective tissue elements (Fig. 4). In this same distribution, the structural integrity of the myocellular elements is preserved. Within the halo zone, a rare rather thickened connective tissue fiber oriented parallel to the endocardium is noted to bridge the necrotic zone to the outer zone where connective tissue elements are still preserved. Among animals in the later stages of disease, the connective tissue abnormalities are more confluent, and transmural in certain areas. Quantitatively, the connective tissue abnormalities more than double the volume of myocardium affected by the cardiomyopathic process. The loss of connective tissue is thought to result in myofiber slippage in the perinecrotic zone, and ultimately to contribute to wall thinning.

Whether the connective tissue abnormalities observed in the cardiomyopathic Syrian hamster are a primary or secondary pathologic effect is still not known. What is clear is that the connective tissue alterations in association with myocellular pathology interact dynamically to result in significant dysfunction and ultimately in the early cardiac demise of these animals.

Pathology of the Cardiac Collagen Matrix in Ischemia

As discussed earlier, the myocardium has only a limited histopathologic response to chronic injury, i.e., hypertrophy and fibrosis. Ischemic insult is no exception. In autopsies from patients who have suffered remote transmural myocardial infarction,

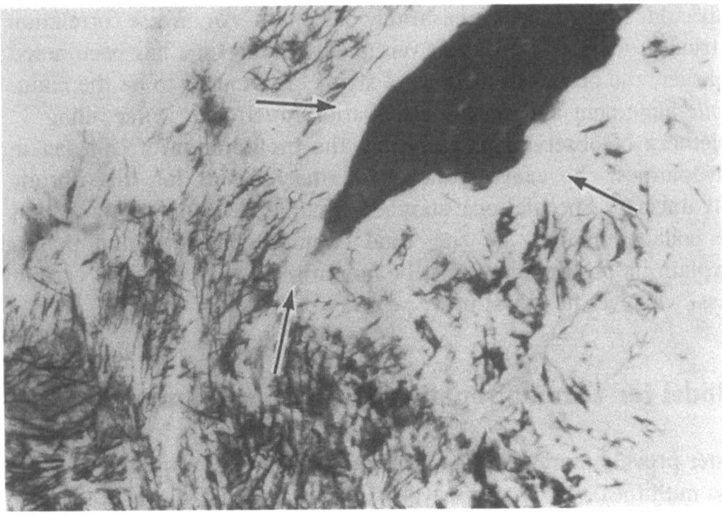

Fig. 4. A section of myocardium impregnated with silver from a cardiomyopathic Syrian hamster. The dark solid mass is a calcified scar due to focal myocardial necrosis. Immediately surrounding the calcified zone, viable myocytes are present (not visible with this technique); however, the silver impregnation demonstrates a clear halo (*arrows*) with the normal myocardium relatively devoid of matrix connective tissue. The loss of the connective tissue may contribute to systolic contractile dysfunction characteristic of this animal model (Del Rio y Hortega stain, × 40)

the gross appearance of a dense fibrous scar in a hypertrophied heart is the typical finding. Similarly, in hearts of patients with severe large vessel arteriosclerotic coronary artery disease in the absence of infarction, an increased interstitial fibrosis is noted, particularly in the subendocardial regions. Focal myocellular 'dropout' often occurs with the subsequent histologic hallmark of replacement fibrosis. Ultrastructural studies have revealed that the cells encased in dense collagen are devoid of intercellular struts. Hence, it is believed that distolic function is compromised when subendocardial fibrosis is present [12].

Ischemic cardiomyopathies in patients with severe coronary artery disease, congestive heart failure, and markedly dilated ventricles in the absence of significant scar formation have been attributed to the long-term consequences of repeated ischemic insults. Recently, a great deal of interest has turned to the pathology of the stunned myocardium: the effects of a brief ischemic insult which may cause symptomatology (e.g., angina) but not necrosis [16].

Ultrastructural studies (Fig. 5) of the myocardium of dogs subjected to repeated brief left anterior descending coronary artery occlusions followed by reperfusion have been performed [17]. Although no hemodynamic changes occur, a significant regional dysfunction results, manifest by decreased systolic shortening and increased systolic bulging of the ischemic area with resultant focal increased myocardial compliance. Ultrastructurally, reversible myocellular ischemic changes are present (for example, chromatin clumping and glycogen depletion), but there is no evidence of irreversible cell death. Most significantly, profound alterations in the collagen

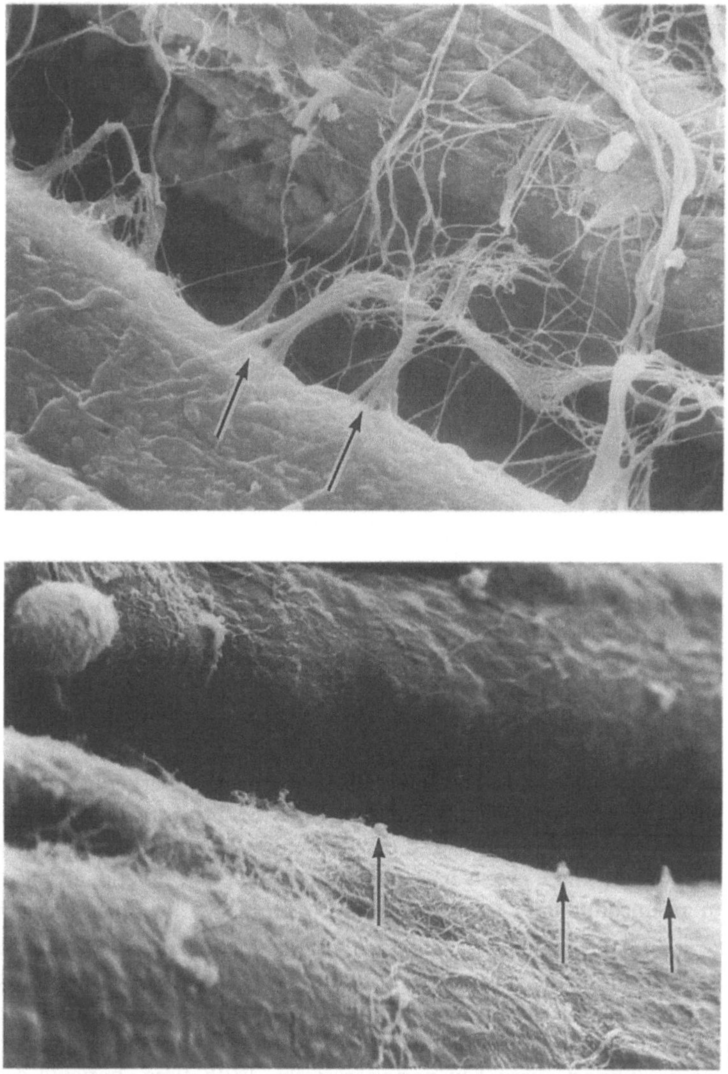

a

b

Fig. 5. a A scanning electron micrograph from a normal canine heart reveals complex connective tissue struts extending between and tethering two myocytes. The struts can be observed to attach to the sarcolemmal surface by root-like projections (*arrows*) (original magnification, × 1980). **b** A scanning electron micrograph from a canine heart after producing 12 repetitive episodes of ischemia and reperfusion with the development of segmental dysfunction and wall thinning characteristic of stunned myocardium. Note that the interstitial space shows no evidence of connective tissue. No struts are present, but small stubs of connective tissue (*arrows*) can be seen on the sarcolemmal surface as remnants of strut attachment sites following collagen degradation. The myocytes are viable, but the connective tissue breakdown may partially account for the loss of segmental contractile function and mural thinning due to slippage and untethering of myocardial cells (original magnification, × 1980).

c

Fig. 5 *(Continued)*
c A silver-impregnated section from a stunned canine heart reveals focal areas of matrix connective tissue loss in the subendocardium. Such silver negative zones correspond to the connective tissue degradation observed with scanning electron microscopy (Del Rio y Hortega stain, × 40)

matrix are observed. The intercellular struts are markedly decreased in number, with the presence of stub-like structures resulting from breakage. The collagen weave is much less dense, and even absent in certain areas. Perimysial coils become unravelled. These events have been interpreted in a functional sense as contributing to an uncoupling of force generation, with resultant impact on performance.

Several days usually elapse which allows functional recovery to occur after myocardial stunning, and this may represent the time course necessary for connective tissue repair [17]; for example, following cardiac surgery, when a stunned myocardium with functional deficits requires mechanical (e.g., intraaortic balloon pump) or inotropic support every day postoperatively. With time, patients can be effectively weaned from these supports [16].

The connective tissue of the procine heart has been studied to investigate acute changes induced by prolonged myocardial ischemia resulting in myocellular necrosis [18]. Twenty minutes following occlusions of the left anterior descending coronary artery, irregularities of the endocardial collagen are observed. After forty minute occlusions, irregularities in the collagen struts bridging capillaries to myocytes and in all of the collagen layers are noted. After two hours, the collagen fibrils lose their anchorage to the basement membrane, and individual perimysial fibers can not be discerned at the ultrastructural level.

As a functional correlate, it is interesting to note that left ventricular compliance decreases in the first hour after infarction [19], and may in part be attributable to the early loss of the mechanical coupling function provided by the cardiac collagen matrix.

Acknowledgments. We sincerely appreciate the technical efforts of Ms. Regina Dominitz who performed the silver impregnation studies. We thank Dr. Meng-jia Zhao for graciously providing the two scanning electron micrographs.

References

1. Robinson TF, Factor SM, Capasso JM, Wittenberg BA, Blumenfeld OO, Seifter S (1977) Morphology, composition, and function of struts between cardiac myocytes of rat and hamster. Cell Tissue Res 249: 247–255
2. Caulfield JB, Borg TK (1979) The collagen network of the heart. Lab Invest 40: 364–372
3. Borg TK, Caulfield JB (1981) The collagen matrix of the heart. Fed Proc 40: 2037–2041
4. Spotnitz HM, Spotnitz WD, Cottrell TS, Spiro D, Sonnenblick EH (1974) Cellular basis for volume related wall thickness changes in the rat left ventricle. J Mol Cell Cardiol 6: 317–331
5. Robinson TF, Factor SM, Sonnenblick EH (1986) The heart as a suction pump. Sci Am 254: 84–91
6. Factor SM, Flomenbaum M, Zhao MJ, Eng C, Robinson TF (1988) The effects of acutely increased ventricular cavity pressure on intrinsic myocardial connective tissue. J Am Coll Cardiol 12: 1582–1589
7. Robinson TF, Cohen-Gould L, Factor SM (1983) Skeletal framework of mammalian heart muscle. Arrangement of inter- and pericellular connective tissue structures. Lab Invest 49: 482–498
8. Tanaka M, Fujiwara H, Onodera T, Wu DJ, Hamashima Y, Kawai C (1986) Quantitative analysis of myocardial fibrosis in normals, hypertensive hearts, and hypertrophic cardiomyopathy. Br Heart J 55: 575–581
9. Hess OM, Schneider J, Koch R, Bamert C, Grimm J, Krayenbuehl HP (1981) Diastolic function and myocardial structure in patients with myocardial hypertrophy. Special reference to normalized viscoelastic data. Circulation 63: 360–371
10. Wegelius O, von Knorring J (1964) The hydroxyproline and hexosamine content in human myocardium at different ages. Acta Med Scand 412 (Suppl): 233–237
11. Pearlman ES, Weber KT, Janicki JS, Pietra GG, Fishman AP (1982) Muscle fiber orientation and connective tissue content in the hypertrophied human heart. Lab Invest 46: 158–164
12. Caulfield JB: Alterations in cardiac collagen with hypertrophy. In: Tarazi RC, Dunbar JB (eds) Perspectives in Cardiovascular Research, 8: 49–57
13. Fouad FM, Slominski M, Tarazi RC (1984) Left ventricular diastolic function in hypertension: relation to left ventricular mass and systolic function. J Am Coll Cardiol 3: 1500–1506
14. Sole MJ, Factor SM (1985) Hamster cardiomyopathy; a genetically transmitted sympathetic dystrophy. In: Beamish RE, Panagia V, Dhalla NS(eds) Pathogenesis of Stress-Induced Heart Disease. Martinius Nijhoff Boston, pp 34–43
15. Cohen-Gould L, Robinson TF, Factor SM (1987) Intrinsic connective tissue abnormalities in the heart muscle of cardiomyopathic Syrian hamsters. Am J Pathol 127: 327–334
16. Braunwald E, Kloner RA (1982) The stunned myocardium: prolonged, postischemic ventricular dysfunction. Circulation 66: 1146–1149
17. Zhao M, Zhang H, Robinson TF, Factor SM, Sonnenblick EH, Eng C (1987) Profound structural alterations of the extracellular collagen matrix in postischemic dysfunctional('stunned') myocardium. J Am Coll Cardiol 10: 1322-1334
18. Sato S, Ashraf M, Millard RW, Fujiwara H, Schwartz A (1983) Connective tissue changes in early ischemia of porcine myocardium: an ultrastructural study. J Mol Cell cardiol 15: 261–275
19. Forrester JS, Diamond G, Parmley WW, Swan HJC (1972) Early increase in left ventricular compliance following myocardial infarction. J Clin Invest 51: 598–602

The Cellular Basis of Stunned Myocardium

Hideo Kusuoka[1], Eduardo Marban[2], and Myron L. Weisfeldt[2]

Summary. Contractility is decreased after ischemia, even when there is no histologic evidence of irreversible injury; myocardium so afflicted has come to be known as 'stunned.' The contractility lesion in stunned myocardium might affect at least one of three major factors in force generation; the $[Ca^{2+}]$ transient, myofilament Ca^{2+}-sensitivity and maximal Ca^{2+}-activated force. Our results indicate that contractile dysfunction in stunned myocardium is characterized by a decline in maximal force and a shift to higher Ca in the myocardial sensitivity to extracellular Ca. The latter may reflect either a decrease in myofilament Ca^{2+}-sensitivity or a decrease in the $[Ca^{2+}]$ transient. Reperfusion with a solution of low Ca concentration or an acidic solution ameliorated functional recovery, indicating that calcium entry upon reperfusion plays a major role in the pathogenesis of myocardial stunning. Recent research suggests that the lower intramyocardial level at ATP may be an epiphenomenon and not a causal factor in stunning. Thus, we favor the hypothesis that calcium is the crucial mediator of reversible post-ischemic dysfunction.

Introduction

Reperfusion after myocardial ischemia of brief duration does not result in myocardial necrosis, but induces prolonged contractile dysfunction which persists for up to two weeks [1]. This phenomenon of reversible myocardial contractile dysfunction following a sublethal period of ischemia has been termed myocardial 'stunning' [12]. Although this review will focus on the cellular features of stunning in a perfused heart model, it is important to note that reversible postischemic dysfunction has a genuine clinical manifestation: patients undergoing thrombolytic therapy for acute myocardial infarction demonstrate a delayed improvement in regional function after the restoration of perfusion [3]. Nevertheless, the physiology of stunned myocardium is not well characterized, and the underlying mechanism of stunning remains a subject of considerable controversy [4].

In thinking about stunned myocardium, we must keep in mind that it is not a static process. Appreciable contractile reserve is preserved; paired pacing or catecho-

[1] The First Department of Medicine, Osaka University School of Medicine, 1-1-50 Fukushima, Fukushima-ku, Osaka, 553 Japan

[2] Division of Cardiology, Department of Medicine, The Johns Hopkins University School of Medicine, Baltimore, MD 21205, USA

lamines induce a robust positive inotropic response [5]. This finding indicates that stunned myocardium, although apparently restricted in its force generation, can still mount an appreciable response to stimuli that are known to affect the delivery of activator calcium to the contractile proteins.

Beyond the characterization of force generation in stunned myocardium remains the question of the underlying mechanism of celllular injury. Two major hypotheses have been put forward: (1) cellular ATP depletion, or (2) 'abnormalities in calcium flux' [2]. Recently, a cause-and-effect relationship between ATP depletion and depressed contractile force has been questioned [6]. The possibility of abnormalities in Ca metabolism in stunned myocardium, has received little attention despite extensive evidence implicating cellular Ca overload in the etiology of irreversible reperfusion injury [7]. However, several lines of evidence have been obtained from the myocardial physiology laboratories at Johns Hopkins and Osaka Universities that give special credence to the hypothesis that celllular calcium overload during ischemia and/or reperfusion produces stunning [8, 9, 10].

Pathophysiology of Stunned Myocardium

We first turn our attention to the question of how excitation-contraction coupling might be impaired in stunned myocardium. The 'contractility lesion' in stunned myocardium might affect one of three major links in force generation, as illustrated in Fig. 1a. The first of these is the intracellular free Ca^{2+} concentration, $[Ca^{2+}]_i$, during each cardiac cycle (the '$[Ca^{2+}]$ transient'), depicted schematically on the *left*. The second major link is the *responsiveness* of the contractile machinery to Ca^{2+}. Myofilament Ca^{2+}-responsiveness itself consists of two parameters [11]: the myofilament *sensitivity* to Ca^{2+}, which describes the range of $[Ca^{2+}]_i$ which activates contraction, and maximal Ca^{2+}-activated force, which determines the amplitude of the contractile response. Both of these are subsumed within the apparent Ca^{2+}-tension relation diagrammed *above* the *arrow* linking $[Ca^{2+}]_i$ and force. These three basic steps could account for the contractile dysfunction of stunned myocardium, as shown in panels b, c and d. A decrease in the amplitude of the $[Ca^{2+}]$ transient, as shown in the *left panel* of b, will, of course, decrease force. Note that the Ca^{2+}-responsiveness remains unchanged. In contrast, panel c illustrates the effect of keeping the $[Ca^{2+}]$ transient unchanged but shifting myofilament Ca^{2+}-sensitivity (as shown by the dashed $[Ca^{2+}]_i$-force relation, with its midpoint shifted to the right relative to control); in this case, less force is generated at any submaximal $[Ca^{2+}]_i$, but maximal force remains the same. A third possibility is shown in d. Here, maximal Ca^{2+}-activated force has been scaled down (as shown by the dashed $[Ca^{2+}]_i$-force relation *above* the *arrow*), but neither the $[Ca^{2+}]$ transient nor the midpoint of the $[Ca^{2+}]_i$-force relation has been changed. This or any of the other changes depicted in b,c,d individually suffices to decrease twitch force; if two or more of these changes were present in combination, twitch force would be decreased even further.

It has previously been impossible to characterize the pathophysiologic basis of stunned myocardium, because of lack of suitable methods in perfused hearts. Recently, a new approach has been described that enables the determination of maximal Ca^{2+}-activated force as reflected by isovolumic pressure in the isolated

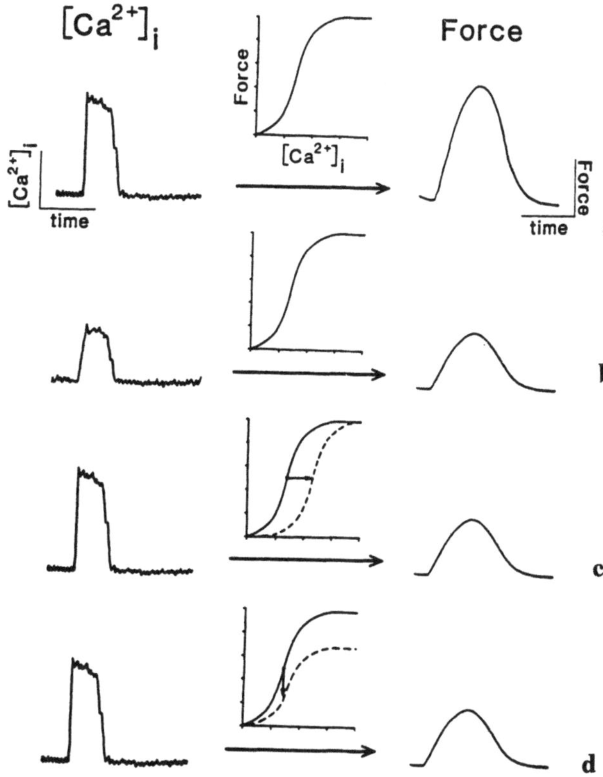

Fig. 1a-d. Schematic representation of $[Ca^{2+}]$ transient (*left*), force (*right*) and the $[Ca^{2+}]_i$-force relation. From [8], by permission of J Clin Invest

perfused heart [12], using tetani elicited by rapid pacing after exposure to ryanodine. We have used this approach to determine functional characteristics of stunned myocardium [8].

We characterized the responsiveness of twitch force to changes in extracellular Ca (Ca_o-responsiveness) before and after stunning by measuring developed pressure at several $[Ca]_o$(0.5–7.5 m*M*) in isovolumic Langendorff-perfused ferret hearts. Stunned myocardium was induced by reperfusion after 15 min of global ischemia at 37°C. Ca_o-responsiveness during twitches represents the net result of the $[Ca^{2+}]$ transient and myofilament Ca^{2+}-sensitivity, as scaled by maximal Ca^{2+}-activated force. At all $[Ca]_o$, developed pressure was depressed in stunned myocardium. Maximal Ca^{2+}-activated pressure (MCAP) measured from tetani was decreased after stunning. Normalization of the relationship between developed pressure and $[Ca]_o$ by corresponding MCAP (Ca_o-sensitivity) revealed a shift to higher $[Ca]_o$ in stunned hearts. These results indicate that contractile dysfunction in stunned myocardium is due to a decline in maximal Ca^{2+}-activated force, and a shift in Ca_o-sensitivity. The latter may reflect either decreased myofilament Ca^{2+}-sensitivity or a decrease in the $[Ca^{2+}]$ transient.

The Role of Cellular Calcium Overload in Stunned Myocardium

Cellular Ca overload is well-recognized as an etiologic factor in irreversible reperfusion injury after a long duration of ischemia [7]. Cellular Ca overload will occur if Ca influx pathways that are inhibited during ischemia are re-activated upon reperfusion. One specific hypothesis [13, 14] involves Na-Ca exchange. This pathway normally extrudes Ca from the cell at the expense of the trans-sarcolemmal Na gradient. Under conditions of high intracellular Na the direction of exchange is reversed, giving Na-induced cellular Ca gain. During ischemia, the Na^+, K^+-ATPase is inhibited by the metabolic suppression, so that intracellular Na accumulates. At the same time Na-Ca exchange is inhibited by the intracellular acidosis. Upon reperfusion, intracellular pH returns to normal very rapidly, removing the inhibition of Na-Ca exchange. When the Na-Ca exchange is reactivated under these conditions, intracellular [Na] is still thought to be elevated, and consequently the Na gradient drives Ca into the cells.

In stunned myocardium, the histologic changes are not pathognomonic for Ca-mediated injury, as they are in reperfusion after longer periods of ischemia. Nevertheless, stunning may share the same basic mechanism, with the major difference being one of degree of injury. It has previously been impossible to show direct evidence of intracellular Ca overload during reperfusion, because of lack of suitable methods to measure $[Ca^{2+}]_i$ in perfused whole heart. It is now clear that $[Ca^{2+}]_i$ increases during 20 min of ischemia at 30°C, and remains elevated for the first 5–10 min of reperfusion [15]. We sought evidence to link the observed rise in $[Ca^{2+}]_i$ to the pathogenesis of stunning, as discussed next.

We reperfused ferret hearts with solutions of varying Ca concentration after 15 min of ischemia at 37°C [8], the same ischemic protocol as in the conventional stunning group. The only variable was the $[Ca]_O$ of the initial reperfusate. We reasoned as follows: if massive Ca influx occurs upon reperfusion and is required for stunning, then reduction of $[Ca]_O$ in the reperfusate will decrease stunning by decreasing the driving force for Ca influx. Hearts were reperfused with low-$[Ca]_O$ solution (0.1 m*M* during the initial 7 min, 0.2 m*M* during 7–14 min 0.5 m*M* during 14–21 min) before returning to the standard perfusate ($[Ca]_O$ = 2 m*M*). Remarkably, these hearts exhibited developed pressure and maximal Ca^{2+}-activated pressure comparable to a nonischemic control and significantly higher than in the stunned group.

Reperfusion with acidic solution represents a different strategy to the low-$[Ca]_O$ reperfusion [10], but shares the goal of blunting Ca overload during reperfusion. Because acidosis is known to antogonize Ca influx and the intracellular binding of Ca, acidosis during reperfusion may prevent Ca overload and ameliorate myocardial stunning. Hearts were subjected to the same ischemic protocol as in the conventional stunning group, and reperfused with acidic solution (pH 6.6 during 0–3 min, pH 7.0 during 4–6 min) before returning to the standard perfusate (pH 7.4). Intramyocardial pH was lower in the acidic group than in the stunned group during early reperfusion. Both Ca_O-responsiveness and maximal Ca^{2+}-activated pressure were greater in the acidic group than in the stunned group.

These results obtained with low-$[Ca]_O$ reperfusion and acidic reperfusion indicate that transient calcium overload during early reperfusion plays a major role in the pathogenesis of myocardial stunning. Then, the next question is whether or not transient Ca overload without ischemia can produce contractile dysfunction which mimics stunned myocardium. The answer is 'yes' [9]. Isolated ferret hearts after three 5-min periods of perfusion with 10 m*M* $[Ca]_O$ without ischemia showed a decrease in both Ca_O-responsiveness and maximal Ca^{2+}-activated pressure virtually indistinguishable from that observed in post-ischemic myocardium.

Pathogenesis of Stunned Myocardium

Among the several possible etiological factors in myocardial stunning, summarized in Table 1, prolonged cellular ATP depletion has been received a great deal of attention because of the close correlation between the recovery of ATP and the recovery of systolic function after regional ischemia in the canine heart [16]. However, the responsiveness of stunned myocardium to catecholamines suggests that the lower level of ATP may be an epiphenomenon and not a causal factor in stunning [5]. In addition, no unique correlation between intramyocardial ATP level ([ATP]) and maximal Ca^{2+}-activated pressure was observed in our experiments [8]. The control hearts show high [ATP] and high maximal Ca^{2+}-activated pressure. Stunned myocardium shows low [ATP] and low maximal Ca^{2+}-activated pressure. These two observations alone are consistent with a cause-and-effect relationship between [ATP] and contractile function. However, the hearts reperfused with low-$[Ca]_O$ solution also show a significantly lower [ATP] than control, but much higher maximal Ca^{2+}-activated pressure than in stunned hearts. It is readily apparent that there is no unique relation between [ATP] and contractile function, indicating that ATP depletion is not the cause of myocardial stunning. A similar result was observed with acidic reperfusion [10]. The lack of a causative role for ATP is further demonstrated by the observation that an increase of [ATP] in stunned myocardium with adminstration of ATP precursors failed to improve contractile function [17].

No significant abnormality in the microcirculation was detectable in our work on stunned myocardium. The contribution of a 'no reflow' phenomenon was considered to be negligible after brief ischemic periods [5, 18]. Increase in oxygen consumption and increased heterogeneity of oxygen extraction were also reported [18]. The free

Table 1. Possible causes of myocardial stunning

1. Transient cellular Ca overload
2. Prolonged depletion of intramyocardial ATP
3. "No reflow" phenomenon
4. Functional denervation of sympathetic nerves
5. Dysfunction of sarcoplasmic reticulum
6. Disturbance in energy utilization by the myofilaments
7. Abnormal excitation
8. Regional increase of load due to dyskinesis of the ventricle
9. Free radicals

energy change of ATP hydrolysis, an index of maximum work available in ATP hydrolysis, is actually increased in stunned myocardium in spite of the decrease in maximal force [19]. These results indicate that an insufficient transfer of energy into contraction or an increased use of energy for noncontractile activities exists in stunned myocardium. Further research to elucidate the intracellular mechanism of this phenomenon is required.

Formation of free radicals during early reperfusion after brief ischemia was directly demonstrated using the electron spin resonance technique [20]. Administration of superoxide dismutase showed a preventive effect on stunning [21]. These results are consistent with the idea that free radicals may play an important role in myocardial stunning, but a direct relation to maximal Ca^{2+}-activated force has not yet been demonstrated. If they are involved, it seems amply clear they must interact with cell Ca either in their generation or removal or further downstream in the cascade that leads to functional cell injury.

Conclusion

A number of questions about pathophysiology and pathogenesis of stunned myocardium still remain unanswered. Stunned myocardium is one of the important topics in clinical cardiology, and models such as ours to produce stunned myocardium under controlled conditions may be useful in quantifying the benefits of reperfusion. 'Percent salvage' is generally expressed as necrotic area over area at risk. Clearly, there is a large 'gray zone' over which the benefits of reperfusion, modified or unmodified, must be assessed in a more discriminating manner.

The lack of a requirement for pretreatment before the ischemic episode is important for realistic clinical strategies to prevent myocardial stunning. From this viewpoint, reperfusion with low-$[Ca]_o$ solution or acidic solution may be applicable during recanalization in patients with acute myocardial infarction to protect myocardium from stunning.

Acknowledgements. We thank Drs. James K. Porterfield and Harlan F. Weisman for their collaboration in the work reviewed here. We also thank Dr. Elena Moscarelli for her helpful comments on the manuscript.

References

1. Weiner JM, Apstein CS, Arthur JH, Pirzada FA, Hood Jr WB (1976) Persistence of myocardial injury following brief periods of coronary occlusion. Cardiovasc Res 10 678–686
2. Braunwald E, Kloner RA (1982) The stunned myocardium: prolonged postischemic ventricular dysfunction. Circulation 60: 1146–1149
3. Satler LF, Kent KM, Fox LM, Goldstein HA, Green CE, Rogers WJ, Pollas RS, Del Negro AA, Pearle DL, Rackley CE (1986) The assessment of contractile reserve after thrombolytic therapy for acute myocardial infarction. Am Heart J 111: 821–825
4. Braunwald E, Kloner RA (1985) Myocardial reperfusion: a double-edged sword? J Clin Invest 76: 1713–1719

5. Becker LC, Levine JH, DiPaula AF, Guarnieri T, Aversano T (1986) Reversal of dysfunction in postischemic stunned myocardium by epinephrine and postextrasystolic potentiation. J Am Coll Cardiol 7: 580-589
6. Neely JR, Grotyohann LW (1984) Role of glycloytic products in damage to ischemic myocardium. Dissociation of adenosine triphosphate levels and recovery of function of reperfused ischemic hearts. Circ Res 55: 816-824
7. Shine KL, Douglas AM (1983) Low calcium reperfusion of ischemic myocardium. J Mol Cell Cardiol 15: 251-260
8. Kusuoka H, Porterfield JK, Weisman HF, Weisfeldt ML, Marban E (1987) Pathophysiology and pathogenesis of stunned myocardium. Depressed Ca^{2+} activation of contraction as a consequence of reperfusion-induced cellular calcium overload in ferret hearts. J Clin Invest 79: 950-961
9. Kitakaze M, Weisman HF, Marban E (1988) Contractile dysfunction and ATP depletion after transient calcium overload in perfused ferret hearts. Circulation 77: 685-695
10. Kitakaze M, Weisfeldt ML, Marban E (1988) Acidosis during early reperfusion prevents myocardial stunning in perfused ferret hearts. J Clin Invest 82: 920-927
11. Kusuoka H, Weisfeldt ML, Zweier JL, Jacobus WE, Marba E (1986) Mechanism of early contractile failure during hypoxia in intact ferret heart: evidence for modulation of maximal Ca^{2+}-activated force by inorganic phosphate. Circ Res 59: 270-282
12. Marban E, Kusuoka H, Yue DT, Weisfeldt ML, Wier WG (1986) Maximal Ca^{2+}-activated force elicited by tetanization of ferret papillary muscle and whole heart. Mechanism and characteristics of steady contractile activation in intact myocardium. Circ Res 59: 262-269
13. Grinwald PM (1982) Calcium uptake during post-ischemic reperfusion in the isolated rat heart: influence of extracellular sodium. J Mol Cell Cardiol 14: 359-365
14. Renlund DG, Gerstenblith G, Lakatta EG, Jacobus WE, Kallman CH, Weisfeldt ML (1984) Perfusate sodium during ischemia modifies post-ischemic functional and metabolic recovery in the rabbit heart. J Mol Cell Cardiol 16: 795-801
15. Kitakaze M, Pike MM, Chacko VP, Marban E (1987) Direct measurement of cystolic free calcium during ischemia and reperfusion in ferret hearts. (abstract) Circulation 76: 380
16. Kloner RA, Deboer LWV, Darsee JR, Ingwall JS, Hale S, Tumas J, Braunwald E (1981) Prolonged abnormalities of myocardium salvaged by reperfusion. Am J Physiol 241: H591-H599
17. Mauser M, Hoffmeister HM, Nienaber C, Schaper W (1985) Influence of ribose, adenosine, and AICAR on the rate of myocardial adenosine triphosphate synthesis during reperfusion after coronary artery occlusion in the dog. Circ Res 56: 220-230
18. Stahl L, Weiss HR, Becker LC (1988) Myocardial oxygen consumption, oxygen supply/demand heterogeneity, and microvascular patency in regionally stunned myocardium. Circulation 77:865-872
19. Kusuoka H, Inoue M, Marban E (1988) Decreased efficiency of energy utilization in stunned myocardium. (abstract) Circulation 78: II-261
20. Zweier JL, Flaherty JT, Weisfeldt ML (1987) Direct measurement of free radical generation following reperfusion of ischemic myocardium. Proc Natl Acad Sci USA 84: 1404-1407
21. Przyklenk K, Koner RA (1986) Superoxide dismutase plus catalase improve contractile function in the canine model of the 'stunned myocardium'. Circ Res 58: 148-156

Isozymic Changes in Human Cardiac Myosins Due to Overload

Yoshio Yazaki, Hidetsugu Tsuchimochi, Masahiko Kurabayashi, and Issei Komuro[1]

Summary. During the process of enlargement of the heart due to overload, a significant reconstitution of the organ including myoctes and intracellular constituents occurs, as well as cellular hypertrophy induced by increased synthesis of cellular components in order to meet an increased demand in work. This adaptational phenomenon can be defined at the molecular level in terms of altered transcription of specific genes.

Cardiac myosin isoforms, through their enzymatic activities, regulate both contractility and energy efficiency for contraction of the muscle. Since analysis of myosin isoforms can provide insight into the process of caridac hypertrophy, we developed monoclonal antibodies specific for two types of cardiac myosin, and characterized these isoforms by an immunofluorescence study. Furthermore, we cloned specific genes encoding the heavy and light chains of these cardiac myosin isoforms from the human cDNA library and demonstrated the regulation of their expression in the overloaded human heart. In this study, we investigated human materials to obtain clinical implications from our observations.

Redistribution of Cardiac Myosin Isozymes in the Human Heart Due to Overload

Enzymatic [1–4], electrophoretic [5, 6], and immunochemical [7, 8] analyses have shown the existence of two distinct cardiac myosin heavy chains, HCα and HCβ. However, it has not been conclusively determined that isoforms are present in human cardiac myosin, since nondenaturing pyrophosphate gel electrophoretic analysis failed to separate individual isozymes of human cardiac myosin. On electrophoretic gels, human cardiac myosin is composed of a single band, which appears to be very close to the rat V3 isomyosin.

In order to discriminate human cardiac HCα and HCβ, we have prepared monoclonal antibodies specific for the heavy chains of either type HCα or HCβ . Anti-myosin monoclonal antibodies were produced by hybridomas obtained by fusion of myeloma cells (P3X63Ag8U1) with isolated spleen cells of BALB/c mice immunized with bovine atrial and human ventricular myosin as HCα or HCβ antigen, respectively, essentially by the method of Köhler and Milstein [9]. Two clones (CMA19 and HMC14) of hybrid cells secreting anti-myosin antibodies were selected for discriminating the antigenic difference between HCα and HCβ by ELISA tests [10, 11].

[1] The Third Department of Internal Medicine, The Faculty of Medicine, University of Tokyo, 7-3-1 Hongo, Bunkyo-ku, Tokyo, 113 Japan

The distribution of myosin isozymes in human atrial and ventricular myocardium was investigated by immunofluorescence assay using these monoclonal antibodies. Human myocardial specimens were obtained from the patients with or without valvular disease during open heart surgery or at autopsy. Cryostat tissue sections were stained with the indirect immunofluorescence procedure.

Our immunofluorescence study using two types of monoclonal antibodies demonstrates (a) the existence of two types of myosin isozymes in the human myocardium, and (b) a striking difference in their distribution between atrial and ventricular myocardium. HCβ is a predominant myosin heavy chain isoform in the ventricle, whereas a main isoform in the atrium is HCα (Fig. 1). Considering that ventricular

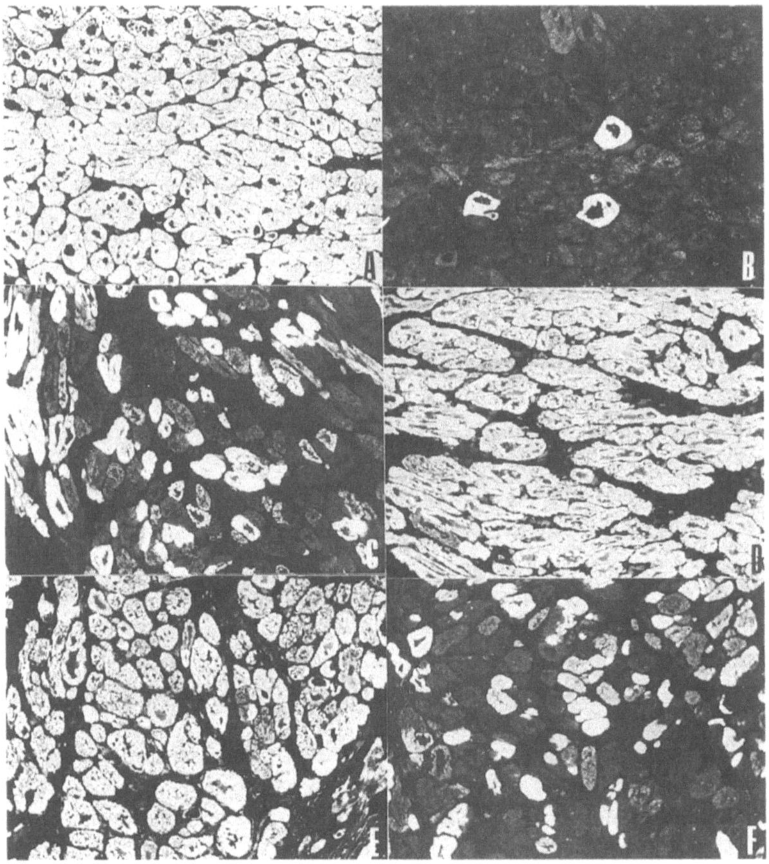

Fig. 1 A Cryostat section of normal human ventricle stained by HMC14. All fibers reacted strongly and uniformly. **B** As in **A**, except that the muscle fibers were stained by CMA19. A small number of fibers were reactive. **C** Cryostat section of normal human atrium stained by HMC14. The staining intensity was highly variable. **D** As in **C**, except that the myofibers were stained by CMA19. Almost all fibers were strongly reactive. **E** Cryostat section of a pressure-overloaded atrium stained by HMC14. The specimen was obtained from a patient with mitral stenosis and regurgitation. Almost all fibers became reactive. **F** As in **E**, except that the myofibers were stained by CMA19. Unreactive fibers were significantly increased [11].

pressure is much higher than atrial pressure, and that HCβ appears to be suited for pressure work, the difference in the distribution of myosin isozymes between atrial and ventricular myocardium is regarded as a physiological adaptation to their loaded hemodynamic work.

Evidence has been accumulated that relative proportions of the two basic isoforms of myosin were changed in rat and rabbit ventricles during development and cardiac overload [7, 8, 12–14]. To confirm whether or not redistribution of these isozymes occurs in the human myocardium subjected to overload, we performed again an immunohistochemical study using monoclonal antibodies, CMA19 and HMC14, on the overloaded human myocardium.

When pressure-overloaded atria, obtained from patients with mitral stenosis and/or regurgitation or tricuspid regurgitation during open heart surgery, were stained by each antibody, a striking reversal of the staining pattern was observed (Fig. 1). The number of the myocytes reactive with HMC14 was increased significantly, while that of the myocytes reactive with CMA19 showed a corresponding decrease. To demonstrate the changes in the amount of each myosin isozyme more quantitatively, we counted the number of the myocytes which reacted with each antibody and the total scores per 1000 myocytes were calculated in each specimen and

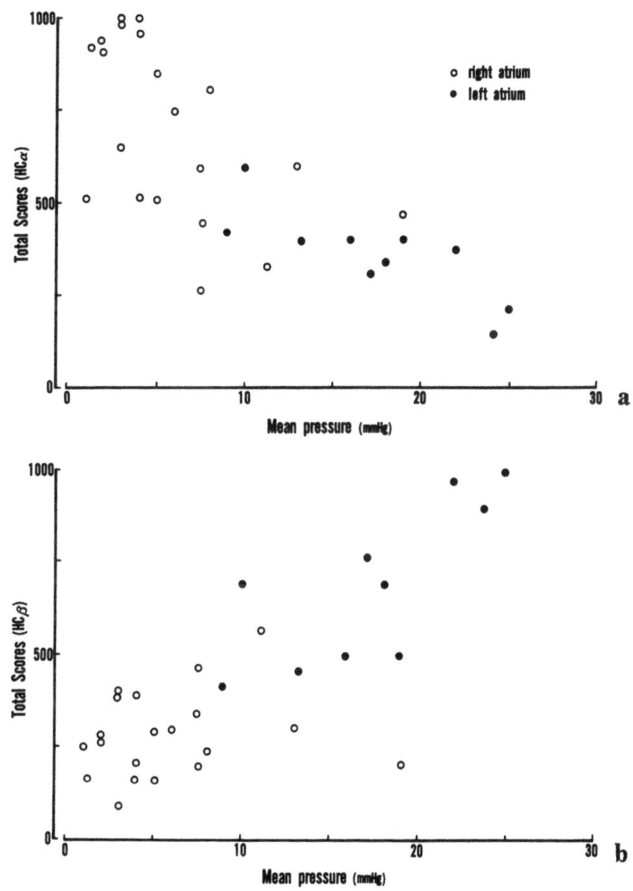

Fig. 2a, b. The total scores per 1 000 myofibers of HCα **a** or HCβ **b** plotted against the mean atrial or pulmonary capillary wedge pressure. ○ right atrium; ● left atrium [11]

plotted against each mean atrial pressure or mean pulmonary wedge pressure (Fig. 2). This observation revealed that the degrees of HCα decrement and HCβ increment correlated well with the mean atrial pressure [11]. We also observed a decrease in the number of the myocytes reactive with CMA19 in the left ventricular specimen obtained from the patients with mitral and/or aortic valvular disease according to the increment of peak left ventricular pressure [15]. These results showed that pressure overload may play a major role in the isozymic transition of cardiac myosin.

The isozymic redistribution in pressure-overloaded myocardium may have a physiological implication. Since the low adenosinetriphosphatase (ATPase) activity related to low velocity of shortening improves the efficiency of contraction for an equivalent amount of work, as previously described [16, 17], HCβ appears to be a physiological mysoin isozyme for performing pressure work. Overloaded atrial myocytes may promote synthesis of HCβ instead of HCα in order to improve the efficacy of contraction. Thus, the isozymic redistribution can be considered as a physiological adaptive mechanism to meet increased pressure overload. This isozymic transition from HCα to HCβ in the overloaded atrium may participate in a compensatory mechanism by which mechanical function of the atrium changes from volume work to pressure work. In the ventricular myocardium, normal mysoin isozyme pattern has proved to be exclusively HCβ-dominant. Therefore, the physiological importance of isozymic redistribution in the human heart induced by overload seems to be greater in the atrium than in the ventricle, since the content of HCα which could be transformed to HCβ was much larger in the atrial myocardium.

Molecular Cloning and Characterization of Human Cardiac α and β Type Myson Heavy and Light Chain cDNA Clones: Regulation of Their Expression During Pressure Overload

To understand the molecular basis of the myosin heavy chain isozyme transition in overloaded human myocardium, we investigated the level of the expression of individual myosin heavy chain mRNA in pressure-overloaded human atrium. For this purpose, we have isolated two cDNA clones encoding either HCα or HCβ from a fetal human heart cDNA library and analyzed the differences in accumulation of these mRNAs in the human heart with Northern blot hybridization using these cDNAs as probes.

Our successful cloning of HCα and HCβ cDNAs has established the existence of the two molecular variants in the human heart [18]. In Fig. 3, the nucleotide

Fig. 3. Nucleotide and deduced amino acid sequences of HCα and HCβ clones. Nucleotides are numbered to the right of each line. *Line 1* amino acid sequence of HCα clone; *Line 2* nucleotide sequence of HCα clone; *Line 3* nucleotide sequence of HCβ clone; *Line 4* amino acid sequence of HCβ clone. Nucleotides and amino acids are written only where different from that of HCα clone. The *dashed line* indicates sequences identical to that of the HCα clone. The underlined 26 bases represent the complementary sequences of the oligonucleotides synthesized for probes. *Wavy lines* indicate the polyadenylation signal sequence

```
Glu Ala Val Asn Ala Lys Cys Ser Ser Leu Glu Lys Thr Lys His Arg Leu Gln Asn Glu Ile Glu Asp Leu Met Val Asp Val Glu Arg
GAG GCT GTT AAT GCC AAA TGC TCC TCA CTG GAG AAG ACC AAG CAC CGG CTA CAG AAT GAG ATA GAG GAC TTG ATG GTG GAC GTA GAG CGC   90
                            --- --G --- --- --- --- --- --- --- --- --- --- --C --- --- --- --- --- --- --- --- ---
                            --- --- --- --- --- --- --- --- --- --- --- --- --- --- --- --- --- --- --- --- --- ---

Ser Asn Ala Ala Ala Ala Ala Leu Asp Lys Lys Gln Arg Asn Phe Asp Lys Ile Leu Ala Glu Trp Lys Gln Lys Tyr Glu Glu Ser Gln
TCC AAT GCT GCT GCT GCT GCT CTG GAC AAG AAG CAG AGG AAC TTT GAC AAG ATC CTG GCC GAG TGG AAG CAG AAG TAT GAG GAG TCG CAG  180
--- --- --- --- --- --- --- --- --- --- --- --- --- --- --C --- --- --- --- --- --- --- --- --- --- --- --- --- --- ---
--- --- --- --- --- --- --- --- --- --- --- --- --- --- --- --- --- --- --- --- --- --- --- --- --- --- --- --- --- ---

Ser Glu Leu Glu Ser Ser Gln Lys Glu Ala Arg Ser Leu Ser Thr Glu Leu Phe Lys Leu Lys Asn Ala Tyr Glu Glu Ser Leu Glu His
TCT GAG CTG GAG TCC TCA CAG AAG GAG GCT CGC TCC CTC AGC ACA GAG CTC TTC AAG CTC AAG AAC GCC TAC GAG GAG TCC CTG GAG CAC  270
--G --- --- --- --- --G --- --- --- --- --- --- --- --- --- --- --- --- --A --- --- --- --- --T --- --- --- --- --A --T
--- --- --- --- --- --- --- --- --- --- --- --- --- --- --- --- --- --- --- --- --- --- --- --- --- --- --- --- --- ---

Leu Glu Thr Phe Lys Arg Glu Asn Lys Asn Leu Gln Glu Glu Ile Ser Asp Leu Thr Glu Gln Leu Gly Glu Gly Gly Lys Asn Val His
CTA GAG ACC TTC AAG CGG GAG AAC AAG AAC CTT CAG GAG GAA ATC TCG GAC CTT ACT GAG CAG CTA GGA GAA GGA GGA AAG AAT GTG CAT  360
--G --- --- --- --- --- --- --- --A --- --G --- --- --G --- --C --- T-G --- --- --- T-G --T TCC A-C --- --- -C- A-C ---
--- --- --- --- --- --- --- --- --- --- --- --- --- --- --- --- --- --- --- --- --- --- --- Ser Arg --- --- Thr Ile ---

Glu Leu Glu Lys Val Arg Asn Gln Leu Glu Val Glu Lys Met Glu Leu Gln Ser Ala Leu Glu Glu Ala Glu Ala Ser Leu Glu His Glu
GAG CTG GAG AAG GTC CGA AAC CAG CTG GAG GTG GAG AAG ATG GAG CTG CAG TCA GCC CTG GAG GAG GCA GAG GCC TCC CTG GAG CAC GAG  450
--- --- --- --- --- --- --G --- --- --- -CC --- --- --- --- --- --- --- --- --- --- --- --C --- --- --- --- --- --- ---
--- --- --- --- --- --- Lys --- --- --- Ala --- --- --- --- --- --- --- --- --- --- --- --- --- --- --- --- --- --- ---

Glu Gly Lys Ile Leu Arg Ala Gln Leu Glu Phe Asn Gln Ile Lys Ala Glu Ile Glu Arg Asn Val Ala Glu Lys Asp Glu Glu Met Glu
GAG GGC AAG ATC CTC CGG GCC CAG CTA GAG TTC AAC CAG ATC AAG GCA GAG ATC GAG CGG AAC GTG GCA GAG AAG GAC GAG GAG ATG GAA  540
--- --- --- --- --- --- --- --- --G --- --- --- --- --- --- --- --- --- --- --- --- *-- --- --- --- --- --- --- --- ---
--- --- --- --- --- --- --- --- --- --- --- --- --- --- --- --- --- --- --- --- --- --- --- --- --- --- --- --- --- ---

Gln Ala Lys Arg Asn His Gln Arg Val Val Asp Ser Leu Gln Thr Ser Leu Asp Ala Glu Thr Arg Ser Arg Asn Glu Val Leu Arg Val
CAG GCC AAG CGC AAC CAC CAG CGG GTG GTG GAC TCG CTG CAG ACC TCC CTG GAT GCA GAG ACA CGC AGC CGC AAC GAG GTC CTG AGG GTG  630
--- --- --- --- --- --- -T- --- --* --- --- --- --- --- --- --- --- --C --- --- --- --- --- --- --- --- -C- --- --- ---
--- --- --- --- --- --- Leu --- --- --- --- --- --- --- --- --- --- --- --- --- --- --- --- --- --- --- Ala --- --- ---

Lys Lys Lys Met Glu Gly Asp Leu Asn Glu Met Glu Ile Gln Leu Ser His Ala Asn Arg Met Ala Ala Glu Ala Gln Lys Gln Val Lys
AAG AAG AAG ATG GAA GGA GAC CTC AAT GAG ATG GAG ATC CAG CTC AGC CAC GCC AAC CGC ATG GCT GCC GAG GCC CAG AAG CAA GTC AAG  720
--- --- --- --- --- --- --- --- --- --- --- --- --- --- --- --- --- --- --- --- --- --C --- --- --- --- --- --- --- ---
--- --- --- --- --- --- --- --- --- --- --- --- --- --- --- --- --- --- --- --- --- --- --- --- --- --- --- --- --- ---

Ser Leu Gln Ser Leu Leu Lys Asp Thr Gln Ile Gln Leu Asp Asp Ala Val Arg Ala Asn Asp Asp Leu Lys Glu Asn Ile Ala Ile Val
AGC CTC CAG AGC TTG CTG AAG GAC ACC CAG ATC CAG CTG GAC GAT GCG GTC CGT GCC AAC GAC GAC CTG AAG GAG AAC ATC GCC ATC GTG  810
--- --- --- --- --- T-- --- --- --- --- --T --- --- --- --- --A --- --- --- --- --- --- --- --- --- --- --- --- --- ---
--- --- --- --- --- --- --- --- --- --- --- --- --- --- --- --- --- --- --- --- --- --- --- --- --- --- --- --- --- ---

Glu Arg Arg Asn Asn Leu Leu Gln Ala Glu Leu Glu Glu Leu Arg Ala Val Val Glu Gln Thr Glu Arg Ser Arg Lys Leu Ala Glu Gln
GAG CGG CGC AAC AAC CTG CTG CAG GCT GAG CTG GAG GAG TTG CGC GCC GTG GTG GAG CAG ACG GAG CGG TCC CGG AAG CTG GCG GAG CAG  900
--- --- --- --- --- --- --- --- --- --- --- --- --- --- --- --- --- --- --- --- --A --- --- --- --- --- --- --- --- ---
--- --- --- --- --- --- --- --- --- --- --- --- --- --- --- --- --- --- --- --- --- --- --- --- --- --- --- --- --- ---

Glu Leu Ile Glu Thr Ser Glu Arg Val Gln Leu Leu His Ser Gln Asn Thr Ser Leu Ile Asn Gln Lys Lys Lys Met Asp Ala Asp Leu
GAG CTG ATT GAG ACT AGT GAG CGC GTG CAG CTG CTG CAT TCC CAG AAC ACC AGC CTC ATC AAC CAG AAG AAG AAG ATG GAT GCT GAC CTG  990
--- --- --- --- --- --- --- --G --- --- --- --- --- --- --- --- --- --- --- --- --- --- --- --- --- --- --C --- --- ---
--- --- --- --- --- --- --- --- --- --- --- --- --- --- --- --- --- --- --- --- --- --- --- --- --- --- --- --- --- ---

Ser Gln Leu Gln Ser Glu Val Glu Glu Ala Val Gln Glu Cys Arg Asn Ala Glu Glu Lys Ala Lys Lys Ala Ile Thr Asp Ala Ala Met
TCC CAG CTC CAG TCG GAA GTG GAG GAG GCA GTG CAG GAG TGC AGA AAC GCC GAG GAG AAG GCC AAG AAG GCC ATC ACG GAT GCC GCC ATG 1080
--- --- --- --- A-T --- --- --- --- --- --- --- --- --- --G --T --T --- --- --- --- --- --- --- --- --* --- --- --- ---
--- --- --- --- Thr --- --- --- --- --- --- --- --- --- --- --- --- --- --- --- --- --- --- --- --- --- --- --- --- ---

Met Ala Glu Glu Leu Lys Lys Glu Gln Asp Thr Ser Ala His Leu Glu Arg Met Lys Lys Asn Met Glu Gln Thr Ile Lys Asp Leu Gln
ATG GCA GAG GAG CTG AAG AAG GAG CAG GAC ACC AGC GCC CAC CTG GAG CGC ATG AAG AAG AAC ATG GAG CAG ACC ATT AAG GAC CTG CAG 1170
--- --- --- --- --- --- --- --- --- --- --- --- --- --- --- --- --- --- --- --- --- --- --A --- --- --- --- --- --- ---
--- --- --- --- --- --- --- --- --- --- --- --- --- --- --- --- --- --- --- --- --- --- --- --- --- --- --- --- --- ---

His Arg Leu Asp Glu Ala Glu Gln Ile Ala Leu Lys Gly Gly Lys Lys Gln Leu Gln Lys Leu Glu Ala Arg Val Arg Glu Leu Glu Gly
CAC CGG CTG GAC GAG GCC GAG CAG ATC GCC CTC AAG GGC GGC AAG AAG CAG CTG CAG AAG CTG GAA GCG CGG GTG CGG GAG CTG GAG GGT 1260
--- --- --- --- --A --- --- --- --- --- --- --- --- --- --- --- --- --- --- --- --- --- --- --- --- --- --- --- --- AA-
--- --- --- --- --- --- --- --- --- --- --- --- --- --- --- --- --- --- --- --- --- --- --- --- --- --- --- --- --- Asn

Glu Leu Glu Ala Glu Gln Lys Arg Asn Ala Glu Ser Val Lys Gly Met Arg Lys Ser Glu Arg Arg Ile Lys Glu Leu Thr Tyr Gln Thr
GAG CTG GAG GCC GAG CAG AAG CGC AAC GCA GAG TCG GTG AAG GGC ATG AGG AAG AGC GAG CGG CGC ATC AAG GAG CTC ACC TAC CAG ACA 1350
--- --- --- --- --- --- --- --- --- --- --- --- --- --- --- --- --- --- --- --- --- --- --- --- --- --- --- --- --- --G
--- --- --- --- --- --- --- --- --- --- --- --- --- --- --- --- --- --- --- --- --- --- --- --- --- --- --- --- --- ---

Glu Glu Asp Lys Lys Asn Leu Leu Arg Leu Gln Asp Leu Val Asn Lys Leu Gln Leu Lys Val Lys Ala Tyr Lys Gly Gln Ala Glu Glu
GAG GAG GAC AAA AAG AAC CTG CTG CGG CTG CAG GAC CTG GTA AAC AAG CTG CAA CTG AAG GTC AAG GCC TAC AAG GGC CAG GCC GAA GAG 1440
--- --- --- -GG --A --- --- --- --- --- --- --- --- --- G-- --- --- --G --A --- --- --- --- --- --- C-- --- --- --G ---
--- --- --- Arg --- --- --- --- --- --- --- --- --- --- Asp --- --- --- --- --- --- --- --- --- --- Arg --- --- --- ---

Ala Glu Glu Arg Ala Asn Thr Asn Leu Ser Lys Phe Arg Lys Val Gln His Glu Leu Asp Glu Ala Glu Glu Arg Ala Asp Ile Ala Glu
GCG GAG GAG CGA GCC AAC ACC AAC CTG TCC AAG TTC CGC AAG GTG CAG CAT GAG CTG GAT GAG GCA GAG GAG CGG GCG GAC ATC GCT GAG 1530
--- --- --- -A- --- --- --- --- --- --- --- --- --- --- --- --C --- --- --- --- --- --- --- --- --- --- --- --- --C ---
--- --- --- Gln --- --- --- --- --- --- --- --- --- --- --- --- --- --- --- --- --- --- --- --- --- --- --- --- --- ---

Ser Gln Val Asn Lys Leu Arg Ala Lys Ser Arg Asp Ile Gly Ala Lys Lys Met His Asp Glu Glu END
TCC CAG GTC AAC AAG CTT CGA GCC AAG AGC CGT GAC ATT GGT GCC AAG AAA ATG CAC GAT GAG GAG TGA  CACTGCCTCGGGAACCTCACTGCTTG 1619
--- --- --- --- --- --G --G --- --- --- --- --- --- --C A-G --- GGC T-- A-T --G --- T-- CTT  TG-CA-A--TT--T-TG-T-A--GCC
--- --- --- --- --- --- --- --- --- --- --- --- --- --- Thr --- Gly Leu Asn Glu --- END

CCAACCTGTAATAAATATGAGTGCC(A)n
TG-GGTGCC-GC---GCCCCA---TGGAGCCTGTGTAACAGCTCGCTTGGGAGGAAGCAGAATAAAGCAATTTCCTTGAAGCCG(A)n
```

sequences of these two types of heavy chain cDNA clones are compared. These cDNAs are quite homologous, exhibiting 96% nucleotide homology within the translated region; they differ in only 77 out of the 1,596 nucleotides compared. In contrast, 3'-untranslated regions show extensive sequence divergence.

We examined the expression of these two types of myosin heavy chain genes in the human heart by Northern blot analysis using these cDNAs as probes. Since the nucleotide sequences of these two cDNAs are highly homologous, it is difficult to identify the specific heavy chain mRNA expressed in the myocardium by Northern blot analysis using the cDNA coding sequence. To overcome this difficulty, synthetic oligonnucleotide probes complementary to the unique 3'-untranslated regions were used in Northern blot analysis.

Hybridization patterns of atrial mRNA from the patients whose atria suffered from pressure overload due to valvular disease or pulmonary hypertension, are shown in Fig. 4. Nine patients were examined [18]. In the control patient who died from breast cancer, HCα mRNA was predominant in both right and left atria. However, in pressure-overloaded atria, accumulation of HCβ mRNA was observed. In Fig. 4, **1** shows the hybridization pattern of atrial mRNA from the patient with mitral regurgitation. Strong hybridization of left atrial mRNA with HCβ probe was

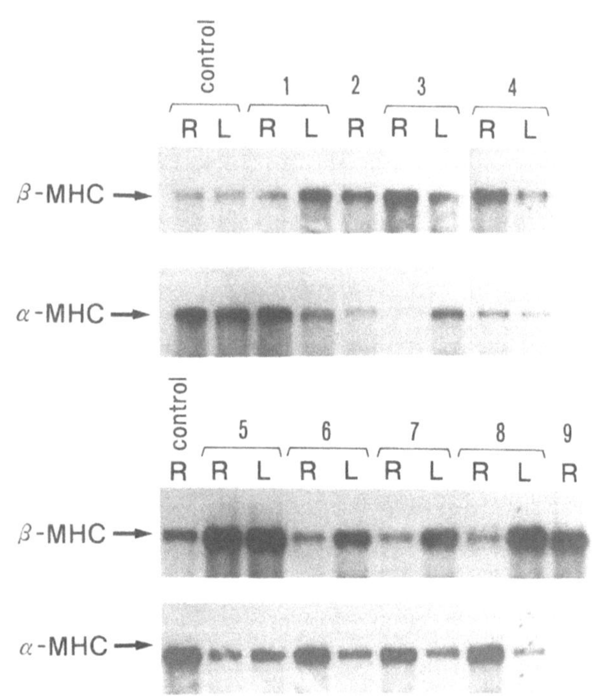

Fig. 4. Detection of HCα and HCβ mRNA in total cellular RNA from normal and diseased human atria. The same RNA blot was sequentially hybridized with labeled oligonucleotides specific for HCβ mRNA, exposure, 24 h; HCβ mRNA, exposure, 24 h. Normal atrial RNA controls are shown in the *left lanes* of the figures. The lanes containing right (*R*) and left (*L*) atrial RNA (10 μg) from patients 1-9 are indicated. The heavy chain mRNA band is marked on the *left* (HCα, HCβ) [18]

observed, whereas diminished hybridization with HCα probe was evident. In this patient, the left atrium suffered from pressure overload. In patient **3**, almost complete reversal of the hybridization pattern was shown in right atrial mRNA. Strong hybridization with HCβ probe but very hybridization with HCα probe was observed. This patient suffered from primary pulmonary hypertension. In this case, the right atrium had been subjected to pressure overload. The degrees of Northern blot hybridization of atrial mRNA with the probes were analyzed by densitometric scanning. The results were summarized in Fig. 5. In overloaded atria, the level of HCβ mRNA was significantly elevated, whereas the level of HCα mRNA was markedly decreased. Our study demonstrated that HCα and HCβ genes in the human atrium respond to hemodynamic overload in the same fashion as seen in the hypertrophied rat ventricle.

Comparison of the results from Northern blot hybridization and immunofluorescence staining reveals that changes in the HCα and HCβ mRNA level are to a large extent reflected in the changes in their respective proteins. Although immunofluorescence staining is not a quantitative determination, translation and post-translational mechanisms, if present, do not appear to play a major role in the production of heavy chain isozyme switches in response to hemodynamic overload.

It is very interesting to determine whether the same isoform transition occurs in cardiac mysoin light chain. Cummins et al., in an electrophoretic study, have

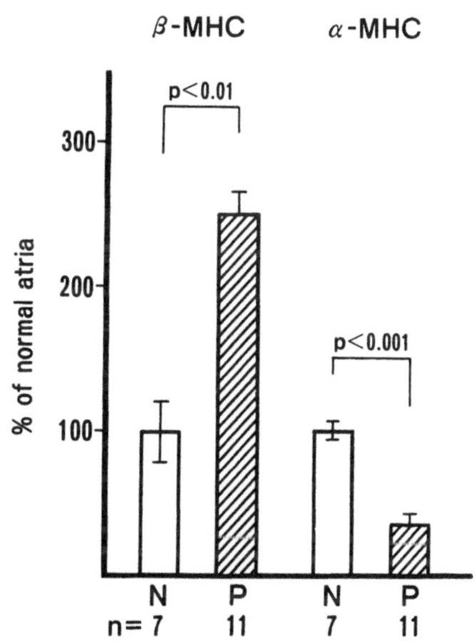

Fig. 5. Steady state mRNA content for HCα and HCβ in normal (*N*) and pressure-overloaded (*P*) atria. Data from each group are presented as mean ± SEM. RNA content in each atrium was quantitated by means of densitometric analysis of autoradiograms of Northern blots hybridized with the appropriate ^{32}P-labeled probes. Results are expressed as percentage of values in normal atria. Student's *t* test indicated the HCβ mRNA level was significantly higher and HCα mRNA level was significantly lower in pressure-overloaded atria compared with normal atria [18]

reported that the ventricular type light chain replaces the atrial type light chain in overloaded human atria [19]. We have isolated essentially full-length cDNA clones encoding atrial or ventricular alkali light chain from a fetal human heart cDNA library and examined the level of each type of myosin light chain mRNA in overloaded atria by Northern blot analysis [20]. Our data demonstrated the increased level of ventricular light chain mRNA from overloaded atria (Fig. 6), suggesting that ventricular myosin light chain gene expression is upregulated by overload, as observed for ventricular heavy chain gene expression. Although the physiological significance of ventricular light chain induction in overloaded atria remains unclear, the concomitant replacement of ventricular type heavy and light chains should work against pressure overload.

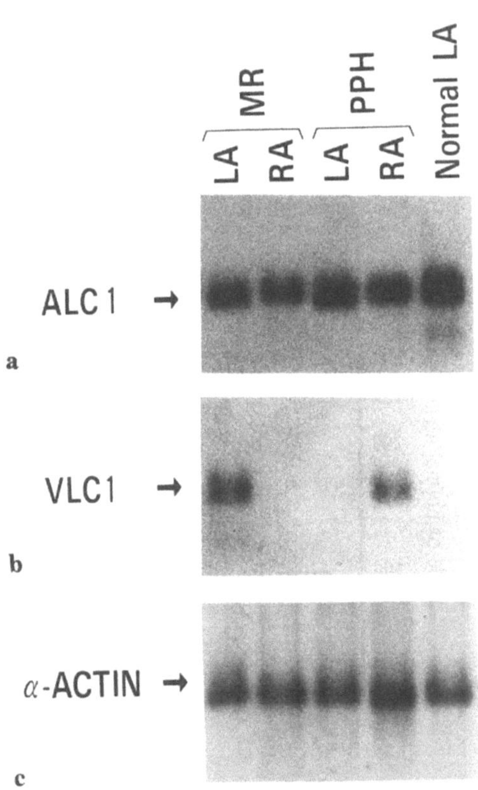

Fig. 6a-c. Detection of atrial and ventricular light chain mRNA ALC1 and VLC1 in total cellular RNA from normal and diseased human atria. The same RNA blot was sequentially hybridized with labeled oligonucleotides specific for ALC1 or VLC1. Normal atrial RNA control is shown in the *right lane* of the figure. The lanes containing left (*LA*) and right (*RA*) atrial RNA (10 μg) from patients with mitral regurgitation (*MR*) or primary pulmonary hypertension (*PPH*) are indicated. The light chain mRNA band is marked on the *left* (ALC1, VLC1). α-Actin band is shown as an internal standard [20]

Conclusion

We demonstrated the distribution of two types of cardiac myosin heavy chains (HCα and HCβ) in the human heart by using monoclonal antibodies. The ventricle comprised mainly HCβ which has low ATPase activity, whereas the atrium was predominantly composed of HCα which has high ATPase activity. We also demonstrated isozymic transition of HCα to HCβ in the human atrium and ventricle due to hemodynamic overload, regarded as a compensatory mechanism to meet an increased demand in work.

To examine the molecular mechanism for the expression of these HCs, we have isolated human HCα and HCβ cDNA clones from a fetal heart cDNA library. Comparison of the nucleotide and amino acid sequence deduced from the DNA between these cDNA clones showed 91% and 96% homology, respectively. Using HCα and HCβ gene-specific sequences, we demonstrated that the transition of HCα to HCβ in the overloaded human heart was induced by the expresson of HCβ-gene.

References

1. Yazaki Y, Raben MS (1974) Cardiac myosin adenosinetriphosphatase of rat and mouse: distinctive enzymatic properties compared with rabbit and dog cardiac myosin. Circ Res 35: 15–23
2. Yazaki Y, Raben MS (1975) Effect of thyroid state on the enzymatic characteristic of cardiac myosin: a difference in behavior of rat and rabbit cardiac myosin. Circ Res 36: 208–215
3. Yazaki Y, Ueda S, Nagai R et al. (1979) Cardiac atrial myosin adenosine triphosphatase of animals and humans: distinctive enzymatic properties compared with cardiac ventricular myosin. Circ Res 45: 522–527
4. Mercadier JJ, Bouveret P, Gorza L et al. (1983) Myosin isozymes in normal and hypertrophied human ventricular myocardium. Circ Res 53: 52–62
5. Hoh JFY, McGrath PA, Hale PJ (1978) Electrophoretic analysis of mutiple forms of rat cardiac myosin: effect of hypophysectomy and thyroxine replacement. J Mol Cell Cardiol 10: 1053–1076
6. Chizzonite RA, Everett AW, Clark WA et al. (1982) Isolation and characterization of two molecular variants of myosin heavy chain from rabbit ventricle: change in their content during normal growth and after treatment with thyroid hormone. J Biol Chem 257: 2056–2065
7. Sartore S, Gorza L, Pierobon-Bormioli S et al. (1981) Myosin types and fiber types in cardiac muscle. I. Ventricular myocardium. J Cell Biol 88: 226–233
8. Clark WA Jr, Chizzonite RA, Everett AW et al. (1982) Species correlations between cardiac isomyosins: a comparison of electrophoretic and immunological properties. J Biol Chem 257: 5449–5454
9. Köhler G, Milstein C (1973) Continuous cultures of fused cells secreting antibody of predefined specificity. Nature 256: 495–497
10. Yazaki Y, Tsuchimochi H, Kuro-o M et al. (1984) Distribution of myosin isozymes in human atrial and ventricular myocardium: comparison in normal and overload heart. Eur Heart J (Suppl F): 103–110
11. Tsuchimochi H, Sugi M, Kuro-o M et al. (1984) Isozymic changes in myosin of human atrial myocardium induced by overload: immunohistochemical study using monoclonal antibodies. J Clin Invest 74: 662–665
12. Chizzonite RA, Everett AW, Prior G et al. (1984) Comparison of myosin heavy chains in atria and ventricles from hyperthyroid, hypothyroid, and euthyroid rabbits. J Biol Chem 259: 15564–15571

13. Mercadier JJ, Lompre AM, Wisnewsky C et al. (1981) Myosin isozymic changes in several models of rat cardiac hypertrophy. Circ Res 49: 525-532
14. Litten RZ, Martin BJ, Low RB et al. (1982) Altered myosin isozyme patterns from pressure-overloaded and thyrotoxic hypertrophied rabbit hearts. Circ Res 50: 856-864
15. Kawana M, Kimata S, Taira A et al. (1986) Isozymic changes in myosin human ventricular myocardium induced by pressure overload. Circulation 74: II-82
16. Alpert NR, Murieri LA (1982) Increased myothermal economy of isometric force generation in compensated cardiac hypertrophy induced by pulmonary artery constriction in the rabbit. Circ Res 50: 491-500
17. Holubarsch C, Goulette RP, Litten RZ (1985) The economy of isometric force development, myosin isozyme pattern, and myofibrillar ATPase activity in normal and hypothyroid rat myocardium. Circ Res 56: 78-86
18. Kurabayashi M, Tsuchimochi H, Komuro I et al. (1988) Molecular cloning and characterization of human cardiac α- and β-form myosin heavy chain cDNA clones: regulation of expression durng development and pressure overload in human atrium. J Clin Invest 82: 524-531
19. Cummins P (1982) Transition in human atrial and ventricular myosin light-chain isozymes in response to cardiac pressure-overload induced hypertrophy. Biochem J 205: 195-204
20. Kurabayashi M, Tsuchimochi H, Komuro I et al. (1988) Molecular cloning and characterization of human atrial and ventricular myosin alkali light chain cDNA clones. J Biol Chem 263: 13930-13936

Chapter B. Mechanical Properties of the Ventricle

Edward L. Yellin

This section progresses from the cellular and ultrastructural level to the macro level of the intact heart. An interesting transition is made by Arts et al., who postulate that control of left ventricular hypertrophy is at the cellular level and need not be fully encoded at the genetic level. They advance this interesting hypothesis on the basis of model studies that require that the process of hypertrophy is controlled by feedback signals related to fiber stress, end-diastolic sarcomere length, and sarcomere shortening during ejection. Sagawa et al. review the phenomenological model of cardiac performance based on the time varying elastance concept, test its assumptions in light of recent findings that the end-systolic pressure–volume relation is nonlinear under conditions of large perturbations in loading, and examine the windkessel model of arterial impedance to derive an approximation of arterial elastance with which to couple the ventricle and aorta. Mirsky uses an entirely different approach based a stress-strain analysis to develop the concept of systolic myocardial stiffness. The concept is validated by comparing contractile states pre- and post-operatively in several patients with valve disease. It will be interesting to see how this new concept is applied in the future.

Glantz and Gilbert present a comprehensive review of the determinants of diastolic function. They point out that diastolic function should no longer be characterized solely by a balance of forces due to interventricular pressure and chamber elasticity; consideration must be given to rate and extent of relaxation, elastic recoil, and pericardial constraints, among others. Nikolic et al. provide experimental evidence for the existence of elastic recoil and for its physiological role in the intact canine left ventricle. This study at the organ level can be shown to be related to the cellular and ultrastructural studies presented in Chapter A (Krueger; van Hoeven and Factor). An additional attempt to relate diastolic function to cellular mechanisms is described in the next paper. Serizawa et al. use a new cardiotonic agent that increases the sensitivity of the contractile elements to Ca^{2+}, in isolated normal and hypoxic rabbit hearts. The data lead them to conclude that relaxation and stiffness can be separated by assuming an exponential decay of left ventricular pressure with a time constant representing relaxation and an asymptote representing stiffness.

Maruyama et al. review the close link between coronary circulation and cardiac function. They focus on the effects of coronary flow on systolic and diastolic function, the effects of systolic and diastolic pressure on flow, and on the effects of loading conditions on the ischemic myocardium. A very important new direction for research is the study of left atrial chamber properties in man by Arakawa et al. It will be shown in the following sections that ventricular filling depends significantly on the atrial pressure, which is in turn a function of the chamber properties. This editor thinks that the role of the left atrium as the source of the upstream component of the driving pressure gradient has been seriously ignored; the Arakawa study points us in the right directin.

These chapters, dealing with mechanical properties, provide a background for the next sections which contain reviews of both systolic and diastolic function in health and disease.

Integrity of Myocardial Fiber Structure Maintained by Fiber Load Induced Local Growth

THEO ARTS[1], PETER H.M. BOVENDEERD[1], FRITS W. PRINZEN[2], and ROBERT S. RENEMAN[2]

Summary. In a mathematical simulation of left ventricular wall mechanics, it is hypothesized that the process of left ventricular hypertrophy is controlled by the feedback signals: fiber stress, end-diastolic sarcomere length, and sarcomere shortening during ejection. In this simulation, a stable final solution is found for the transmural course of fiber orientation, in which cardiac muscle is loaded homogeneously within $\pm 2\%$. The solution is self-correcting for deviations in the anatomical structure. The transmural course of fiber orientation thus found is not significantly different from anatomical findings reported in many species. Thus, according to the simulation, the global, complicated cardiac structure can be maintained by control signals on the cellular level.

Introduction

The heart is a muscular pump. Commonly, the weight of the heart is closely related to the amount of pumping work needed. Comparing different species, smaller animals have smaller hearts. Despite the differences in size, the pattern of fiber orientation seems to be quite invariant.

Within a few weeks after a chronic change in hemodynamic load, the size of the heart adapts to the amount of hemodynamic work demanded. An increase in hemodynamic load may be introduced by hypertension, by disease of a cardiac valve, or by an exercise training program. Despite a change in the size of the myocardial wall, the fiber structure within the wall seems to be maintained. When the cause of hypertrophy is removed, the whole process appears to reverse. This seemingly obvious finding requires a high degree of growth control. How does the heart know which parts have to gain weight, and how much? To simplify matters, the present study is confined to the control mechanism of hypertrophy of the left ventricle.

Fiber Stress as Feedback for Myocardial Hypertrophy

When the left ventricle is subject to a supranormal intracavitary pressure, due, for instance, to aortic valve stenosis or hypertension, the wall thickens considerably,

Departments of Biophysics[1] and Physiology[2], University of Limburg, 6200 MD Maastricht, The Netherlands whereas the internal size of the cavity changes relatively little. However, when stroke volume increases while systolic pressure is unaffected, as in the case of mitral regurgitation, the sizes of wall and cavity both increase [2]. The character of these changes point to control of myocardial growth by fiber stress. After all, in previous model studies [3-5], it was found that the ratio of myocardial fiber stress (S_{mf}) to left ventricular pressure (P_{lv}) is practically linear with the ratio of ventricular cavity volume (V_{lv}) to wall volume (V_{wall}) according to

$$S_{mf}/P_{lv}=1+3\ V_{lv}/V_{wall}. \tag{1}$$

Using the model given by Eq. 1, growth of the myocardium may be induced by fiber stress. Consider an elevated left ventricular pressure. As a result, fiber stress increases above its set-point value. Thus, an increase of muscle mass is induced resulting in a decrease of fiber stress. The negative feedback guarantees stable control. So, macroscopically, muscle fiber stress provides an appropriate means to control myocardial mass.

In order to further investigate the stress feedback hypothesis, a section of left ventricular wall (Fig. 1) is simulated by a cylinder segment [3, 6]. The cylinder segment may shorten circumferentially, shorten from base to apex, and shear in a way which is associated with torsion of the left ventricle around the base to apex axis. The myocardial fibers are considered to be embedded in soft tissue. In the wall of the left ventricle, the equilibrium of stresses and external forces is satisfied. In a first approximation, the transmural course of fiber orientation is taken to be linear (Fig. 2). At the epicardium, midwall, and endocardium the fibers are arranged in a left-handed helix, circumferentially, and in a right-handed helix, respectively. End-diastolic sarcomere length is set homogeneously across the wall.

Fig. 3 shows the transmural course of fiber stress as calculated on the basis of a linear transmural course of fiber orientation. In the subepicardial layers fiber stress appears to be minimum, and in the inner half fiber stress is maximum. In the stress

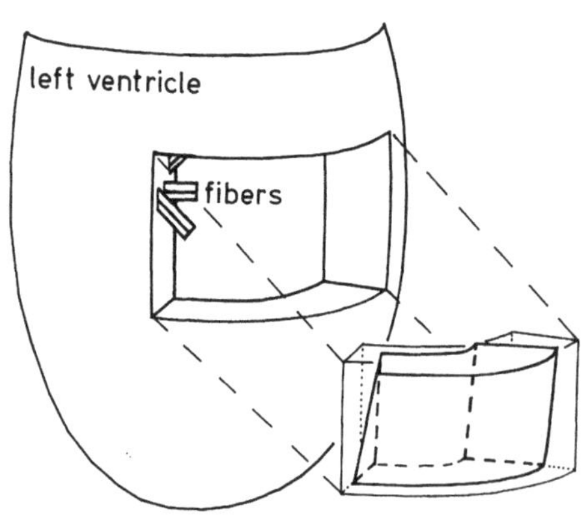

Fig. 1. Schematic representation of a segment of the cardiac left ventricular wall by a cylindric segment. Fiber orientations are indicated. The segment deforms due to circumferential and axial shortening and torsion around the base to apex axis. The wall segment is in mechanical equilibrium with its environment

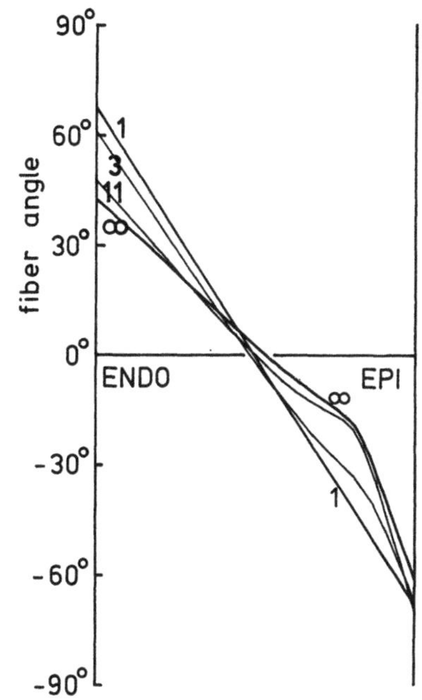

Fig. 2. Simulation of the development of the transmural course of fiber orientation in the left ventricular wall departing from the initial linear course (*1*). The number represents the iteration number. The final course is stable. *ENDO*, endocardium; *EPI*, epicardium

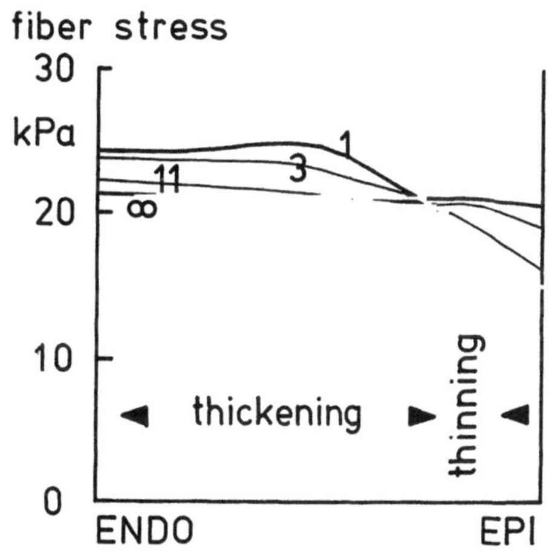

Fig. 3. Transmural distribution of systolic myocardial fiber stress associated with the transmural course of fiber orientation as indicated in Fig. 2. The final, stable solution shows a homogeneous distribution of fiber stress

feedback hypothesis, the layers with low stress degenerate, whereas the layers with high stress hypertrophy. As a result, the transmural course of fiber orientation modifies. Using this new distribution, the transmural distribution of fiber stress is calculated again. The new transmural distribution of fiber stress is much more homogeneous, and so far, fiber stress seems to be a good candidate for control of myocardial growth.

This simulation is a typical example of an apparently stable behavior, while in fact the system is not stable at all. In the first iterations, a fast convergence to a stable solution is found, but after several (more than 10) iterations, small irregularities in

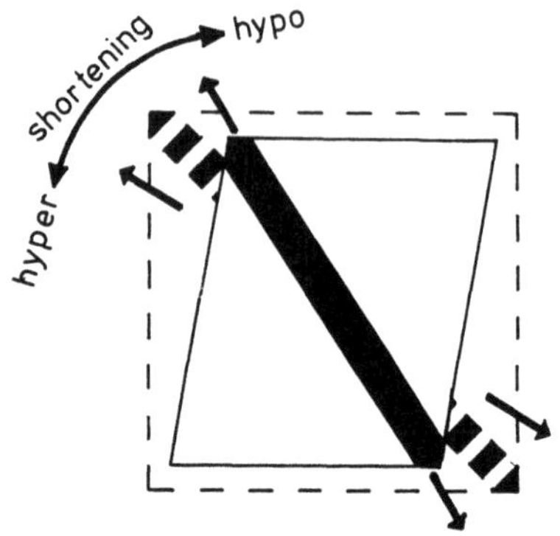

Fig. 4. Cause of local instability when controlling hypertrophy by fiber stress. The *diagonal bar* represents the local average of fiber orientation. The *parallelogram* indicates the square after deformation. Fibers deviating from the main fiber orientation sense a different degree of shortening, causing differences in stress. *hypo*, subnormal shortening, hypertrophy; *hyper*, supranormal shortening, degeneration

the transmural course of fiber orientation tend to grow. These irregularities finally produce a wave pattern in the transmural course of fiber orientation with an ever increasing amplitude. The mechanism of this instability is elucidated by Fig. 4. The *solid bar* represents the average fiber orientation in a certain layer. The instability is associated with the inability of the stress control mechanism to correct for small local deviations in fiber orientation. When some fibers are directed to orientations with less shortening, the fibers will bear a supranormal load. Those fibers will gain in importance by the fiber stress feedback mechanism. On the other hand, fibers alligned with directions with less shortening will degenerate. As a result, the fiber orientation will rotate continuously to directions with less shortening. Because of this rotation, the global solution is influenced, and finally no stable solution can be found. So, the hypertrophying mechanism cannot be controlled by fiber stress and end-diastolic sarcomere length alone.

The Possible Role of Local Dispersion of Fiber Orientation

Upon taking a closer look at the fiber structure of the myocardium, it can be seen that the fibers are not arranged in parallel as perfectly as those in skeletal muscle. The dispersion around the average fiber orientation is approximately $\pm 10°$. In the various layers of the myocardium, shortening during the ejection phase depends on the actual fiber orientation with respect to the local deformation field. So, in a certain area, fibers may sense the direction where fiber shortening is optimal. By a local feedback mechanism they may be directed along this orientation. The precise mechanism is not known, but the signals needed for direction feedback are available on the local level and may be sensed by the local cells. The set-point of optimum fiber shortening may be a muscular property.

In the new hypothesis for the control mechanism of hypertrophy end-diastolic sarcomere length is controlled directly; sarcomere shortening during the ejection

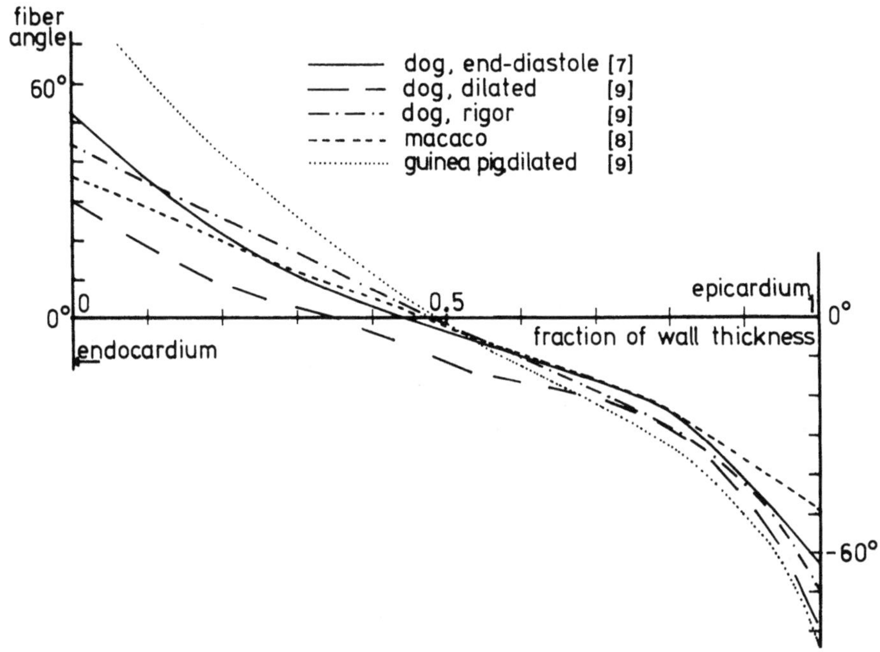

Fig. 5. Comparison of the calculated transmural course of fiber orientation in the wall of the left ventricle with experimental results obtained by various investigators

phase is controlled by finding the optimal fiber orientation, and the cross-sectional area of the myocardial fiber is controlled by fiber stress. When using this model set-up, a linear transmural course of fiber orientation develops a new, stable distribution (Fig. 2). Small perturbations in the transmural distribution of fiber orientation are corrected for automatically. The final solution is independent of the starting conditions for the transmural course of fiber orientation as long as the deviations are not too severe. The convergence occurs within a certain range of starting values for the fiber orientation. In the final solution, the transmural course of fiber stress is homogeneous within $\pm 2\%$ (Fig. 3). End-diastolic sarcomere length and sarcomere shortening are also homogeneous.

Discussion

In the present study, control of the hypertrophying process of cardiac muscle is investigated using a model of myocardial mechanics as presented earlier. In a simple model relating global myocardial fiber stress to left ventricular pressure, myocardial growth could be controlled by feedback on fiber stress. However, on the local level, this feedback mechanism appeared to be unstable. In a second, more complicated set-up, growth is controlled by fiber stress, end-diastolic sarcomere length, and fiber shortening. This mechanism appeared to be stable. The transmural course of fiber orientation thus found appears to be close to the distributions (Fig. 5) reported by

Streeter and Hanna, [7], Ross and Streeter [8], and Hort [9]. All measurements as well as the model show a convex curvature of the fiber angle in the subepicardial layers. The model is based on local control of muscle mechanics. Fiber stress as well as fiber strain are homogeneous. The transmural course of fiber orientation as calculated on this base is not significantly different from experimental results on this course. So, in the real left ventricle, the transmural distribution of mechanical load may be about as homogeneous as in the model.

The present model of myocardial growth control is based on feedback signals, obtained in the cellular environment. So the complicated structure of the left ventricle may be just a result of this feedback mechanism. The structure itself is not necessarily recorded in full detail in the genetic code of the organism. A rough draft of the cardiac architecture may be sufficient to start within the stable range of convergence to the final solution. An argument for this hypothesis may be found in the fact that hearts of many species have a similarly complicated structure, close to the species-independent mathematical solution.

References

1. Pearlman ES, Weber KT, Kanicki JS, Pietra GG, Fishman AP (1982) Muscle fiber orientation and connective tissue content in the hypertrophied human heart. Lab Invest 46: 158-168
2. Cooper GIV (1987) Cardiocyte adaptation to chronically altered load. Annu Rev Physio 49: 501-518
3. Arts T, Veenstra PC, Reneman RS (1982) Epicardial deformation and left ventricular wall mechanics during ejection in the dog. Am J Physiol 243: H379-H390
4. Arts T, Reneman RS (1987) Conversion of fiber stress to global left ventricular pump work. In: S Sideman (ed) Activation, metabolism, and perfusion of the heart. Martinus Nijhoff, pp 353-364
5. Regen DM (1984) Myocardial stress equations: fiber stresses of the prolate spheriod. J Theor Biol 109: 191-215
6. Arts T, Veenstra PC, Reneman RS (1979) A model of the mechanics of the left ventricle. Ann Biomed Eng 7: 299-318
7. Streeter DD, Hanna WT (1973) Engineering mechanics for successive states in canine left ventricular myocardium: II. Fiber angle and sarcomere length. Circ Res 33: 656-664
8. Ross MA, Streeter DD (1975) Non-uniform subendocardial fiber orientation in the macaque left ventricle. Eur J Cardiol 3:229-247
9. Hort W (1960) Makroskopische und mikrometrische Untersuchungen am Myocard verschieden stark gefuellter linker Kammern. Virchows Arch [A] 33: 523-564

Contractility and Pump Function of In Vivo Left Ventricle and Its Coupling with Arterial Load: Testing the Assumptions

Kiichi Sagawa, David A. Kass, Seiryo Sugiura, Daniel Burkhoff, and Joe Alexander[1]

Summary. In the studies reviewed, we tested the assumptions inherent in applying the ventricular and vascular elastance concepts to in vivo physiology. The greatest difference between in vivo and isolated heart data appeared to be in the frequently nonlinear shape of the end-systolic pressure-volume relationship (ESPVR). This suggests that linear model parameters may be of only limited value for predictive or modeling purposes. Extreme changes in ejection history in isolated hearts revealed an important *positive* (as well as negative) influence of ejection; however, in vivo where afterload alterations and ejection fraction changes are likely to be much less marked, the relative impact of these factors remains unclear. Finally, while substantial differences exist between real arterial and 3-element Windkessel model impedances, when they are loaded on a simple elastance model of ventricular contraction, we predicted good agreement on many important hemodynamic coupling variables. Furthermore, the effective arterial elastance E_a appeared to be a reasonable representation of in vivo aortic impedance, and thus may be a useful afterload parameter to be coupled with end-systolic elastance(E_{es}).

Modifications of the coupling equations derived previously for stroke work and myocardial efficiency will likely need to be made to reflect the nonlinear ESPVR. However, as we noted, this may not substantially alter several previously proposed principles and conclusions based on linear models.

Introduction

Early studies of pressure-volume relations in isolated blood perfused mammalian hearts revealed the end-systolic pressure-volume relationship (ESPVR) to be linear over a wide load range, with a slope (end-systolic elastance, E_{es} that varies with contractile state [1, 2]. Characterization of aortic input impedance by the effective arterial elastance (E_a) provided a means to couple the ventricle to the vasculature, and predict stroke volume and stroke work for any given end-diastolic volume [3, 4]. These concepts have been increasingly applied to more intact physiologic animal studies, as well as in man [5–7].

Recent studies have revealed differences between isolated, denervated left ventricles ejecting into a computer simulated 3-element Windkessel afterload impedance model, and an intact heart ejecting into a normal arterial tree. In addition to disparities of structure, chemical environment, contractile state, innervation, and load-

[1] Department of Biomedical Engineering, School of Medicine, The Johns Hopkins University, Baltimore, MD 21205, USA

ing, isolated heart data are often obtained in steady state, while in vivo data are more frequently generated under transient conditions. This difference could potentiate the influence of contraction history on the ESPVR. For a pertinent application of the pressure-volume relationship framework in vivo, critical examination is required of these differences and the assumptions underlying the use of E_{es} and E_a.

In this report, we address four issues. Two regard potential factors affecting the characterization of the ESPVR by a single end-systolic elastance (E_{es}), and two examine the use of E_a to assess in vivo afterload impedance. Specifically we have asked:

1. Is the in vivo ESPVR linear, and if not, then how might curvilinearity influence analysis of ventriculo-vascular coupling?
2. What is the short-versus long-term effect of ejection history on end-systolic pressure?
3. Is the 3-element Windkessel model of the arterial input impedance reasonable, and how might more complex loading models be tested?
4. Can E_a be determined in vivo from pressure-volume relations?

ESPVR Nonlinearity and Its Implications

Although most isolated heart studies throughout the 1970s and 1980s demonstrated linear ESPVRs, more and more investigations have revealed significant nonlinearity of ESPVR [8, 9]. This likely relates to improved experimental conditions and the more frequent use of in situ preparations with typically higher contractile states than in isolated hearts. In a study by Burkhoff et al. [8], ESPVRs were determined in blood-perfused, isolated canine left ventricles, with contractile state varied over a wide range via pharmacologic agents, altered coronary perfusion, and extrasystolic premature and potentiated beats. Within a narrower range of inotropic states typical of isolated hearts defined by end-systolic elastance (E_{es}) of 3.5 through 7 mmHg/ml, ESPVRs were nearly linear. However, when contractile state was more strongly augmented, the relations clearly demonstrated curvilinearity concave to the volume axis. At much lower inotropic states, curvilinearity became convex (Fig. 1a).

Similar curvilinearity has recently been demonstrated in ejecting in vivo ventricles [9] (Fig. 1b). In these studies, ventricular inotropic state was independently assessed using the relatively load-independent slope of the stroke work - end-diastolic volume relationship (preload recruitable stroke work - PRSW). The extent of ESPVR nonlinearity (defined by the second degree term of a quadratic model) was found to consistently vary as a function of inotropic state (Fig. 2). The curvilinearity markedly increased at high contractilities.

ESPVR curvilinearity and its apparent dependence on contractile state is in concordance with mechanisms of length-dependent calcium activation demonstrated in intact and skinned papillary muscles [10, 11]. This phenomenon has been ascribed to alterations in both the affinity to and availability of Ca^{2+} to myofilaments as a function of muscle and/or sarcomere length.

To what extent does the nonlinearity of ESPVR alter the conceptual framework of ventricular performance and vascular coupling based on the linear ESPVR models? The slope of an ESPVR, or end-systolic elastance (E_{es}), becomes a function of the

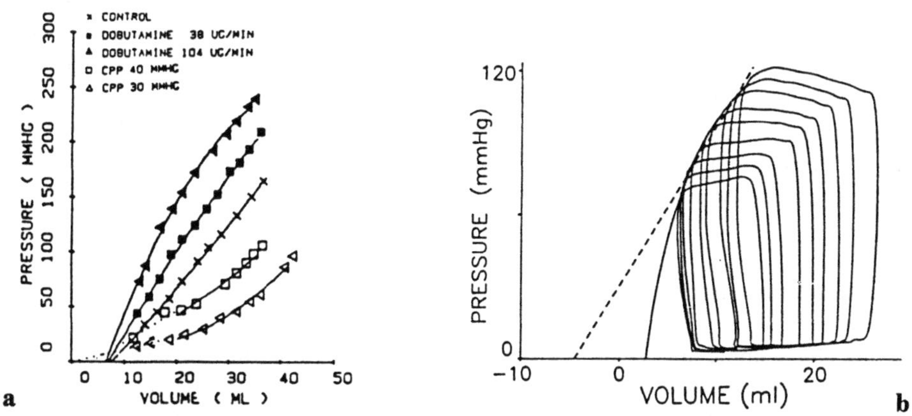

Fig. 1a. end-systolic pressure-volume relationship (ESPVR) curvilinearity in isolated canine hearts. Five different contractile states varied by dobutamine and reduced coronary perfusion pressure are shown. Data are fit to a quadratic ($P_{es} = aV_{es}^2 + bV_{es} + c$). P_{es}, end-systolic pressure; V_{es}, end-systolic volume. ESPVR shape was dependent on contractile state. **b** ESPVR nonlinearity in situ. The *dashed line* is a linear approximation, while the *solid line* is the best fit parabola

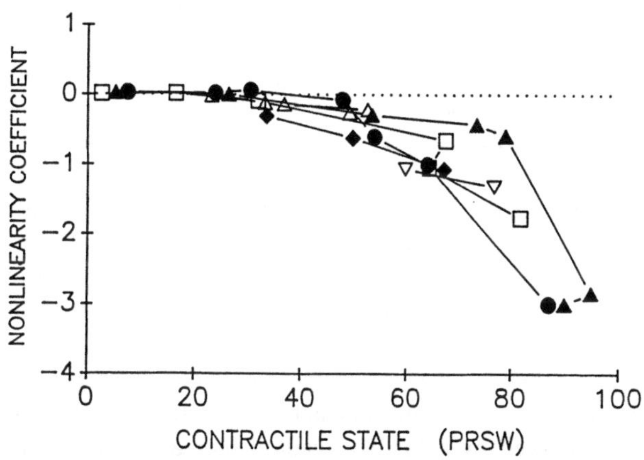

Fig. 2. Dependence of in vivo ESPVR curvilinearity on contractile state. Contractility is indexed by the slope of the relation between stroke work and end-diastolic volume (preload recruitable stroke work *PRSW*). Nonlinearity is indexed by the quadratic term of a parabolic model. With increasing contratile state, ESPVR nonlinearity increased

operating loading range. Thus, extrapolations outside this range, or comparisons of ESPVRs over markedly differing load ranges must be made very cautiously.

The ESPVR, whether linear or not, still represents the limits of systolic pump performance, and if characterized by an appropriate model, can still be used to predict pump performance when coupled to a vascular loading system. For example, the prediction that maximal stroke work (SW) is achieved when E_{es} is matched to the effective arterial elastance (E_a) (i.e., $E_{es} = E_a$) can be mathematically shown to hold true for nonlinear ESPVRs. In the nonlinear case, SW is maximal when the local slope of the ESPVR at a given end-systolic volume [$E_{es}(V_{es})$] equals the arterial elastance E_a. One important difference, however, is that with a linear ESPVR, the ejection fraction at the optimal coupling is $\leq 50\%$, whereas with a curvilinear ESPVR concave to the volume axis, the optimal EF can be 67%, more typical of normal man, without an increase in the E_{es}/E_a ratio. Further studies will be needed to test these predictions and to evaluate the role of ESPVR nonlinearity in various model predictions regarding myocardial efficiency.

Influence of Ejection History

While the simple model of ventricular contraction as a time-varying elastance implied that pressure could be predicted at any moment by the instantaneous volume, many studies have found that additional factors needed to be considered [12, 13]. Both the decoupling effect of instantaneous shortening velocity and the deactivating effect of muscle shortening during ejection have been found to reduce systolic pressure. We have recently reported, however, that ejection can have a *positive* effect, in that the end-systolic pressure (P_{es}) of an ejecting beat can exceed the pressure attained by an isovolumic beat at the same end-systolic volume. In a study by Hunter et al.[14], isolated, blood-perfused left ventricles were loaded with a simulated arterial system. Steady-state ejecting beats were abruptly switched to an isovolumic contraction at an end-systolic volume (V_{es}) similar to that attained during ejection. When the starting ejection fraction (EF) was less than 50%, P_{es} of the ejecting beat exceeded that of the isovolumic beat (by 10 mmHg with EFs $\approx$ 30%), while with EFs larger than 50%, the negative influence of ejection gradually prevailed.

These results have recently been extended to include not just the immediate but also the long-term response to altered contraction mode [15]. End-systolic volume was clamped at 20–25 ml, and stroke volume (SV) varied from 5 to 25–30 ml. In hearts with small SVs and EFs, the initial beat after changing from ejecting (EJ) to isovolumic (ISO) contraction showed a fall in P_{es}, (as Hunter et al. found), and an even further decline was observed over the ensuing 3 min before steady-state was re-established. Switching contraction back to ejection mode led to an immediate rise in P_{es}, followed by a further gradual increase to the initial baseline value (Fig. 3A). With larger SVs (and EFs), switching from EJ to ISO showed a *rise* in P_{es}, which was followed by a gradual decline. The steady-state isovolumic P_{es} could still be higher than, equal to, or lower than that of the initial EJ beat depending on how large the increase in SV and EF is. With the return to ejection mode, an acute reduction in P_{es} was observed, followed by a gradual rise until the initial baseline was reattained (Fig. 3B).

These studies demonstrated that the net effects of ejection on end-systolic pressure is a balance between its positive and negative influences. In ejecting contractions at low EFs, a positive influence can prevail. This positive influence may be ascribed to length-dependent increase in calcium affinity of and availability to myofilaments due to the larger volumes (both at end-diastole and averaged throughout the cardiac cycle) of the ejecting beats [16, 17]. The effect of length on the affinity is instan-

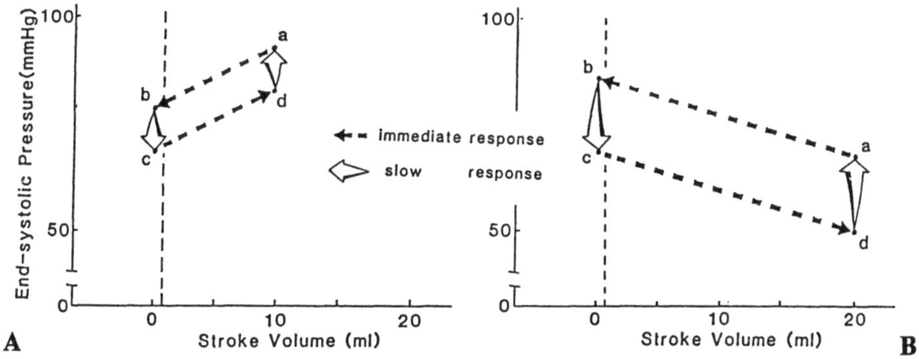

Fig. 3A, B. Schematic diagrams showing end-systolic pressure (Pes) response to altering contraction mode. In both **A** and **B**, mode is altered from ejecting (*a*) to isovolumic (*b*) at similar end-systolic volume. The longer term isovolumic response (*c*) precedes a rapid switch back to ejecting mode (*d*). **A** shows the response with a small stroke volume, **B** for a large stroke volume

taneous, but the effect on the availability develops slowly [17]. The positive influence is counterbalanced by the negative effects of shortening that increase with speed and stroke volume. Thus, in beats with low SV, the positive influence of a contraction history at higher volumes predominates, and P_{es} is reduced with isovolumic contraction immediately by the decreased affinity and subsequently by reduced calcium availability. With greater shortening, the deactivation effect of ejection dominates and P_{es} rises with isovolumic contractions. However, the subsequent loss of length-dependent activation supersedes this initial effect, and P_{es} decreases with time towards, or even below, the initial ejecting end-systolic pressure.

Although these findings were obtained in isolated hearts, they may well have implications for in vivo pressure-volume relations. Many in vivo relations have been obtained by transient (beat by beat) alterations in loading; most commonly by preload reduction (via inferior vena caval occlusion) or by afterload alteration (by pharmacologic agents or aortic occlusion). Studies have shown that ejection fraction is much less sensitive to preload change than to afterload alteration. Thus, it is more likely that ESPVRs obtained with afterload manipulation will be influenced by different ejection fractions than those obtained by preload manipulation. Recent in vivo animal data show predominately a negative effect of shortening [18], while preliminary human data have not shown much of an influence of ejection history [19]. The full extent to which the behavior observed in the isolated heart studies affects in vivo pressure-volume relations remains unclear.

Three-Element Windkessel vs In Vivo Vasculature

A large body of information regarding ventricular pump function and the ESPVR has derived from studies performed in isolated, isovolumically contracting hearts. Yet, there has always been an appreciation and interest in understanding the behavior of the heart coupled to an arterial impedance, and certainly in vivo, the latter data is of paramount importance. The 3-element Windkessel model has been extensively used to characterize arterial loading because of its simplicity and consequent ease in use for real time loading by on-line solution of simple differential equations. While the Windkessel embodies many prominent features of real impedance, there are important differences. We have recently examined the theoretical impact of several of these differences on ventricular pump performance parameters such as stroke volume, stroke work, pressure-volume area (PVA) (related to myocardial oxygen consumption [MVO_2], and blood pressure [21].

Real aortic input impedance was repeatedly determined in 5 open-chest dogs under a wide range of steady-state inotropic and loading conditions using random pacing (white noise) analysis. Each spectrum was then analyzed to provide parameter values (R_a, R_c, and C_a) for the best fit Windkessel. R_c was determined using the mean of the high frequency terms, and C_a by the technique of Liu et al. [20]. As a first step to understand the impact of the differences between real impedance and the Windkessel model on cardiac performance, both the real and Windkessel impe-

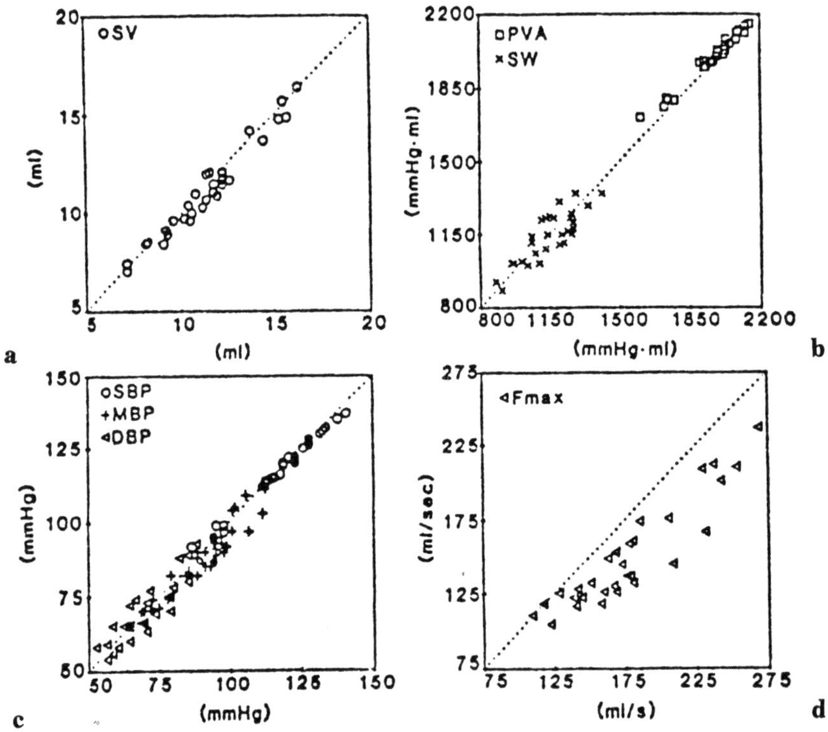

Fig. 4a-d. Comparison of model predicted results for **a** (*SV*) stroke volume **b** (SW) stroke work **c** arterial pressure and **d** aortic flow using real aortic impedance loading versus a 3-element Windkessel model. Both best fit models were coupled to a simple time-varying elastance afterload model of the ventricle. Pump performance parameters were very similar with both impedance loads, with the exception of aortic flow, which was lower with the Windkessel than with real impedance loading. *PVA*, pressure-volume area; *SBP*, systolic bloodpressure; *MBP*, mean blood pressure; *DBP*, diastolic blood pressure; *Fmax*, peak aortic flow dance spectra were coupled with a ventricular model using a linear time-varying elastance that was insensitive to loading conditions. The simplicity of the ventricular model permitted clear understanding of causality. Coupling with the real impedance spectrum was performed by convolution of the history of aortic flow with the impulse response of the arterial system, the latter being calculated by inverse Fourier transform of the impedance spectra.

Paired model predictions of several hemodynamic variables are shown in Fig. 4. In each panel, prediction with the real input impedance is plotted on the *abscissa* and that with the Windkessel on the *ordinate*. As indicated by the distribution of data close to the identity line, there was little influence of the impedance difference on stroke volume, stroke work, PVA, or systolic, diastolic, and mean arterial pressures. The only exception was aortic flow, which was significantly lower (15%) with the Windkessel model than with real impedance loading. Thus, for the purpose of predicting stroke work, stroke volume, PVA, or mechanical efficiency ($SW/MVO_2 = SW/[a\mathrm{PVA} + b]$) with a time-varying elastance model, the use of a 3-element Windkessel model appears satisfactory. Future work will test the effects of real aortic input impedances on beating hearts by real-time convolution. In this way we can better assess the influence of reflected waves on ventricular pump performance.

E_a as an Index of In Vivo Aortic Impedance

It is very cumbersome to couple the frequency-domain representation of arterial impedance with ventricular end-systolic elastance (E_{es}). Sunagawa et al.[3, 4] proposed the concept of an effective arterial elastance (E_a), by which the arterial input impedance (modeled as a Winkessel) could be represented by a single parameter. E_a is related to the Windkessel parameters (R_a, R_c, C_a), and the ejection (t_s) and arterial diastolic (t_d) time intervals as follows:

$$E_a = (R_a + R_c) / \{t_s + \tau[1 - e^{(-t_d/\tau)}]\} \tag{1}$$

where $\tau = R_a \cdot C_a$.

From this definition, it could be shown that E_a could be approximated by:

$$E_a \approx MAP/SV \tag{2}$$

where MAP is mean arterial pressure. By substituting end-systolic pressure (P_{es}) for MAP, E_a could further be simplified:

$$E_a \approx P_{es}/SV. \tag{3}$$

This version of E_a is readily available from end-systolic and diastolic pressure-volume data, thus markedly simplifying coupling calculations. Studies in the isolated canine hearts have confirmed the equivalence of the two expressions (Eqs. 1, 3), and the ability to predict stroke volume (and/or stroke work) from knowledge of the ventricular E_{es} and vascular impedance model parameters [4]. However, whether the most simplified version of E_a (Eq. 3) can adequately represent in vivo aortic input impedance and whether it is influenced by changes in mean arterial pressure (MAP) remained to be tested.

Open-chest, autonomically blocked (hexamethonium) dogs were instrumented with an aortic flow probe and an aortic catheter-tip micromanometer. A left atrial cannula connected to a reservoir enabled alteration of MAP. Aortic input impedance was characterized by Fourier spectral analysis of randomly paced beats, from which R_a, R_c, and C_a parameters were identified (as previously described). Aortic flow was integrated to obtain stroke volume. E_a was calculated using both Eqs. 1 and 2 and the results compared (Fig. 5). The data showed the approximation of E_a by P_{es}/SV was quite good: $P_{es}/SV = 1.05\ E_a + 0.26$ ($r^2 = 0.988$).

We next assessed in vivo whether E_a varied with P_{es} when cardiac output was changed over a wide range of ventricular preloads. Six anesthetized dogs were instrumented with a volume (conductance) catheter and left ventricle micromanometer for instantaneous pressure-volume data, and preload was altered by transient left atrial hemorrhage. E_a for each beat was approximated by P_{es}/SV, where P_{es} was defined at the point of maximal elastance $[p/(V\text{-}V_o)]$ on each pressure-volume loop. Data were assessed before and after intravenous propranalol (2 mg/kg) to block the sympathetic cardiac reflexes. Fig. 6 shows representative data for unblocked and blocked groups. Data are normalized to the P_{es} and E_a of the first (high volume-load) beat. With intact reflexes, preload reduction clearly led to a significant increase in E_a with successive beats. After beta blockade, however, similar preload reduction led to only small alteration in E_a. Even with active reflexes, E_a change over the first 20% reduction in P_{es} was small. Thus, with adequate blockade, P_{es}/SV approximates a consistent E_a over a fair range of MAPs, although some changes may be seen with extreme alterations in loading.

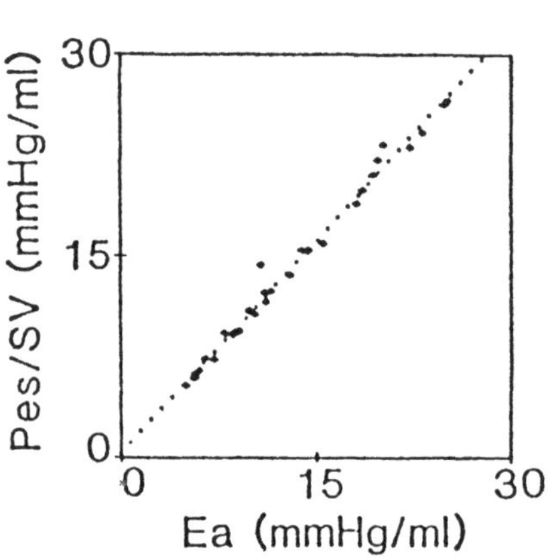

Fig. 5. Comparison of effective arterial elastance (*Ea*) determined from Windkessel model parameters (Eq. 1 of text) versus the more simplified form, Ea ≈ Pes/SV. *Pes*, end-systolic pressure; *SV*, stroke volume. The data demonstrate a good correlation between the two forms, showing that the Pes/SV approximation is quite good in vivo

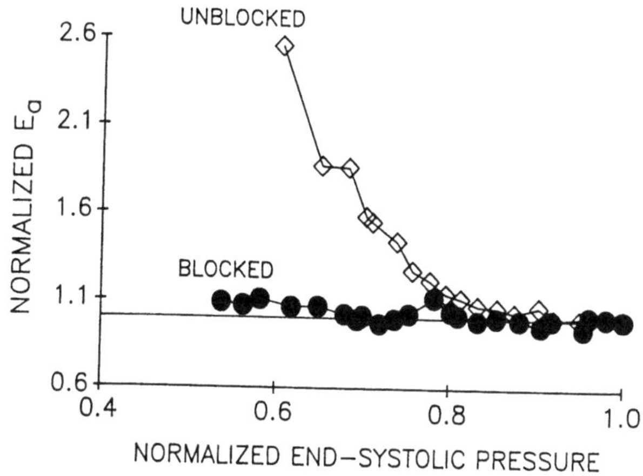

Fig. 6. Influence of altered loading in arterial elastance (E_a) in vivo. Multiple pressure volume loops were determined during preload reduction. End-systolic pressure (P_{es}) and E_a were normalized to their respective starting values. With intact reflexes (*diamonds*), load reduction by ≈20% altered E_a little; thereafter, reflex activation substantially increased afterload with a greater than two fold increase in E_a. With autonomic blockade (*solid circles*), E_a was only little changed over a similar loading range

Conclusion

We have demonstrated that under physiological conditions the hemodynamic properties of the systemic arterial tree appear to be represented well (for most purposes) by a 3-element windkessel model. This model can be simplified further to a single number, the effective arterial elastance (E_a), which allows for analytical solution of aorto-ventricular coupling equations [3, 4, 22]. The E_a representation appears to be valid in the intact animal. On the other hand, the ESPVR is neither linear nor independent of the mode of contraction. Thus, modifications of the coupling equations derived previously for stroke work [4] and myocardial efficiency [22] will likely need to be made to reflect these shortcomings. However, as we noted above, these may not substantially alter several previously proposed principles and conclusions based on linear models.

References

1. Suga H, Sagawa K (1974) Instantaneous pressure-volume relationships and their ratio in the excised, supported canine left ventricle. Circ Res 35: 117
2. Suga H, Sagawa K, Shoukas AA (1973) Load independence of the instantaneous pressure-volume ratio of the canine left ventricle and effects of epinephrine and heart rate on the ratio. Circ Res 32: 314
3. Sunagawa K, Maughan WL, Burkhoff D, Sagawa K (1983) Left ventricular interaction with arterial load studied in isolated canine ventricle. Am J Physiol 245 (Heart Circ Physiol 14): H773

4. Sunagawa K, Maughan WL, Sagawa K (1985) Optimal arterial resistance for the maximal work studied in isolated canine left ventricle. Circ Res 56: 586-595
5. Mehmel HC, Stockinsk B, Ruffnamm K, Olshuausen KV, Schuler G, Kobler W (1981) The linearity of the end-systolic pressure-volume relationship in man and its sensitivity for assessment of left ventricular function. Circulation 63: 1216
6. McKay RG, Aroesty JM, Heller GV, Royal HD, Warren WE, Grossman W (1986) Assessment of the end-systolic pressure-volume relationship in human beings with the use of a time varying elastance model. Circulation 74: 97
7. Kass DA, Midei M, Graves W, Brinker JA, Maughan WL (1988) Use of a conductance (volume) catheter and inferior vena caval occlusion for rapid determination of pressure-volume relationships in man. Cathet Cardiovasc Diagn 15: 192
8. Burkhoff D, Sugiura S, Yue DT, Sagawa K (1987) Contractility-dependent curvilinearity of end-systolic pressure-volume relations. Am J Physiol 252 (Heart Circ Physiol 21): H1218
9. Kass DA, Beyar R, Heard M, Maughan WL, Sagawa K (1987) Curvilinear ESPVR in situ can yield negative estimated volume intercept (Vo) and Vo change with inotropic state. (abstract) Circulation 76 (Suppl IV): IV-428
10. Hibberd MG, Jewell BR (1982) Calcium and length dependent force production in rat ventricular muscle. J Physiol 329: 527
11. Kentish JC, ter Keurs HEDJ, Ricciardi L, Bucx JJ, Nobel MIM (1986) Comparison between sarcomere length-force relations of intact and skinned trabeculae from rat right ventricle. Circ Res 58: 755
12. Suga H, Sagawa K, Demer L (1979) Determinants of instantaneous pressure in canine left ventricle: time and volume specification. Circ Res 46: 238
13. Schroff SG, Janicki JS, Weber KT (1985) Evidence and quantification of left ventricular systolic resistance. Am J Physiol 249: (Heart Circ Physiol 18): H353
14. Hunter WC, Burkhoff D, Oikawa R, Sagawa K (1985) Evidence for mechanisms opposing deactivation during ejection in canine left ventricles. (abstract) Fed Proc 44: 1736
15. Sugiura S, Hunter WC, Sagawa K (1989) Long term versus intrabeat history of ejection as determinants of canine ventricular end-systolic pressure. Circ Res 64: 255
16. Allen DG, Kurihara S (1982) The effect of muscle length on intracellular calcium transients in mammalian cardiac muscle. J Physiol 327: 79-94
17. Parmley WW, Chuck L (1973) Length-dependent changes in myocardial contractile state. Am J Physiol 224: 1195
18. Baan J, van der Velde E (1988) Sensitivity of left ventricular end-systolic pressure-volume relation to the type of loading intervention in dogs. Circ Res 62: 1247-1258
19. Kass DA, Maughan WL (1988) From 'Emax' to pressure-volume relations: a broader view. Circulation 77: 1203
20. Liu Z, Brin KP, Yin FCP (1986) Estimation of total arterial compliance: an improved method and evaluation of current methods. Am J Physiol 251: (Heart Circ Physiol 20) H588-H600
21. Burkhoff D, Alexander J, Schipke J (1988) Assessment of windkessel as a model of aortic input impedance. Am J Physiol 255 (Heart Circ Physiol 24): H742-H753
22. Burkhoff D, Sagawa K (1986) Ventricular efficiency predicted by an analytical model. Am J Physiol 250 (Regulatory Integrative Comp Physiol 19): R1021-R1027

The Concept of Systolic Myocardial Stiffness with Applications to the Assessment of Myocardial Contractility in Health and Disease*

Israel Mirsky[1]

Summary. This review article briefly describes the new concept of systolic myocardial stiffness and its application to assessing the myocardial contractile state in patients with aortic and mitral valve disease. The physiologic studies demonstrate linearity of the end-systolic stress-strain relation (ESSSR) which results in nonlinearity of the end-systolic pressure-volume relation. If V_0 (volume at zero stress), is not markedly altered by drug interventions, changes in the slope of the ESSSR reflect changes in the inotropic state. The concept also permits the development of the entire ejection fraction-afterload relation (for a given preload) thus enabling contractile states to be compared between ventricles.

An approximate method based on a single beat analysis to assess myocardial contractility is applied to clinical data obtained from patients with aortic and mitral valve disease. The contractile state based on conventional methods shows marked discrepancies compared with those assessed on the basis of the systolic stiffness concept. The value of this concept is demonstrated by comparing contractile states pre- and postoperatively in several patients.

Introduction

The elegant concept of ventricular elastance developed by Suga and Sagawa [1] continues to have wide use by many cardiologists in spite of a number of controversies, and these have been adequately addressed by the recent article of Kass and Maughan [2].

In contrast to the ventricular elastance concept which attempts to assess ventricular contractility, the concept of systolic myocardial stiffness [3] permits the assessment of myocardial contractility via the ejection fraction-afterload relation, which is corrected for preload.

This review article describes briefly the applications of this concept [3] and demonstrates (a) how linearity of the end-systolic stress-strain relation results in a new approach for assessing myocardial contractility per se and for assessing changes in the inotropic state and (b) that myocardial contractility may be assessed in aortic valve disease without the use of multiple drug interventions to alter load, and comparisons of contractile states can be made between ventricles of different size and shape.

* This work was supported by grant HL 34596 from the National Heart, Lung and Blood Institute, USA.

[1] Harvard Medical School and Brigham and Women's Hospital, Boston, MA 02115, USA

Terminology

It is appropriate first to define precisely terms employed in this review.

Stress difference (σ). The total stress difference $\sigma = \sigma_\theta - \sigma_r$ is the difference of the circumferential (σ_θ) and radial stress (σ_r) components, which are averaged over the entire cross section. The associated *strain difference* (ϵ), $\epsilon = e_\theta - e_r$, is the difference of the midwall circumferential (e_θ) and radial (e_r) strain components at the equator of an ellipsoid, the assumed geometry of the left ventricle (LV).

Circumferential midwall natural strain (e_θ), $e_\theta = \log (D_m/D_{om})$, is the instantaneous midwall minor diameter of the LV, and D_{om} is the midwall minor diameter at zero stress.

Average systolic myocardial stiffness (E_{av}), $E_{av} = \sigma/\epsilon = \sigma/Km\ e_\theta$ where $Km = (2/3)(2 + D_m^2/L_m^2)$ ave and Lm is the midwall major axis [4].

End-systole. This is the time at which systolic myocardial stiffness attains its maximum value.

Maximum ventricular elastance (E_{es}) $E_{es} = \max\ [P/(V\text{-}V_{op})]$ where V_{op} is the LV volume at zero pressure ($P = 0$).

Physiologic Studies

Animal Preparation

Adult mongrel dogs were anesthetized with intravenous sodium pentobarbitol and prepared with pacing wires, high fidelity micromanometer, and three pairs of ultrasonic crystals for measurements of long and short axes and wall thickness (Fig. 1). Heart rate was controlled during the interventions at approx 120 beats per minute (bpm). Load was altered by inferior vena cava (IVC) cuff occlusion, nitroprusside, atropine, angiotensin II, propranolol, and various combinations with propranolol.

End-Systolic Stress-Strain Relations

Employing the Rankin model [4, 5] for the evaluation of stress difference, average systolic myocardial stiffness (E_{av}), $E_{av} = \sigma/[Km \log (D_m/D_{om}]$ was evaluated iteratively in the following manner:

1. A value for D_{om} was first assumed, and for each intervention, E_{av} was evaluated from the onset of systole until it attained a maximum; by definition, this is end-systole.
2. The stress vs log D_m points at these maxima were plotted and a new value for D_{om} was obtained by appropriate extrapolation to zero stress (Fig. 2a, b).
3. This iterative procedure was continued until D_{om} converged to a constant value. Linearity of the stress vs log D_m relation was demonstrated and the associated end-systolic stress-strain relation (σ_{es} vs ϵ_{es}) was represented in the form (Fig. 2c),

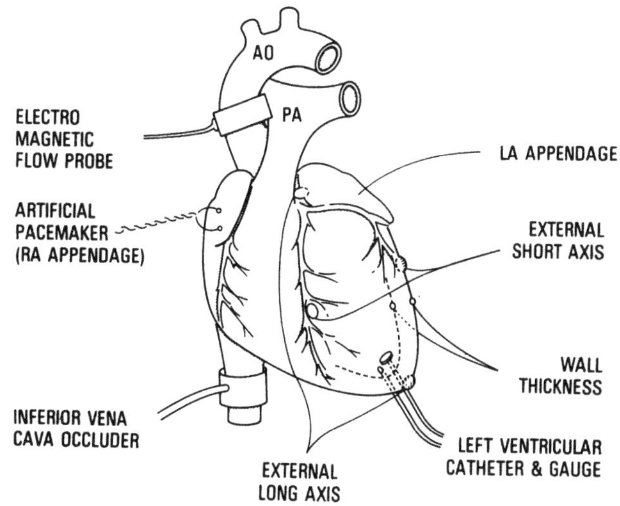

Fig. 1. Instrumentation of the canine heart. *AO*, aorta; *PA*, pulmonary artery; *RA*, right atrium; *LA*, left atrium. From [3] with permission of the American Heart Association

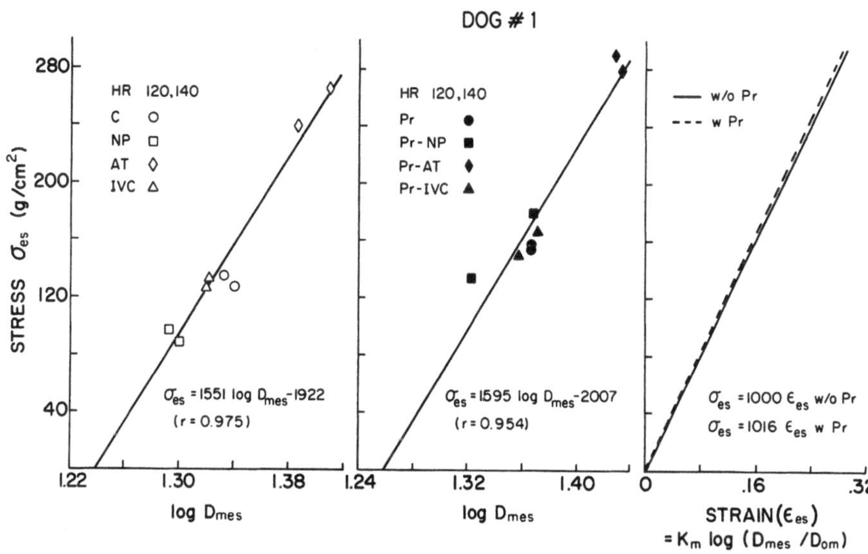

a b, c

Fig. 2a-c. End-systolic stress vs log D_{mes} and associated stress-strain relations **a** without and **b** with propranolol. Afterload was altered from control (*C*) by nitroprusside (*NP*), angiotensin II (*AT*), and inferior vena cava occlusion (*IVC*), and in combination with propranolol (*Pr*). D_{mes}, end systolic midwall diameter; D_{om}, midwall diameter at zero stress. The end-systolic stress-strain relations **c** are linear and imply that end-systolic stiffness $(E_{av})_{es}$ is independent of end-systolic stress (σ_{es}). From [3] with permission of the American Heart Association

$$\sigma_{es} = (\max E_{av}) \cdot \epsilon_{es} \tag{1}$$

Thus max E_{av} is the slope of the linear stress-strain relation and is analogous to E_{max} (the slope of the end-systolic pressure-volume relation [ESPVR]) based on the ventricular elastance concept [1].

End-Systolic Pressure-Volume Relations Based on Maximum Stiffness and Maximum Elastance Concepts

Stiffness Concept

The end-systolic pressure-volume relation (Fig. 3) was derived analytically from the end-systolic stress-strain relation with the result [3]

$$P_{es} = (Km\ \gamma\ \max E_{av}/G) \log (V_{es}/V_{om}) \tag{2}$$

where use has been made of the relations

$$Dm_{es} = A\ V_{es}^{\gamma}; \quad \sigma_{es}/P_{es} = G = \alpha + \beta\ V_{es}.$$

Ventricular Elastance Concept

The ESPVR evaluation follows closely that described for the development of the end-systole stress relation and may be written in the form

$$P_{es} = E_{max} (V_{es} - V_{op})$$

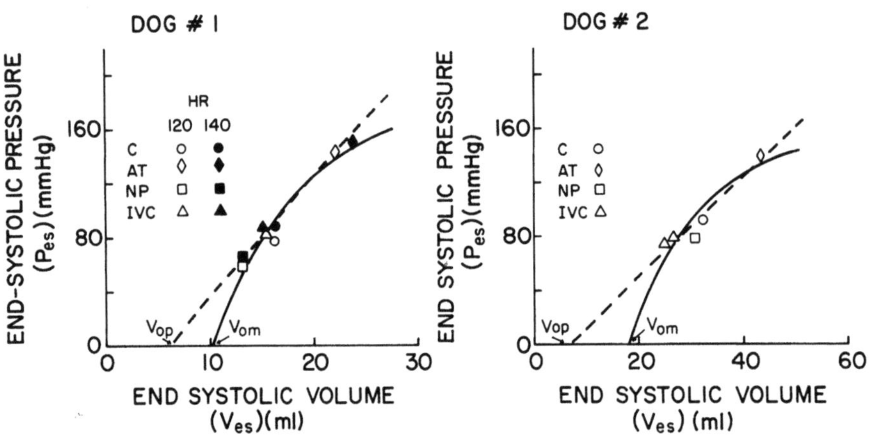

Fig. 3. End-systolic pressure-volume relations (ESPVR) based on maximum elastance and maximum stiffness concepts. Note the nonlinearity of the ESPVR based on the stiffness concept (*solid curve*) in contrast to the linear relation (*dotted line*) based on the elastance concept. The isolated points represent the measured pressure and volumes at the time of maximum ventricular elastance. V_{om} and V_{op} are zero stress volumes based on stiffness and elastance concepts, respectively. *C*, control; *AT*, angiotensin II; *NP*, nitroprusside, *IVC*, inferior vena cava occlusion. permission of the American Heart Association

where V_{op} = volume at zero pressure.
Note the linearity of this relationship in contrast to the above nonlinear relationship Eq. 2 obtained from the stiffness concept.

Ejection Fraction-Afterload Relation Based on the Stiffness Concept
In earlier studies [3] the ejection fraction-afterload relation (EF_{es} vs σ_{aft}) was obtained in the form (Km γ max E_{av}) log $[(1\text{-}EF_{es})V_{ed}/V_{om}] = G\ \sigma_{aft}/(G - 0.5)$ (3) where the afterload σ_{aft} is defined as the circumferential stress (σ_θ) at end systole and V_{ed} is the end-diastolic volume corresponding to a prescribed preload σ_{ed} (end-diastolic stress). Fig. 4 displays these relationships based on the stiffness and ventricular elastance concepts.

Clinical Studies

The concept of systolic myocardial stiffness has been applied to the clinical setting of aortic valve disease in an approximate manner and eliminates the need to alter load by multiple drug interventions [6].

Patients

Of the 70 patients who underwent diagnostic right and left heart catheterization, 26 patients had aortic stenosis (AS), 18 patients had aortic insufficiency (AI), and 26

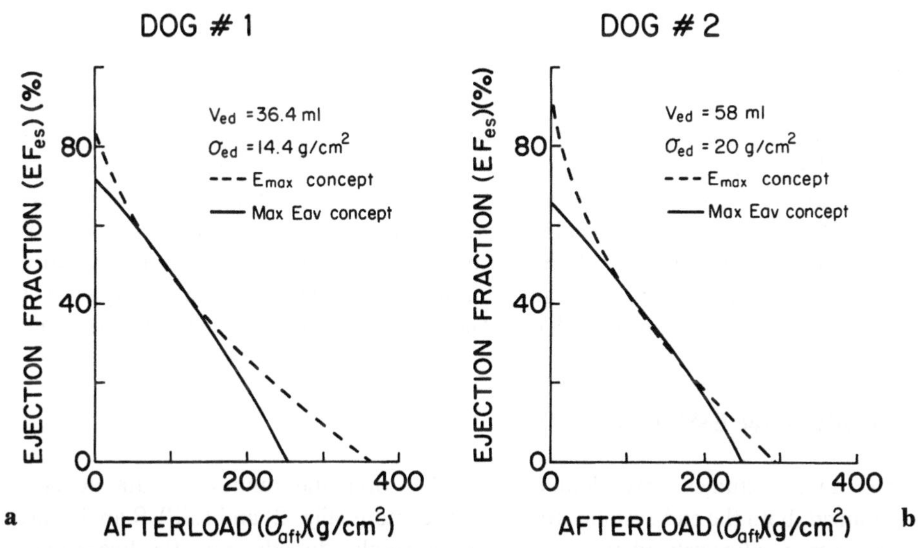

Fig. 4a, b. Ejection fraction vs afterload (EF_{es} vs σ_{aft}) relationships at constant preload based on the stiffness (*solid curve*) and elastance (*dotted curve*) concepts **a** without and **b** with propranolol. These relationships differ markedly at low and high afterload, but are in good agreement over the physiological ranges of ejection fraction and afterload. V_{ed}, volume at end diastole; σ_{ed}, stress at end diastole. From [3] with permission of the American Heart Association patients had chronic mitral regurgitation (MR). Nine patients, evaluated for symptoms of chest pain but who were found to be without significant coronary artery or cardiovascular disease, served as controls.

Patients with AS were divided into two groups on the basis of LV mass index (LVMI), namely AS1: LVMI $< 172\ g/m^2$; AS2: LVMI $\geq 172\ g/m^2$. The MR patients were divided into two groups, since seventeen were in normal sinus rhythm (MR1) and nine were in atrial fibrillation (MR2). In addition, 21 patients with either AS or AI were also evaluated approximately 19 months after successful aortic valve replacement.

End Systolic Stress-Strain, Pressure-Volume, and Ejection Fraction-Afterload Relations

As noted in the animal studies described earlier, the end systolic stress-strain (ESSSR) and end-systolic pressure-volume (ESPVR) relations are governed by the two parameters V_{om} (volume at zero stress) and max E_{av}.

By curve-fitting the single beat stress-volume relations (Fig. 5) in the form $\sigma, \sigma_\theta = A - B\, V^{-\alpha}$, V_{om} was obtained by extrapolation to zero stress with the result $A - B\, V_{om}^{-\alpha} = 0$. Once V_{om} was determined, max E_{av} was evaluated directly from the relation

$$\max E_{av} = \sigma_{es} / [\mathrm{Km}\ \ \gamma\ \ \log (V_{es}/V_{om})].$$

Finally, the analysis for the ESPVR and EF_{es} vs σ_{aft} relations is similar to that described for the animal studies.

Discussion

Ever since Suga and Sagawa [1] proposed their elegant concept of ESPVR, many attempts have been made to apply the concept in the clinical setting, with only moderate success. This lack of success is attributed to (a) the assumption of linearity of the ESPVR outside the physiological range thus bringing into question the constancy of E_{max}, (b) the normalization of the parameter E_{max}, and (c) the resulting negative values often obtained for V_o in many clinical and animal studies as a consequence of the linearity assumption of ESPVR. These problems are now discussed in more detail.

Linearity of the ESSSR and ESPVR

The animal studies have demonstrated that over the ranges of data actually obtained, both the end-systolic stress-strain relations (Fig. 2) and ESPVR are linear. However, nonlinearity of the entire ESPVR results directly from the linear end-systolic stress-strain relation. How then does one reconcile the apparent contradiction? Physiologic considerations dictate that V_{om} cannot attain negative values, which only occur when linear extrapolation of the ESPVR has been performed [7–9]. Furthermore, nonlinearity of the ESPVR at high end-systolic volumes has now been

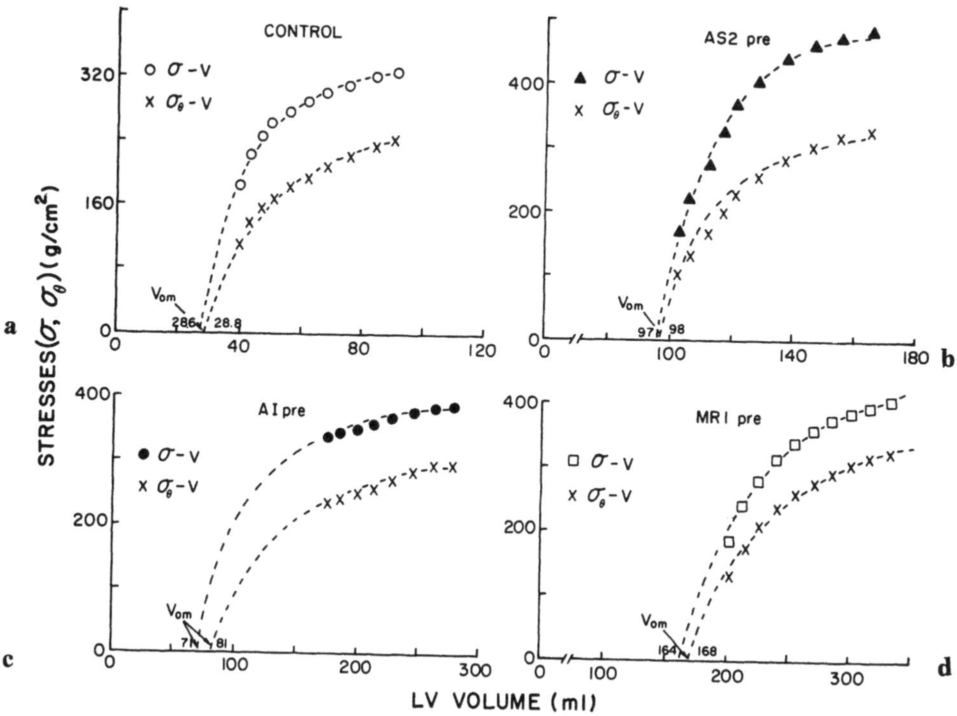

Fig. 5a-d. Estimation of zero stress volume (V_{om}) from systolic stress-volume relations. Stress difference-volume (σ-V) and circumferential stress-volume(σ_θ – V) data are curve-fitted from peak stress to the first minimum volume point in the form σ, $\sigma_\theta = A - B\,V^{-\alpha}$ and V_{om} is determined from the equation $A - B\,V_{om}{}^{-\alpha} = 0$. **a-d** are representative of patients from **a** control, **b** preoperative aortic stenosis (*AS*2 pre), **c** preoperative aortic insufficiency (*AI pre*), and **d** preoperative mitral regurgitation (*MR*1 *pre*) groups. *LV*, left ventricle. From [6] with permission of the American Heart Association

demonstrated by Suga and Sagawa and their associates [2, 10, 11] and in the present animal (Fig. 3) and clinical studies (Fig. 6). In addition, changes in the slope of ESSSR reflect changes in the inotropic state if V_{om} is not markedly altered by drug interventions.

Ejection Fraction-Afterload Relationship as an Assessment of the Myocardial Contractile State

The ejection fraction-afterload relationship has been widely used by cardiologists for assessing the myocardial contractile state. However, its use has been limited in the animal and clinical setting, since ejection fraction is preload dependent; hence, comparisons between ventricles were not possible.

The present concept of systolic myocardial stiffness not only permits comparisons to be made at common levels of preload and afterload, but also enables one to develop the entire ejection fraction-afterload relationship ($EF_{es} - \sigma_{aft}$). Fig. 4 displays for two dog experiments these relationships, which are based on both the

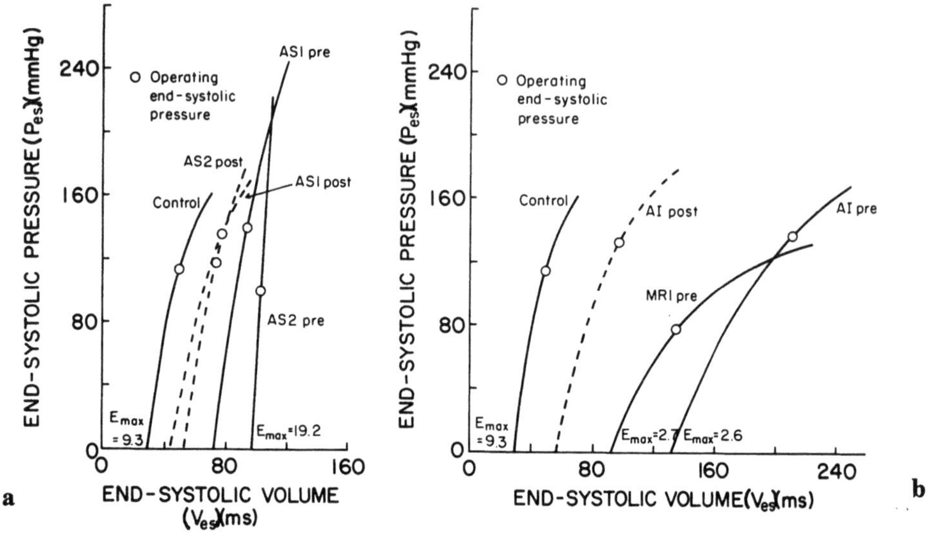

Fig. 6a, b. End-systolic pressure-volume relation (ESPVR) in pressure overload and volume overload-induced hypertrophy. Note the marked nonlinearity of the ESPVR in aortic insufficiency (*AI*) and mitral regurtitation (*MR*) patients. **a** Clinical usefulness of the parameter E_{max} (slope of ESPVR at zero pressure) is questionable since $E_{max} = 19.2$ in *AS2* (aortic stenosis, group 2) patients with depressed contractility compared to control $E_{max} = 9.3$. **b** On the other hand, AI and MR patients operated at normal levels of contractility in the presence of low E_{max} values. From [6] with permission of the American Heart Association

maximum elastance and maximum stiffness concepts. Although the morphology of these curves differ at low and high afterload, there is good agreement over the physiologic ranges of ejection fraction and afterload. It is of interest to note, however, that the convexity of the curve based on the present concept is similar to the pump function curve (mean systolic pressure vs cardiac output) obtained by Elzinga and Westerhof [12]. Further studies are needed to resolve the differences that occur at the high and low afterloads.

With regard to the assessment of myocardial contractility in the clinical setting, three methods have been employed, and the results are displayed in Fig. 7. The conventional approach is to plot single ejection fraction-afterload [13] or shortening-afterload points [14] for each animal or patient, conduct a linear regression analysis, and apply the criterion that myocardial contractile state is depressed if EF or shortening (for a given afterload) lies below the confidence band mean — 2 SDs.

Two alternative methods based on the present concept are displayed in Fig. 7b, c. In Fig. 7b, normality was assessed when $EFc \geq 57.7\%$, where EF_c is the ejection fraction evaluated at control mean levels of preload and afterload ($\sigma_{ed} = 27.6$ g/cm^2, $\sigma_{aft} = 146$ g/cm^2). Fig. 7c is based on a method similar to one proposed by Wisenbaugh [15] who evaluated the ejection fraction (EF_p) at control mean preload and *operating* afterload. As one can observe, there are discrepancies in the assessment of contractile state between all three methods. Further studies both pre- and postoperative are required in order to assess the reliability and sensitivity of these three approaches.

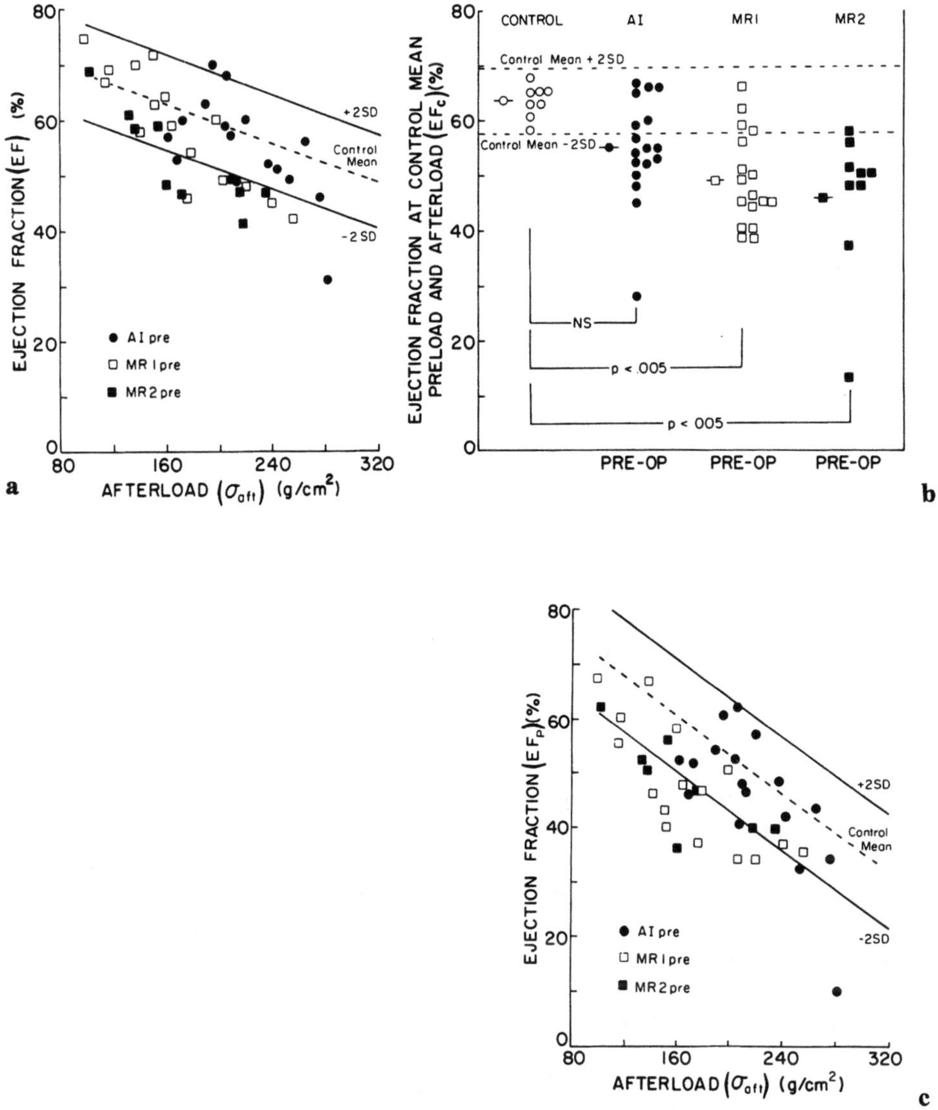

Fig. 7a-c. Assessment of myocardial contractility in aortic insufficiency (*AI*) and in mitral regurgitation (*MR*) based on **a** conventional and **b, c** new methods. **a** A linear inverse relation between ejection fraction (*EF*) (at end ejection) and end-systolic stress (σ_{aft}) is obtained for control group under operating conditions. Myocardial contractility is normal if EF ≥ control mean — 2 SD. This method assessed normal contractility in 13 of 17 AI, 11 of 17 MR1, and 4 of 9 MR2 patients compared with 6 of 17 AI, 4 of 17 MR1, and 1 of 9 MR2 patients based on method **6** and 13 of 17 AI, 8 of 17 MR1, and 4 of 9 MR2 patients based on **c** method. Note the marked discrepancies in the assessment of contractility by **a** and **c** methods compared with **b** method. *PRE-OP*, preoperative levels. From [6] with permission of the American Heart Association

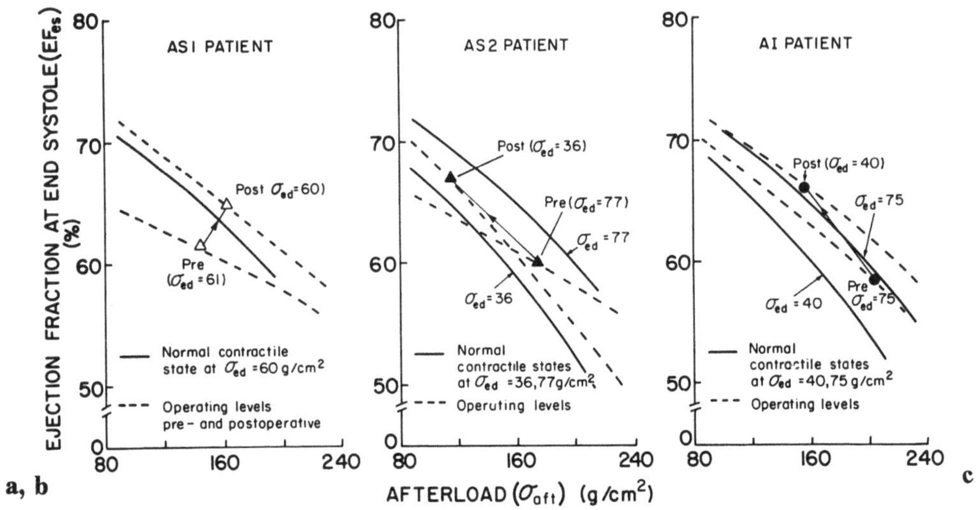

Fig. 8a-c. Preoperative (*Pre*) and postoperative (*Post*) EF_{es} vs σ^{aft} relations in aortic stenosis (*AS*) and aortic insufficiency (*AI*). Preoperatively, contractile state relations in AS1, AS2, and AI patients at operating levels are mildly depressed but return to normal postoperatively. *Solid curves* represent the control relations at specified preload levels. From [6] with permission of the American Heart Association

Fig. 8 displays the preoperative and postoperative contractile states (as assessed by EF_{es} vs σ_{aft} relations at operating preload levels) for representative patients of the AS and AI groups, a result not attainable by current conventional methods. In both AS1 and AS2 groups (Fig. 8a, b), the contractile states were improved postoperatively from preoperative abnormal levels. On the other hand, in the AI group (Fig. 8c), the contractile states remained within normal limits both preoperatively and postoperatively.

Limitations of the Analyses

There are several limitations to the present analyses, and these must be addressed.

1. The estimation of V_{om} is always fraught with error and will require more detailed study in future investigations, particularly in patients with AI. However, the present method always yields positive values for V_{om}, in contrast to negative values obtained occasionally by linear extrapolation [7–9].
2. The theoretical model assumes that stress is a function of strain (volume) alone. Future studies need to examine first-order (viscous) and second-order (inertial) effects [16, 17]. This will require the measurement of volumes at more frequent intervals than those used in the present studies.
3. End-systolic stiffness $(E_{av})_{es}$ has been shown to be load independent in the normal conscious dog (3) and in mild perinephritric hypertension [18], but has yet to be validated in severe pressure and volume overload induced hypertrophy. However, if load dependence of $(E_{av})_{es}$ is demonstrated, the present analyses may still be conducted after making several modifications.

Acknowledgments. The author is indebted to Drs. KL Peterson, HP Krayenbuehl, and associates for their collaborative efforts in the animal and clinical studies described in this review article.

References

1. Suga H, Sagawa K (1974) Instantaneous pressure-volume relationships and their role in the excised supported canine left ventricle. Circ Res 35: 117–126
2. Kass DA, Maughan WL (1988) From 'Emax' to pressure-volume relations: a broader view. Circulation 77: 1203–1212
3. Mirsky I, Tajimi T, Peterson KL (1987) The development of the entire end-systolic pressure-volume and ejection fraction-afterload relations: A new concept of systolic myocardial stiffness. Circulation 76: 343–356
4. Mirsky I, Rankin JS (1979) The effects of geometry, elasticity and external pressures on the diastolic pressure-volume and stiffness-stress relations. How important is the pericardium? Circ Res 44: 601–611
5. Rankin JS, McHale PA, Arentzen CE, Ling D, Greenfield JC Jr, Anderson RW (1976) The three-dimensional dynamic geometry of the left ventricle in the conscious dog. Circ Res 39: 304–313
6. Mirsky I, Corin WJ, Murakami T, Grimm J, Hess OM, Krayenbuehl HP (1988) Correction for preload in assessment of myocardial contractility in aortic and mitral valve disease: An application of the concept of systolic myocardial stiffness. Circulation 78: 68–80
7. Grossman W, Braunwald E, Mann T, McLaurin LP, Green LH (1977) Contractile state of the left ventricle in man as evaluated from end-systolic pressure-volume relations. Circulation 56: 845–852
8. Mehmel HC, Stockins B, Ruffman K, Olhausen K, Schuler G, Kubler W (1981) The linearity of the end-systolic pressure-volume relationship in man and its sensitivity for assessment of left ventricular function. Circulation 63: 1216–1222
9. Kaseda S, Tomoike H, Ogata I, Nakamura M (1985) End-systolic pressure-volume, pressure-length, and stress-strain relations in canine hearts. Am J Physiol 249: H648–H654
10. Sagawa K, Maughan WL, Burkhoff D, Yue D (1985) The similarity of ventricular pressure-volume relation (abstract) and muscle force-length relation. Cardiovasc Res 19: 523
11. Suga H, Yamada O, Goto Y, Igarashi Y (1986) Peak isovolumic pressure-volume relation of puppy left ventricle. Am J Physiol 250: H167–H172
12. Elzinga G, Westerhof N (1979) How to quantify pump function of the heart: the value of variables derived from measurements on isolated muscle. Circ Res 44: 303–308
13. Mirsky I, Pfeffer JM, Pfeffer MA, Braunwald E (1983) The contractile state as the major determinant in the evolution of left ventricular dysfunction in the spontaneously hypertensive rat. Circ Res 53: 767–778
14. Colan SD, Borow KM, Neumann A (1984) Left ventricular end-systolic wall stress-velocity of fiber shortening relation: A load-independent index of myocardial contractility. J Am Coll Cardiol 4: 715–724
15. Wisenbaugh T (1988) Does normal pump function belie muscle dysfunction in patients with chronic severe mitral regurgitation? Circulation 77: 515–525
16. Hunter WC, Janicki JS, Weber KT, Noordergraaf A (1983) Systolic mechanical properties of the left ventricle: Effects of volume and contractile state. Circ Res 52: 319–327
17. Shroff SG, Janicki JS, Weber KT (1985) Evidence and quantitation of left ventricular systolic resistance. Am J Physiol 249: H358–H370
18. Aoyagi T, Fujii AM, Gelpi RJ, Hittinger L, Crocker VM, Mirsky I (in Press) Application of the systolic stiffness concept to assess myocardial function in developing hypertension. Jpn Heart J

Determinants of Diastolic Function*

Stanton A. Glantz and John C. Gilbert[1]

Summary. The original concept of diastolic function of the left ventricle was that the pressure within the left ventricle is determined by the balance between the forces due to pressures within the ventricular cavity that expand the ventricle and forces due to elasticity of the myocardium that resist this expansion. While this concept remains the centrepiece of our understanding of the diastolic pressure-volume relationship, it is now clear that several other factors play a role. During early diastole changes in the rate of active relaxation of the myocardium and so-called diastolic suction are important. During late diastole (and at end-diastole) the extent of relaxation, mechanical interaction between the ventricles modulated by the pericadium, and the pericardium itself are important (Fig. 1). Other factors — pulmonary-cardiac contact pressure, viscoelasticity of the myocardium, and engorgement of the coronary vasculature — are less important in determining the left ventricular diastolic pressure-volume relationship.

Geometry, Muscle Elasticity, and the Basic Pressure-Volume Relationship

The diastolic pressure-volume relationship results from forces acting both *on* and *within* the myocardium. The forces acting on the myocardium, in turn, result from the pressure within the left ventricle and the other cardiac chambers and from the constraining forces exerted on the epicardium by the pericardium and the lungs. The forces acting within the myocardium result from the fact that the myocardium is an elastic material so that the muscle itself develops a resisting force as it is extended or compressed.

The *Laplace Law* relates wall *stress*, σ, to pressure and geometry, i.e., radius and wall thickness in an idealized thin spherical shell:

$$\sigma = pr/2h.$$

The relationship between wall stress and ventricular volume depends on the elastic properties of the myocardium. As in most biological materials, muscle becomes

* This work is a condensed version of [1] which was supported by NHLBI Grant HL-25689. Dr Gilbert was supported by NIH Training Grant HL-07192.

[1] Cardiovascular Research Institute and Department of Medicine, University of California, San Francisco, CA 94143, USA

stiffer as it is extended, i.e., it requires greater increments of force to produce each additional increment of length. *Strain*, ϵ, is a normalized, dimensionless measure of the deformation of a solid such as the myocardium. *Volume strain* is defined as increase over equilibrium volume divided by the equilibrium volume $\epsilon = (V - V_o)/V_o$. The *equilibrium volume*, V_o, is the volume that a shell (such as the left ventricle) exhibits when subjected to no transmural pressure. The pressure the ventricle exhibits depends on the strain of the myocardium and the intrinsic elastic stiffness of the myocardium.

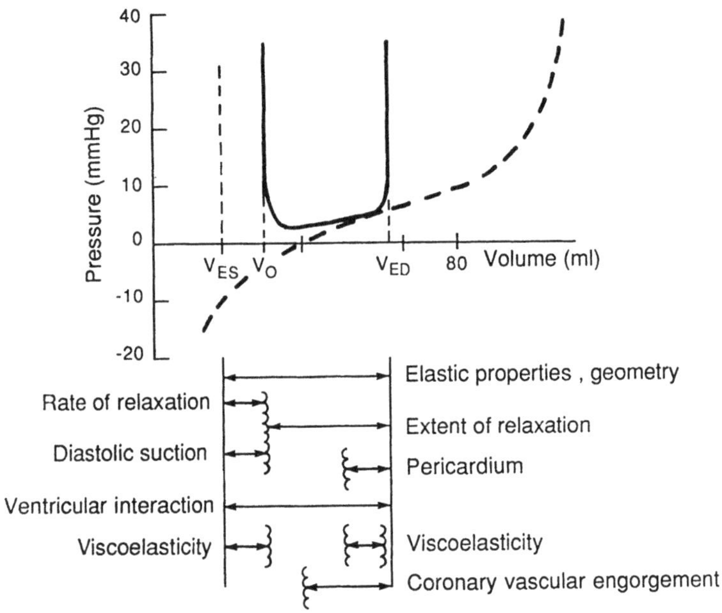

Fig. 1. Several different factors affect the diastolic pressure during diastole, with different factors being important at different times. The myocardium's elasticity and geometry (size and wall thickness) are important throughout diastole; the other effects, superimposed on the pressure due to the myocardial elasticity. During early diastole, active relaxation and the recoil effect of elastic energy stored in the myocardium from the previous systole (so-called 'diastolic suction') determine the atrioventricular pressure gradient and ventricular filling rate. The *dashed line* shows the purely elastic pressure-volume curve of the myocardial shell in the absence of the component due to active relaxation from the previous beat; the negative pressures here reflect the elastic energy stored in the myocardium from the previous systole, which combines with the relaxation component to determine the observed pressure. Late in diastole, ventricular interaction and the pericardium become important. Viscoelastic properties of the myocardium play a small role during rapid filling and during atrial contraction, and coronary vascular engorgement has a small effect late in diastole. The different factors are listed below the curve, with an indication of when they are important in determining the diastolic pressure. *Wavy lines* indicate the approximate boundary when each factor is important. V_{ES}, end-systolic volume; V_{ED}, eqilibrium volume. [1] with permission

Effects of Hypertrophy

The left ventricular responds to a chronic volume or pressure overload by changing its geometry in different ways [2]. Chronic volume overload appears to increase the size of the ventricle without altering the intrinsic elastic properties of the myocardium [3–5]. Chronic pressure overload, on the other hand, causes the ventricular wall to thicken without substantially increasing ventricular diastolic volume [6].

Immediately after the left ventricle is subjected to a volume overload, the myocardium is strained more than usual during diastole because the difference between the volume and equilibrium volume, V_0, is increased. As a result of this increased strain, diastolic wall stress and end-diastolic pressure increase. Over time, cardiac myocytes respond by elongating and thickening in a way that effectively increases V_0 with little or no increase in wall thickness. This increase in V_0 means that the larger volume now corresponds to a lower strain, and thus leads to correspondingly lower wall stress and diastolic pressure. The left ventricle appears to stop growing when it has increased V_0 to the point that end-diastolic wall stress has returned to values present before imposition of the volume overload [7]. This so-called eccentric hypertrophy also occurs in response to physical training [3, 8] and in people with aortic or mitral insufficiency [9].

Subjecting the left ventricle to a pressure overload requires the myocardium to develop greater stress than normal during systole due to the increased afterload. The left ventricle responds to this greater stress by increasing its wall thickness which allows the ventricle to generate increased systolic pressure at a lower level of average wall stress (i.e., force development in individual sarcomeres) than would occur in the absence of hypertrophy. In addition to these purely geometric effects, the structure of the cardiac interstitium changes to increase the amount of collagen present and thus increase the elastic stiffness [5, 10, 11]. The net effect of these changes is to lead to higher diastolic pressure at any given volume. This upward shift in the diastolic pressure-volume relationship accounts in part for the increase in diastolic pressure that accompanies pressure overload and the potential for patients with pressure overload to develop congestive heart failure.

Relaxation

The rate and extent of relaxation are two different quantities, each of which influences the left ventricular diastolic pressure-volume relationship. The extent of relaxation is the end state of the muscle after active relaxation has completed for a given beat. It is the more important of the two, because it determines the myocardium's equilibrium length, which, in turn, determines the left ventricle's equilibrium volumes, which lead to larger strains and higher diastolic pressure at a given brium volumes, which lead to larger strains and higher diastolic pressures at a given end-diastolic volume. Higher pressures at the same volume result in an upward shift in the diastolic pressure-volume relationship. Thus, the extent of relaxation is important at the end of diastole.

The rate of relaxation affects the diastolic pressure-volume relationship early in diastole, because it affects the atrioventricular pressure gradient and therefore affects filling rate during the rapid filling phase of diastole. The rate of relaxation generally does not influence end-diastole directly, because the relaxation process is over by the time diastole ends.

Effects of Ischemia on the Extent of Relaxation

Slowed calcium uptake by the sarcoplasmic reticulum in response to ischemia has been proposed to explain the shifts in the diastolic pressure-volume curve after cardiac pacing induced angina in patients with ischemic heart disease. This hypothesis led several investigators to measure the rate of relaxation in patients following production of ischemia [12–17]. Unfortunately, although people routinely observed slowed relaxation (i.e., a larger value of T, the time constant of isovolumic pressure fall), the reduction in relaxation rate secondary to ischemia was not large enough to produce the shifts in the diastolic pressure-volume curve that had been observed at end-diastole. Moreover, animal studies of the effects of global ischemia and hypoxia on myocardial stiffness [18, 19] failed to demonstrate that ischemia can cause changes in myocardial stiffness.

However, a new experimental preparation which more closely modelled the situation that existed in the clinical studies, clarified the link between ischemia and shifts in the diastolic pressure-volume curve [20, 21]. In this preparation, occluders were placed on the circumflex and left anterior descending coronary arteries of open-chest, open-pericardium dogs and then tightened until flow in these two arteries was reduced by approximately 50%. The left ventricular diastolic pressure-volume curve was not affected by this decrease in flow. The hearts were then paced at approximately twice the resting heart rate. Immediately after pacing was stopped, the pressure-volume curve shifted upward, but returned to normal within 1 min. Thus, in contrast to previous animal studies, reversible shifts in the diastolic pressure-volume relationship were produced that mimicked the shifts observed in patients with coronary artery disease.

Differences from previous experimental studies led Serizawa et al. [20] to propose that there are actually two types of ischemia in the intact heart: "supply-induced" and "demand-induced." During supply-induced ischemia the normally beating heart is underperfused (or perfused with a deoxygenated perfusate), which reduces systolic work, oxygen demand, and the amount of metabolic waste products produced. This is the situation that prevailed during the earlier animal studies. Under these conditions, a low coronary flow is sufficient both to supply oxygen and to remove metabolites; consequently, there are little or no changes in the diastolic properties of the myocardium. In contrast, during demand-induced ischemia, such as that during pacing after coronary artery occlusion, myocardial metabolic demands increase while the supply of oxygen and the capacity to remove waste products are limited by the occlusion. As a result, the sarcoplasmic reticulum takes up calcium more slowly, and more actin-myosin cross-bridges remain intact at the end of relaxation [12, 15, 22–24].

Although the distinction between supply-induced and demand-induced ischemia provides an appealing explanation for the differences observed in different experimental models of cardiac ischemia, it still does not completely explain the mechanism for upward shifts in the diastolic pressure-volume curve in people with coronary artery disease. However, with the advent of percutaneous transluminal coronary angioplasty, the clinical analog of animal experiments in which a coronary artery was transiently occluded became possible. Using this technique, Serruys et al. [25] produced transient upward shifts in the left ventricular diastolic pressure-volume curve in patients with coronary artery disease who underwent angioplasty. The diastolic pressure-volume curve returned to pre-occlusion values after the occlusion ended. The precise mechanisms for these shifts remain to be defined, but these results do indicate that the situation in sick people is more complex than that which occurs in acute animal models of transient ischemia.

Hypertrophy and Relaxation

Whereas so-called physiological hypertrophy, which accompanies increased physical activity, and volume overload hypertrophy do not appear to affect left ventricular relaxation [8, 26], pressure overload hypertrophy [27, 28] and hypertrophic cardiomyopathy [29–31] reduce both the rate and extent of relaxation. The reduced rate and extent of relaxation combine to reduce significantly the rate of left ventricular filling during early diastole. These changes in relaxation appear to be due to changes in the way the sarcoplasmic reticulum handles calcium; the whole process slows in the presence of pressure overload hypertrophy [32] or hypertrophic cardiomyopathy [33, 34].

Myocardial ischemia also plays a role in modulating left ventricular relaxation in the presence of pressure overload hypertrophy, even in the absence of coronary artery disease. Pressure overload hypertrophy is associated with a substantial decrease in capillary density [5] and coronary vasodilator reserve [35]. Thus, in dogs with pressure overload hypertrophy, pacing tachycardia (which increases myocardial oxygen demand) can reduce subendocardial perfusion and produce metabolic evidence of ischemia [36]. Studies of patients with aortic stenosis also reveal hemodynamic and metabolic evidence of ischemia in response to increased demands on the heart [37, 38], as well as histologic evidence of interstitial fibrosis compatible with ischemic injury [39].

The mechanics of pressure overload hypertrophy also aggravate the problems associated with ensuring adequate perfusion of the subendocardium. In thick-walled shells, the stress is highest on the inside wall, and the gradient of stresses across the wall increases as the wall thickens. Thus, as the myocardium thickens to reduce the average stress in the wall, the stress near the endocardium may increase, putting additional demands on the cells located near the endocardium. The reduction of capillary density and coronary reserve in this region that accompanies pressure overload hypertrophy further reduces the capacity to deliver oxygen and remove waste products in this region, particularly when wall stresses increase in response to increased demands put on the heart as a whole.

Implications for Systolic Function

Coronary artery disease is a regional disease in which the regions of the heart that are served by obstructed vessels are affected while the regions served by unobstructed vessels are not affected. Thus, if changes in the extent of relaxation shorten equilibrium lengths of the ischemic muscle, this shorter equilibrium length would cause the left ventricular diastolic pressure-volume curve to shift upward during ischemia. In this situation, it should be possible to show a stiffening of the muscle in regions served by obstructed arteries and normal diastolic properties in the regions perfused by patent arteries. Angiographic studies of regional wall motion in patients with coronary artery disease have demonstrated this point [40]. In regions served by an obstructed coronary artery, the diastolic pressure-segment length curve shifted upward after pacing-induced angina. In parts of the heart perfused by normal coronary arteries, the resulting increase in diastolic pressure led to higher stresses and, therefore, greater strain (extension) of the normal myocardium, which moved up a single diastolic pressure-segment length curve (Fig. 2). This observation confirmed that there are regional changes in muscle stiffness during ischemia in these patients.

Sasayama et al.'s [40] results also have important implications for how the heart maintains overall systolic function in the face of the reduced systolic function of the ischemic myocardium. As shown in Fig. 2, the muscle outside the ischemic region is stretched longer than normal in response to the increase in diastolic pressure due to the stiffening of the ischemic myocardium. Because muscle develops more systolic force when it is stretched (according to the Frank-Starling mechanism), the surrounding normal muscle will increase systolic performance and help compensate for the loss of systolic function in the ischemic region.

This work is the clinical analog of experimental studies on dogs whose circumflex coronary arteries were permanently occluded, then studied in a conscious state after recovering from the surgery [41]. As expected, in the region served by an occluded

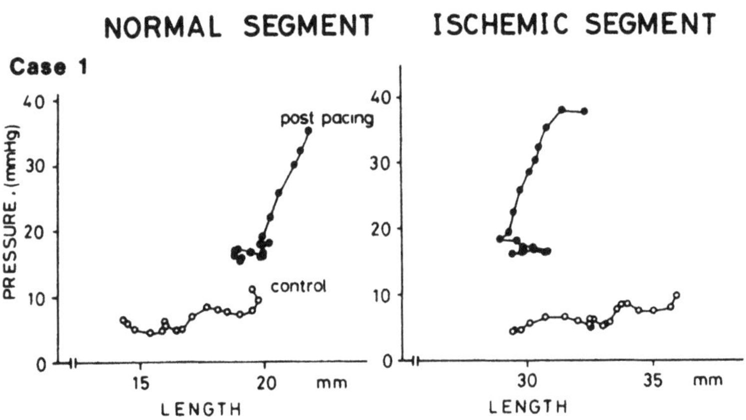

Fig. 2. Angiographic studies of regional wall motion demonstrate the regional effects of ischemia after pacing-induced angina in patients with coronary artery disease. The ischemic segment stiffened, resulting in a higher left ventricular pressure for any given segment length (*right panel*). In the normal segments, this increased pressure caused the muscle to be stretched to longer lengths along a single pressure-length curve (*left panel*). Due to the Frank-Starling mechanism, this increased end-diastolic stretch will make the normal muscle develop more force, partially compensating for the stiffer muscle in the ischemic region. [40] with permission

coronary artery both the end-diastolic segment length and the systolic shortening decreased below pre-occlusion values with a concomitant reduction in systolic function. In contrast, segment lengths increased in the normal myocardium far from the ischemic region of the left ventricle, resulting in an increase in systolic function after 4 weeks. These results can be explained using basic mechanics and the Frank-Starling mechanism without resorting to a "zone of injury" near the ischemic area or compensatory hyperfunction remove from the injured area. Because all the myocardium must remain connected, the normal myocardium near the ischemic region is mechanically constrained by its attachments to the (stiffer) ischemic myocardium [42]. Thus, the stiffer myocardium cannot be stretched to normal end-diastolic lengths, the sarcomeres cannot move up as far on their individual Frank-Starling curves, and the muscle near the ischemic region develops less systolic force. However, for the left ventricle to fill to its normal end-diastolic volume, muscle farther from the stiff region must stretch more than normal and will therefore develop greater force during systole. Thus, compensatory lengthening in the normal muscle due to the mechanical forces acting in the myocardial shell tends to reduce the global severity of regional ischemia on systolic function.

Ventricular Diastolic Suction

Diastolic elastic recoil contributes to the left ventricular diastolic pressure-volume relationship by affecting rapid diastolic filling. Normally, the forces generated during systole compress the muscle fibers to a length shorter than their equilibrium length and in so doing store energy in the muscle fibers and the surrounding collagen matrix as in a compressed spring. The muscle does not need to be compressed very much to generate large recoil forces, because the pressure-volume curve for the left ventricle bends down sharply at negative strains during compression (Fig. 3). These forces decrease the pressure in the left ventricle during early diastole and, therefore, increase the atrioventricular pressure gradient, which in turn increases left ventricular filling rate [43–45]. Under some experimental conditions negative ventricular pressures can be observed as evidence for this elastic recoil, and because a negative ventricular pressure during diastole would tend to suck blood into the ventricle from the atrium, the condition is often called *ventricular diastolic suction*.

Diastolic suction helps the heart respond to increased demands when contractility increases. Increasing contractility leads to smaller end-systolic volumes, stronger recoil forces, a larger atrioventricular pressure gradient [43, 46] and, thus, more rapid early diastolic filling [47]. This increase in filling rate maintains or even increases [48] end-diastolic volume and so maintains end-diastolic sarcomere stretch and, therefore, systolic function.

Ventricular Interaction

The chambers of the heart interact mechanically with each other, particularly during diastole, when the pressures (and, hence, wall stresses) are low. Acute increases in right ventricular volume cause increases in left ventricular diastolic pressure.

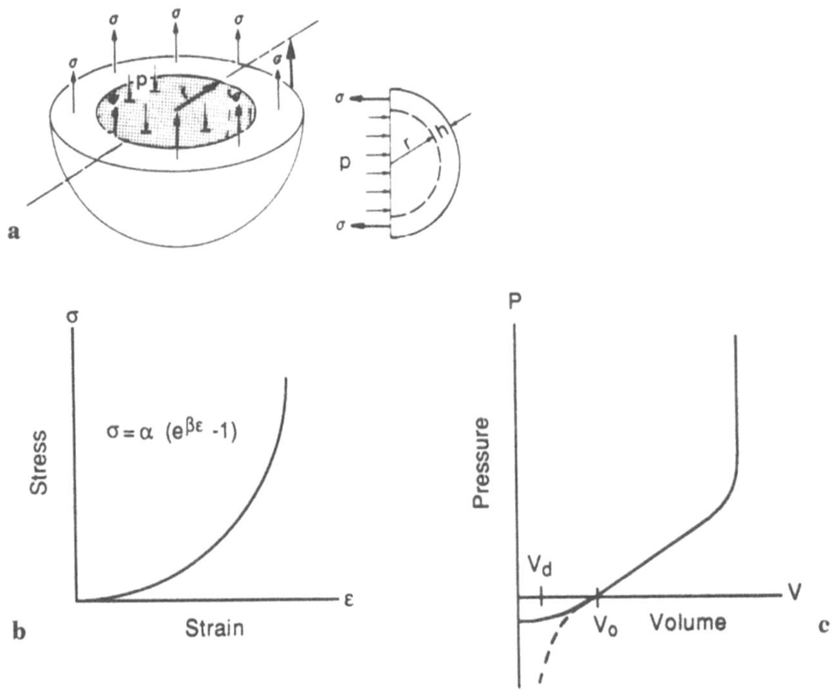

Fig. 3a-c. The elastic component of the diastolic pressure-volume curve is determined by **a** the balance of the force (pressure) in the ventricular chamber tending to push the chamber apart and force (stress) in the wall holding it together, combined with **b** the relationship between stress and strain (distension) of the myocardium. These elements are combined to derive **c** the pressure-volume relationship for the left ventricle shown by the *solid line*. The relationship expressed in **c** provides a good description of the left ventricle's pressure-volume characteristics above the equilibrium volume (V_o), but not below V_o, when the myocardial wall is in compression and storing elastic energy (*dashed line*). The ability of the myocardium to store elastic energy in compression when the end-systolic volume is below the equilibrium volume gives rise to the phenomenon of so-called diastolic suction. σ, stress; p, pressure; r radius of curvature (to the endocardium); h, wall thickness; α and β, elasticity parameters; ϵ, Lagrangian strain; V_d, minimum possible end systolic volume. [1] with permission

Similarly, increases in left ventricular volume increase right ventricular diastolic pressure. Thus, in addition to ischemia, which alters relaxation, ventricular interaction also shifts the left ventricular diastolic pressure-volume curve in response to changes in the vascular loading conditions the left ventricle faces. The degree of direct ventricular interaction and its importance to the left ventricular diastolic pressure-volume relationship and the overall function of the heart depend on the mechanical forces (stresses) to which the muscle is subject and on the elasticity of the myocardium. The pericardium, which encloses the entire heart, strengthens mechanical ventricular interaction.

Ventricular interaction can explain the acute shifts upward or downward in the pressure-volume relationship in patients given the vasoconstrictor angiotensin or the vasodilator nitroprusside [49]. This observation, confirmed by others [50-52], was particularly important because it was the first evidence that hemodynamic changes other than ischemia alter the diastolic pressure-volume relationship.

Direct experimental evidence supports the hypothesis that ventricular interaction, modulated by a constraining pericadium, was the mechanism for the shifts in the diastolic pressure-volume curve in response to vosodilation or vasoconstriction. Shirato et al. [53] administered nitroprusside to conscious dogs instrumented with ultrasonic crystals to determine ventricular dimensions. After giving the vasodilator nitroprusside, the left ventricular pressure-dimension curve shifted downward. These shifts disappeared when the pericardium was removed, thus demonstrating the important role the pericardium plays in modulating the upward and downward shifts in the diastolic pressure-volume relationship after acute administration of vasodilators or vasoconstrictors [54]. The observation that the pericardium modulates interventricular interaction implies that there must be a pericardial constraint and, therefore, that the elastic properties of the pericardium are imposed on the elastic properties of the chambers.

Direct Versus Series Interaction

There are two types of ventricular interaction: direct interaction — via mechanical forces across the septum — and series interaction, because right ventricular output becomes left ventricular input after passing through the pulmonary circulation. Determining the relative roles of direct versus series ventricular interaction in the intact circulatory system is difficult and has led different investigators to reach different conclusions [55, 56]. Whereas Olsen et al. [55] interpreted their data to indicate that direct interaction was the less important effect, Visner et al. [56] reached the opposite conclusion. To resolve this conflict, Slinker and Glantz [57] used a new approach that took advantage of the time lag between a change in right ventricular output and left ventricular input to separate the series (delayed) from the direct (immedate) ventricular interaction. After transiently constricting the pulmonary artery or venae cavae, they measured left- and right-ventricular volumes and pressures over several seconds and found that direct interaction was about one-half as important as series interaction in determining left ventricular end-diastolic size with the pericardium on. Removing the pericardium decreased direct interaction to about one-fifth as important as the series effect, supporting the view that the pericardium modulates ventricular interaction.

The relationship of direct to series interaction also depends on the stiffness of the ventricular walls. In hypertrophied hearts with thickened ventricular walls (resulting from 3 months of renal hypertension), direct interaction at end-diastole was only about one-tenth as important as series interaction in determining left ventricular size with the pericardium on [58]. Removing the pericardium had very little effect on the relative importance of direct versus series interaction in these hypertrophied hearts, because the thicker myocardium (and particularly the intraventricular septum) carried a relatively greater proportion of the wall stress than the normal hearts.

Implications for Systolic Function

Diastolic ventricular interaction is important for maintaining systolic function in response to changes in loading conditions, such as those associated with vasodilation.

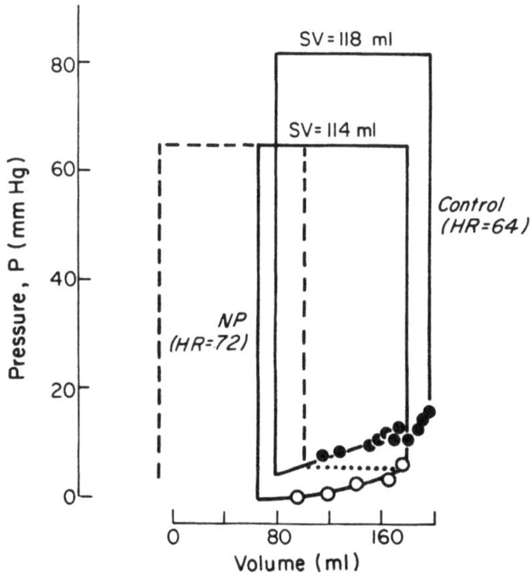

Fig. 4. Data from one patient [49] showing that shifts in the left ventricular diastolic pressure-volume curve help maintain stroke volume (*SV*) in response to changes in loading conditions after administration of the vasodilator sodium nitroprusside (*NP*). During the control condition, the end-diastolic pressure was 16 mmHg, and the end-diastolic volume was 194 ml, with a stroke volume of 118 ml. After administering nitroprusside, the diastolic pressure-volume curve shifted downward, so that the end-diastolic pressure fell to 9 mmHg, but, because the curve shifted, end-diastolic volume fell only slightly (to 176 ml), and stroke volume remained essentially constant at 114 mmHg. Now, suppose that no ventricular interaction and no pericardium existed and the left ventricular diastolic pressure volume curve moved down a unique pressure-volume curve and the end-diastolic volume fell as end-diastolic pressure fell, then the end-diastolic volume would drop to about 94 ml, making it impossible to maintain stroke volume near its original value of 118 ml. However, since the entire pressure-volume curve shifted downward, stroke volume was maintained in the presence of a lowered end-diastolic pressure. Thus, ventricular interaction is important for maintaining systolic function. [59] with permission

For instance, the downward shift in the left ventricular diastolic pressure-volume curve, such as that which occurs after acute administration of the vasodilator nitroprusside, is a critical factor in maintaining stroke volume [59] (Fig. 4).

The Pericardium

The pericardium surrounds the heart, reducing friction between the heart and the surrounding tissues and providing a barrier against infection. Its attachments to the mediastinum help restrain the heart from excessive motion within the chest when body position changes. The pericardium plays an important hemodynamic role by constraining the heart as it expands. Because it constrains the whole heart and is stiffer than the myocardium, the pericardium increases the mechanical interaction between the four cardiac chambers. Thus, an increase in the size of one ventricle causes the pressure-volume relationship of the other ventricle to shift upward due to pericardial constraint.

How the Pericardium Affects the Left Ventricular Diastolic Pressure-Volume Relationship

The key to assessing how much the pericardium influences the left ventricular diastolic pressure-volume relationship under physiological conditions is determining where the pericardium normally operates on its pressure-volume curve. Answering this question is difficult and has been the subject of considerable debate. Several investigators have assessed the effect of the pericardium by comparing the left ventricular diastolic pressure-volume relationships before and after removing the pericardium [53, 57, 58, 60–64]. At high left ventricular filling pressures (above about 10 mmHg), pericardial stiffness clearly increases left ventricular diastolic pressures [18, 60, 63, 65–68]. This effect becomes more pronounced as left ventricular filling pressure and end-diastolic volume increase. In contrast, Stokland et al. [64] reported only a slight (4%) increase in left ventricular myocardial chord lengths after they increased blood volume and opened the pericardium of open-chest dogs, and Tyson et al. [69] reported little effect of the pericardium in conscious, chronically instrumented dogs with low diastolic pressures. Therefore, although the constraining effect of the pericardium contributes significantly to the upward shift of the left ventricular pressure-volume curve under conditions of acute cardiac dilation, there is only minimal effect at low diastolic pressures.

Pericardial Adaptation to Ventricular Hypertrophy

The pericardium adapts to chronic increases in heart size by increasing its mass (size) [70], and as the pericardium expands over time, its effect on the left ventricular diastolic pressure-volume relationship decreases [71]. Shortly after creating experimental volume overload, the heart expands against the pericardium, which leads to an increase in diastolic left ventricular pressure. In contrast, later, after left ventricular mass increased, neither left ventricular end-diastolic dimension nor right ventricular end-diastolic pressure changed when the pericardium was removed, indicating that the pericardium had expanded and was exerting less constraining force on the heart. Thus, when the pericardium hypertrophies along with the heart, the pericardium plays a smaller role in determining the left ventricular diastolic pressure-volume relationship than in normal hearts [58, 71–73]. The influence of the pericardium decreases, because as the heart increases in size, the pericardium is stressed chronically. The pericardium responds to this increased stress by increasing its size. When this happens, the enlarged pericardium would operate on a flatter portion of its stress-strain curve; the constraining forces are smaller, and the influence of the pericardium on the heart is less than its influence before the hypertrophy.

Another effect of myocardial hypertrophy is that the relative importance of the pericardium and of direct ventricular interaction is less when the ventricular wall thickens. As the ventricular walls thicken in response to the pressure overload, the thicker walls become less distensible, and more of the ventricular pressure is balanced by the elasticity of the myocardium as opposed to the pericardium [57, 58].

Implications for Systolic Function

By restricting the expansion of the heart, and thus restricting extension of the sarcomeres that make up the myocardium, the pericardium can affect left ventricular systolic function via the Frank-Starling mechanism under certain pathological conditions. For example, right ventricular infraction can precipitate a low cardiac output syndrome. Although a reduction in right ventricular systolic function plays a role by reducing right ventricular output and left ventricular input, ventricular interaction, modulated by the pericardium, is also important. Goldstein et al. [74, 75] produced right ventricular infarction in dogs by infusing mercury into the coronary circulation. Immediately after the infarct, the right ventricle attempted to dilate within the pericardium, increasing pericardial pressure, increasing left ventricular diastolic pressure, and inhibiting filling. The associated lower left ventricular end-diastolic volumes impaired systolic function because of the Frank-Starling mechanism. When the pericardium was opened and the right ventricle was free to dilate without competing with the left ventricle within the pericardium, the left ventricle could expand and increase end-diastolic volume, stroke volume, and cardiac output.

Pulmonary-Cardiac Contact Pressure

Just as the ventricles and the pericardium interact mechanically with each other because they are in physical contact, so do the heart and lungs because they are in physical contact. Like the pericardium, the lungs affect the left ventricle's diastolic pressure-volume curve by exerting a normal stress on the heart. The contact pressure, which is on the order of a few millimeters of mercury, has little direct mechanical effect during systole. However, during diastole, when pressures are low, inflating the lungs can increase pressure outside the heart and lower the left ventricular transmural pressure at any given absolute left ventricular pressure. This lower transmural pressure can reduce end-diastolic volume and reduce cardiac output because of the Frank-Starling mechanism. This mechanical interaction between the heart and lungs can explain the reduction in cardiac output that occurs during ventilation with positive end-expiratory pressure (PEEP) [76, 77].

Viscoelastic Properties of the Myocardium

Like most biological materials, the myocardium is viscoelastic so that stresses in the heart wall depend on the rate at which it deforms (strain rate, ϵ) as well as on the magnitude of the deformation (strain, ϵ). This dependence means that the diastolic pressure depends on the filling rate as well as on the ventricular volume and that the diastolic pressure at a given volume increases as filling rate increase [78]. Because viscous forces increase with filling rate, these forces are greatest during the rapid filling phase in early diastole and atrial systole. However, elastic recoil or restoring forces are also most significant during rapid filling. Therefore, it is difficult to separate changes in the diastolic pressure-volume relationship during early diastole due to myocardial viscoelasticity from the changes due to active relaxation. Although there is no doubt that viscous forces contribute to diastolic pressure, particularly during rapid filling and atrial systole, this contribution is probably small.

Coronary Vascular Engorgement

Increasing coronary artery pressure has been hypothesized to stiffen the network of coronary arteries that permeate the myocardium and so stiffen the left ventricular wall, analogous to the way that erectile tissue stiffens when it is engorged with blood. This resultant stiffening of the left ventricular wall would be reflected in the diastolic pressure-volume relationship just as if the muscle itself became stiffer. The effect of coronary vascular engorgement on ventricular stiffness would be most evident late in diastole, when the coronary arteries are maximally perfused. Although myocardial stiffening as a result of vascular engorgement has been demonstrated in isolated heart preparations, the effect is small, so it probably does not contribute appreciably to the diastolic pressure-volume relationship.

Discussion

It is now well established that the traditional view of the diastolic left ventricle as an isolated shell in which filling pressure and volume are uniquely related is incorrect. The diastolic pressure-volume relationship changes in response to changes in the heart's operating environment and depends on the balance between the pressure in the left ventricle and the stresses in the myocardium. These stresses, in turn, depend on the size and thickness of the ventricle and the elasticity of the myocardium. In addition to these factors, there are two major categories of factors that can acutely change the diastolic pressure-volume curve.

The first category of factors includes those that alter the extent of relaxation from the previous systole, such as demand-induced ischemia. This type of ischemia impairs active uptake of calcium from the cytosol which yields a higher resting concentration of calcium in the sarcomeres during diastole and a smaller apparent equilibrium volume for the ventricle. This smaller equilibrium volume means that the strains (and, hence, stresses) within the wall of the heart will be higher for any given volume of the left ventricle than before ischemia; so, there will be a higher pressure at any given volume and an upward stift of the pressure-volume curve.

The second category of factors that alter the diastolic pressure-volume curve includes those that alter the external mechanical environment of the left ventricle, in particular, the other cardiac chambers, the pericardium, and contact pressure with the lungs. Interventions such as vasodilation and vasoconstriction alter the forces that these other components exert on the left ventricle. Since the pressure within the left ventricular cavity represents the net effect of the forces within the myocardium and those outside it, changes in these external forces are reflected in pressure changes within the left ventricle at any given volume and, therefore, shift the diastolic pressure-volume curve.

These changes in the diastolic pressure-volume curve have several important implications for systolic function. First, when the end-systolic volume is less than the left ventricle's equilibrium volume, the myocardial wall is put into compression and stores elastic energy during systole (just as compressing a spring stores energy). This stored energy is released in the form of an elastic recoil when the myocardium relaxes and acts to reduce the ventricular pressure, thus, increasing the atrioventricular pressure gradient during early diastole. Increasing the atrioventricular pressure gradient causes the ventricle to fill more rapidly which helps maintain end-diastolic volume and systolic function, via the Frank-Starling mechanism. Second, shifts in the diastolic pressure-volume relationship in response to vascular loading changes are essential in helping the left ventricle maintain stroke volume during vasodilation, because these shifts maintain end-diastolic volume even though left ventricular end-diastolic pressure falls during vasodilation. Finally, the fact that filling pressure and volume are not uniquely related means that end-diastolic pressure cannot be used as a surrogate for end-diastolic volume when assessing systolic function.

References

1. Gilbert JG, Glantz SA (to be published) Determinants of left ventricular filling and the diastolic pressure-volume relationship. Circ Res
2. Peterson KL, Tsuji J, Johnson A, DiDonna J, LeWinter M (1978) Diastolic left ventricular pressure-volume and stress-strain relations in patients with valvular aortic stenosis and left ventricular hypertrophy. Circulation 58: 77-89
3. Rippe JM, Pape LA, Alpert JA, Ockene IS, Paraskos JA, Kotilainen P, Anas J, Webster W (1982) Studies of systolic mechanics and diastolic behavior on the left ventricle in the trained racing greyhound. Basic Res Cardiol 77: 619-644
4. Nomura S (1986) Diastolic property of left ventricle under experimental volume overload. Jpn Circ J 50: 426-432
5. Michel JB, Salzmann JL, Ossondo Nlom M, Bruneval P, Barres D, Camilleri JP (1986) Morphometric analysis of collagen network and plasma perfused capillary bed in the myocardium of rats during evolution of cardiac hypertrophy. Basic Res Cardiol 81: 142-154
6. Grossman W, Jones D, McLaurin LP (1975) Wall stress and patterns of hypertrophy in the human left ventricle. J Clin Invest 56: 56-64
7. Florenzano F, Glantz SA (1987) Left-ventricular mechanical adaptations to chronic aortic regurgitation in intact dogs. Am J Physiol 252: H969-H984
8. Fagard R, van den Broekle C, Bielen E, Vabhees L, Amery A (1987) Assessment of stiffness of the hypertrophied left ventricle of bicyclists using left ventricular inflow Doppler velocimetry. J Am Coll Cardiol 9: 1250-1254
9. Grossman W, McLaurin LP, Stefadouros MA (1974) Left ventricular stiffness associated with chronic pressure and volume overloads in man. Circ Res 35: 793-800
10. Thiedemann KU, Holubarsch C, Medugorac I, Jacob R (1983) Connective tissue (content and myocardial stiffness in pressure overload hypertrophy: A combined study of morphologic, morphometric, biochemical, and mechanical parameters. Basic Res Cardiol 78: 140-155
11. Weber KT, Janicki JS, Shroff SG, Pick R, Chen RM, Bashey RI (1988) Collagen remodeling of pressure-overloaded, hypertrophied nonhuman primate myocardium. Circ Res 62: 757-765
12. McLaurin LP, Rolett EL, Grossman W (1973) Impaired left ventricular relaxation during pacing-induced ischemia. Am J Cardiol 32: 751-757
13. Mann T, Brodie BR, Grossman W, McLaurin LT (1977) Effect of angina on the left ventricular diastolic pressure-volume relationship. Circulation 55: 761-766

14. Grossman W, Mann JT (1978) Evidence for impaired left ventricular relaxation during acute ischemia in man. Euro J Cardiol 7 (Suppl): 239-249
15. Bourdillon PD, Paulus WJ, Serizawa T, Grossman W (1986) Effects of verapamil on regional myocardial diastolic function in pacing-induced ischemia in dogs. Am J Physiol 251: H834-H840
16. Sasayama S, Nonogi H, Fujita M, Sakurai T, Wakabayashi A, Kawai C, Eiho S, Kuwahara M (1984) Analysis Of asychronous wall motion by regional pressure-length loops in patients with coronary artery disease. J Am Coll Cardiol 4: 259-267
17. Carroll JD, Lang RM, Neumann AL, Borow KM, Rajfer SI (1986) The differential effects or positive inotropic and vasodilator therapy on diastolic properties in patients with congestive cardiomyopathy. Circulation 74: 815-825
18. Palacios I, Johnson RA, Newell JB, Powell WJ Jr (1976) Left ventricular end-diastolic pressure-volume relationship with experimental acute global ischemia. Circulation 53: 428-436
19. Smiseth OA, Kingma I, Refsum H, Smith ER, Tyberg JV (1985) The pericardial hypothesis: A mechanism of acute shifts of the left ventricular diastolic pressure-volume relation. Clin Physiol 5: 403-415
20. Serizawa T, Carabello BA, Grossman W (1980) Effect of pacing-induced ischemia on left ventricular diastolic pressure-volume relations in dogs with coronary stenoses. Circ Res 46: 430-439
21. Paulus WJ, Serizawa T, Grossman W (1982) Altered left ventricular diastolic properties during pacing-induced ischemia in dogs with coronary stenosis potentiation by caffeine. Circ Res 50: 218-227
22. Nayler WG, Williams A (1978) Relaxation in heart muscle: some morphological and biochemical considerations. Eur J Cardiol 7 (Suppl): 35-50
23. Nayler WG, Poole-Wilson PA, Williams A (1979) Hypoxia and calcium. J Mol Cell Cardiol 11: 683-706
24. Carroll JD, Hess OM, Hirzel HO, Krayenbuehl HP (1983) Exercise-induced ischemia: the influence of altered relaxation on early diastolic pressures. Circulation 67(3): 521-528
25. Serruys PW, Wijns W, Piscione F, de Feyter P, Hugenholtz PG (1988) Ejection, filling, and diastasis during transluminal occlusion in man: consideration on global and regional left ventricular function. In: Grossman W, Lorell BH (eds) Diastolic relaxation of the heart. Martinus Nijhoff, Boston, pp 255-280
26. Granger CB, Karimeddini MK, Smith VE, Shapiro HR, Katz AM, Riba AL (1985) Rapid ventricular filling in left ventricular hypertrophy: I. Physiologic hypertrophy. J Am Coll Cardiol 5: 861-862
27. Hanrath P, Mathey DG, Siegert R, Bleified W (1980) Left ventricular relaxation and filling pattern in different forms of left ventricular hypertrophy: an echocardiographic study. Am J Cardiol 45: 15-23
28. Smith VE, Schulman P, Karimeddini MK, White WB, Merran MK, Katz AM (1985) Rapid ventricular filling in left ventricular hypertrophy: II. Pathologic hypertrophy. J Am Coll Cardiol 5: 869-874
29. Lorell BH, Paulus WJ, Grossman W, Wynne J, Cohen PF (1982) Modification of abnormal left ventricular diastolic properties by nifedipine in patients with hypertrophic cardiomyopathy. Circulation 65: 499-507
30. Alvares RF, Shaver JA, Gamble WH, Goodwin JF (1984) Isovolumic relaxation period in hypertrophic cardiomyopathy. J Am Coll Cardiol 3: 71-81
31. Betocchi S, Bonow RO, Bacharach SL, Rosing DR, Maron BJ, Green MV (1986) Isovolumic relaxation period in hypertrophic cardiomyopathy: assessment by radionuclide angiography. J Am Coll Cardiol 7: 74-81
32. Gwathmey JK, Morgan JP (1985) Altered calcium handling in experimental pressure-overload hypertrophy in the ferret. Circ Res 57: 836-843
33. Morgan JP, Morgan KG (1984) Calcium and cardiovascular function: intracellular calcium leavels during contraction and relaxation of mammalian cardiac and vascular smooth muscle as detected aequorin. Am J Med 77 (Suppl 5A): 33-46

34. Gwathmey JK, Copelas L, MacKinnon R, Schoen FJ, Feldman MD, Grossman W, Morgan JP (1987) Abnormal intracellular calcium handling in myocardium from patients with end-stage heart failure. Circ Res 61: 70-76
35. Alyono D, Anderson RW, Parrish DG, Dai X, Bache RJ (1986) Alterations of myocardial blood flow associated with experimental canine left ventricular hypertrophy secondary to valvular aortic stenosis. Circ Res 58: 47-57
36. Bache RJ, Arentzen CE, Simon AB, Vrobel TR (1984) Abnormalities in myocardial perfusion during tachycardia in dogs with left ventricular hypertrophy: metabolic evidence for myocardial ischemia. Circulation 69: 409-417
37. Fifter MA, Bourdillon PD, Lorell BH (1986) Altered left ventricular diastolic properties during pacing-induced angina in patients with aortic stenosis. Circulation 74: 675-683
38. Diver DJ, Royal HD, Aroesty JM, McKay RG, Ferguson JJ, Warren SE, Lorell BH (1988) Diastolic function in patients with aortic stenosis: influence of left ventricular load reduction. J Am Coll Cardiol 12: 642-648
39. Hess OM, Ritter M, Schneider J, Grimm J, Turina M, Krayenbuehl HP (1984) Diastolic stiffness and myocardial structure in aortic valve disease before and after valve replacement. Circulation 69: 855-865
40. Sasayama S, Nonogi H, Miyazaki S, Sakurai T, Kawai C, Eiho S, Kuwahara M (1985) Changes in diastolic properties of the regional myocardium during pacing-induced ischemia in human subjects. J Am Coll Cardiol 5: 599-606
41. Theroux P, Ross J Jr, Franklin D, Covell JW, Bloor CM, Sasayama S (1977) Regional myocardial function and dimensions early and late after myocardial infaction in the unanesthetized dog. Circ Res 40: 158-165
42. Janz RF, Waldron RJ (1978) Predicted effect of chronic apical aneurysms on the passive stiffness of the human left venticle. Circ Res 42(2): 255-263
43. Hori M, Yellin EL, Sonnenblick EH (1982) Left ventricular diastolic suction as a mechanism of ventricular filling. Jpn Circ J 46: 124-129
44. Yellin EL, Hori M, Yoran C, Sonnenblick EH, Gabbay S, Frater RWM (1986) Left ventricular relaxation in the filling and nonfilling intact canine heart. Am J Physiol 250: H620-H629
45. Nikolic' S, Yellin EL, Tamura K, Vetter H, Tamura T, Meisner JS, Frater RWM (1988) Passive properties of canine left ventricle: diastolic stiffness and resting forces. Circ Res 62: 1210-1222
46. Zile MR, Blaustein AS, Gaasch WH (1985) In the normal left ventricle catecholamine induced changes in filling rate are mediated through change in end systolic size. (abstract) Circulation 72 (Suppl III): 88
47. Sonnenblick EH (1980) The structural basis and importance of restoring forces and elastic recoil for the filling of the heart. Eur Heart J 1 (Suppl A): 107-110
48. Suga H, Goto Y, Igarashi Y, Yamada O, Nozawa T, Yasumura Y (1986) Ventricular suction under zero source pressure for filling. Am J Physiol 251: H47-H55
49. Alderman EL, Glantz SA (1976) Acute hemodynamic interventions shift the diastolic pressure-volume curve in man. Circulation 54: 662-671
50. Parmley WW, Chuck L, Chatterjee K, Klausner SC, Glantz SA, Ratshin RA (1976) Acute changes in the diastoic pressure-volume relationship of the left ventricle. Eur J Cardiol 4 (Suppl): 105-120
51. Brodie BR, Grossman W, Mann T, McLaurin LP (1977) Effects of sodium nitroprusside on left ventricular diastolic pressure-volume relations. J Clin Invest 59: 59-68
52. Ludbrook PA, Byrne JD, Kurnik PB, McKnight RC (1977) Influence of reduction of preload and afterload by nitroglycerin on left ventricular diastoic pressure-volume relations and relaxation in man. Circulation 56: 937-943
53. Shirato K, Shabetai R, Bhargava V, Franklin D, Ross J Jr (1978) Alteration of the left ventricular diastolic pressure-segment length relation produce by the pericardium: effects of cardiac distension and afterload reduction in conscious dogs. Circulation 57: 1191-1198
54. Ross J (1979) Acute displacement of the diastolic pressure-volume curve of the left ventricle: role of the pericardium and the right ventricle. Circulation 59: 32-37 (editorial)

55. Olsen CO, Tyson GS, Maier GW, Spratt JA, Davis JW, Rankin JS (1983) Dynamic ventricular interaction in the conscious dog. Circ Res 52: 85–104
56. Visner MS, Arentzen CE, O'Connor MJ, Larson EV, Anderson RW (1983) Alterations in left ventricular three-dimensional dynamic geometry and systolic function during acute right ventricular hypertension in the conscious dog. Circulation 67: 353–365
57. Slinker BK, Glantz SA (1986) End-systolic and end-diastolic ventricular interaction. Am J Physiol 251: H1062–H1075
58. Slinker BK, Chagas ACP, Glantz SA (1987) the importance of direct ventricular interaction decreases in chronic pressure overload hypertrophy in the dog. Am J Physiol 253: H347–H357
59. Tyberg JV, Misbach GA, Parmley WW, Glantz SA (1987) Effects of the pericardium on left ventricular performance. In: Baan J, Arntzenius AC, Yellin EL (eds) Cardiac dynamics, Martinus Nijhoff. The Hague, pp 159–168
60. Glantz SA, Misbach GA, Moores WY, Mathey DG, Lekven J, Stowe DF, Parmley WW, Tyberg JV (1978) The pericardium substantially affects the left ventricular diastolic pressure-volume relationship in the dog. Circ Res 42: 433–441
61. Little WC, Badke FR, O'Rourke RA (1984) Effect of right ventricular pressure on the end-diastolic left ventricular pressure-volume relationship before and after chronic right ventricular pressure overload in dogs without pericardia. Circ Res 54: 719–730
62. Maruyama Y, Ashikawa K, Isoyama S, Kanatsuka H, Ino-Oka E, Takishima T (1982) Mechanical interactions between four heart chambers with and without the pericardium in canine hearts. Circ Res 50: 86–100
63. Spadaro J, Bing OHL, Gassch WH, Laraia P, Franklin A, Weintraub RM (1982) Effects of perfusion pressure on myocardial performance, metabolism wall thickness, and compliance: comprison of the beating and fibrillating heart. J Thorac Cardiovasc Surg 84: 398–405
64. Stokland O, Miller MM, Lekven J, Ilebekk A (1980) The significance of the intact pericardium for cardiac performance in the dog. Circ Res 47: 27–32
65. Janicki JS, Weber KT (1980) the pericardium and ventricular interaction, distensibility, and function. Am J Physiol 238: H494–H503
66. Robotham JL, Mitzner W (1979) A model of the effects of respiration on left ventricular performance. J Appl Physiol 46: 411–418
67. Robotham JL, Rabson J, Permutt S, Bromberger-Barnea B (1979) Left ventricular hemodynamics during respiration. J Appl Physiol 47: 1295–1303
68. Smiseth OS, Refsum H, Tyberg JV (1984) Pericardial pressure assessed by right atrial pressure: a basis for calculation of left ventricular transmural pressure. Am Heart J 108: 603–605
69. Tyson GS Jr, Maier GW, Olsen CO, Davis JW, Rankin JS (1984) Pericardial influences on ventricular filling in the conscious dog: an analysis based on pericardial pressure. Circ Res 54: 173–184
70. Freeman GL, LeWinter MM (1984) Pericardial adaptations during chronic cardiac dilation in dogs. Circ Res 54: 294–300
71. LeWinter MM, Pavelec R (1982) Influence of the pericardium on left ventricular end-diastolic pressure-segment relations during early and later stages of experimental chronic volume overload in dogs. Circ Res 50 (4): 501–509
72. Lee JM, Boughner DR (1981) Tissue mechanics of canine pericardium in different test environments. Circ Res 49: 533–544
73. Lee MC, LeWinter MM, Freeman GL, Shabetai R, Fung YC (1985) Biaxial mechanical properties of the pericardium in normal and volume overload dogs. Am J Physiol 249: H222–H230
74. Goldstein JA, Vlahakes GJ, Verrier ED, Shiller NB, Botvinick E, Tyberg JV, Parmley WW, Chatterjee K (1983) Volume loading improves low cardiac output in experimental right ventricular infarction. J Am Coll Cardiol 2: 270–277
75. Goldstein JA, Vlahakes GJ, Verrier ED, Shiller NB, Tyberg JV, Ports TA, Parmley WW, Chatterjee K (1982) The role of right ventricular systolic dysfunction and elevated intra-

pericardial pressure in the genesis of low output in experimental right ventricular infarction. Circulation 65: 513–522
76. Fewell JE, Abendschein DR, Murray JF, Rapaport E (1980) Continuous positive-pressure ventilation decreases right and left ventricular end-diastolic volumes in the dog. Circ Res 46: 125–132
77. Fewell JE, Abendschein DR, Carlson CJ, Rapaport E, Murray JF (1980) Mechanism of decreased right and left ventricular end-diastolic volumes during continuous positive-pressure ventilation in dogs. Circ Res 47: 467–472
78. Rankin JS, Arentzen CE, McHale PA, Ling D, Anderson RW (1977) Viscoelastic properties of the diastolic left ventricle in the conscious dog. Circ Res 41: 37–45

Equilibrium Volume and Passive Pressure-Volume Relationship in the Intact Canine Left Ventricle

SRDJAN NIKOLIĆ, KOICHI TAMURA, TAKAKO TAMURA, and EDWARD L. YELLIN[1]

Summary. In this chapter we have described a unique method of left ventricular volume clamping designed to quantify the passive properties of the intact ventricle. We prevented complete (end-systolic clamping) or partial filling at different times in diastole. The ventricle thus relaxed completely at different volumes, and we generated pressure-volume coordinates for the passive ventricle that included negative, as welll as positive, values of pressure. We then determined the equilibrium volume, that is, volume at zero transmural pressure, in the working ventricle. We characterized the passive pressure-volume relation with a logarithmic approach that is physically more realistic than the traditional exponential. Finally, we discussed the importance of the concepts of equilibrium volume an restoring forces for diastolic mechanics.

Introduction

In order to understand left ventricular diastolic function and filling dynamics, we must first characterize the properties of the fully relaxed ventricle *at all volumes*. These properties have traditionally been determined for the end-diastolic volume. Since the normal, in situ, ventricle starts to fill while it is still relaxing, it has not been possible to achieve small volumes, except by vena caval occlusions. But this technique is still unable to separate the effects of filling from the effects of relaxation. To overcome this problem, we present here a method of volume clamping the ventricle to prevent filling, thereby allowing the ventricle to completely relax even when it has contracted from normal end-diastolic volumes to end-systolic volumes below the equilibrium volume.

Charactrization of the Passive P-V Relation

We define the *equilibrium volume* (V_0) as the volume of the resting ventricle at zero transmural pressure (Fig. 1). Discussed first by Brecher [1] and investigated further by Tyberg et al. [2] and others [3–5], this concept provides the mechanical model that forms the physiological basis of diastolic suction. Ventricular contraction to an

[1] Department of Cardiothoracic Surgery, Pysiology and Biophysics, and Anesthesiology, Albert Einstein College of Medicine, Bronx, NY 10461, USA

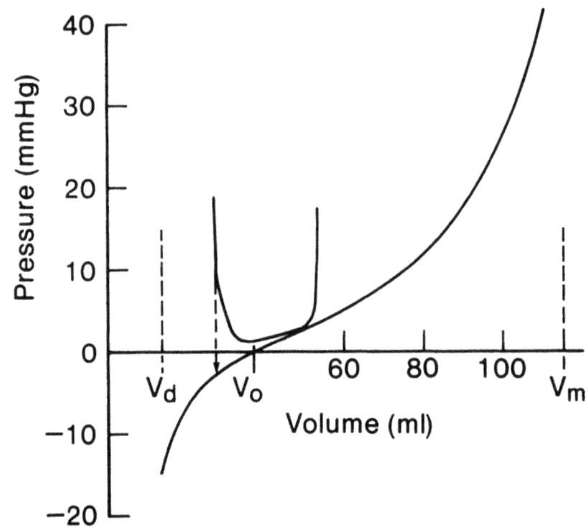

Fig. 1. Representative plot of a measured diastolic pressure-volume (P-V) loop in a normally filling ventricle, superimposed on a passive P-V relation. V_o, volume at zero transmural pressure; V_d and V_m, minimum and maximum volumes. The *arrow* describes the trajectory of a completely isovolumic (i.e., non-filling) relaxation to a negative pressure

end-systolic volume below the equilibrium volume stores energy in the ventricular walls; this stored energy provides an internal restoring force that facilitates filling at low pressures, and that may influence the ventricular relaxation process.

This concept can be clarified by considering a chamber that has actively contracted to below its equilibrium volume and is clamped at that volume. During relaxation, active wall stresses decrease, and when the internal elastic forces tending to restore the chamber to its initial configuration are greater than the decaying active wall forces, there must be a concurrent decrease of chamber pressure to maintain a balance of forces. Thus, the extenal pressure becomes greater than the internal pressure, i.e., the transmural pressure is negative. When the external pressure is atmospheric, as in the open-chest preparation, or below atmospheric, as in the closed-chest condition, the pressure in a chamber clamped below its equlibrium volume will be negative (Fig. 1). Diastolic suction is a concept that has been supported by many experiments [1-6]. We therefore conclude that a complete description of the passive diastolic properties of the left ventricle must include its behavior both above and below its equilibrium volume.

We characterize the positive portion of the P–V relation with a logarithmic approach [7-9]:

$$P_p = -S_p \ln[(V_m - V)/(V_m - V_o)] \qquad (1)$$

where V_m is the maximum attainable volume of the ventricle, S_p is a material property (as described later), and the subscript "$_p$" refers to the positive portion of the P–V relation. The logarithmic approach to the P–V relation is the equivalent of an exponential V–P relation, and is heuristically desirable because it scales the passive ventricle to function within the natural limits of its rest and yield volumes. Differentiating Eq. 1 and rearranging:

$$S_p = (dP_p/dV)\,(V_m - V) \tag{2}$$

Thus, S_p is seen to be a constant, defined as the chamber stiffness normalized by scaling the instantaneous volume to the operating range of the ventricle . Because S_p has the units of pressure, it is reasonable to conclude that it is related to the wall stress, modified by a geometric factor.

A similar approach is followed for the negative portion of the P-V relation:

$$P_n = S_n \ln[(V - V_d)/(V_o - V_d)] \tag{3}$$

$$S_n = (dP_n/dV)\,(V - V_d) \tag{4}$$

where the subscript "n" denotes the negative portion of the P – V relation. Here again, S_n is seen to be a constant with the units of stress and is defined by the local stiffness normalized by the operating range of the ventricle — between the equilibrium volume and a nonzero minimum volume, V_d.

Experimental Methods and Results

Ventricular diastolic volume was controlled with a remote controlled mitral valve (RCMV) described in detail elsewhere [6]. Briefly, a modified ball-in-cage prosthetic valve was implanted in series with an electromagnetic flow probe in the mitral anulus in 8 adult anesthetized mongrel dogs. A controller was triggered by a suitable physiological signal, and after a programmed delay, it rapidly closed the valve. The valve can be closed during systole (end-systolic volume clamping) or at any other time after filling has started (Fig. 2).

Changes in volume were calculated from the integrals of aortic and mitral flow. In essence, we assumed that the end-systolic volume remained constant throughout each run and subsequent volumes were determined from inflow and outflow. It should be noted that we did not measure absolute ventricular volumes, so that all volume measurements are referenced to an unknown end-systolic volume (V_{es}) that is constant for each run. It was thus important that there should be no change in hemodynamic state during a run. This was tested by examining the control beats before each occlusion and accepting only those runs in which there was no change in the control pressures and flows.

Determination of Equilibrium Volume and P-V Relations

Fig. 2a shows a typical mitral flow waveform, with the numbers (1-6) indicating increasing times during diastole when the mitral valve was occluded. Panel b shows the left ventricular volume, relative to end systole, at each of these times. The high-gain left ventricular pressure traces for each of these occlusions is shown in panel c. The solid trace is the control with point 7 being the control left ventricular end-diastolic pressure (LVEDP). The broken traces indicate the time variation of pressure after the mitral valve occlusion. The *numbered points* in panel c indicate the minimum pressure reached during the occlusion.

Points 1–4 in panel c are evidence of elastic recoil and diastolic suction. They occur at volumes below the equilibrium volume, hence the fall to a negative pressure (see Fig. 1); and they mark the end of relaxation, because volume is constant and pressure has stopped falling. Conversely, points 5 and 6 are greater than the equilibrium volume, so that volume clamping at those time did not lead to negative pressure. Since pressure continued to fall to point 5, we conclude that relaxation had not yet ended at the time of occlusion; and by the same reasoning, point 6 occurred after

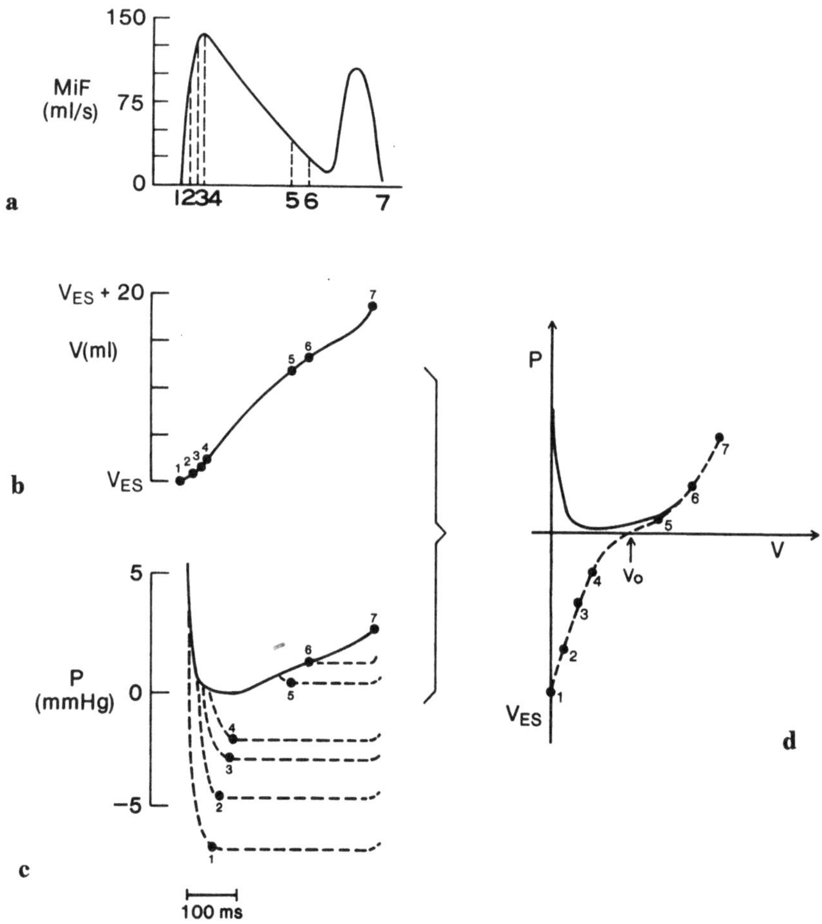

Fig. 2a-d. Description of the volume clamping method of determining the passive-volume relation. The control mitral flow (*MiF*), ventricular volume (*V*), and pressure (*P*), traces and drawn as *solid lines* in **a-c**. The numbers 1–6 in **a** mark the times of mitral valve occlusion. Number 7 is the end-diastolic point. Integration of the flow curve to the time of occlusion gives the change in ventricular volume, relative to the control end-systolic volume. The broken lines in **c** denote the pressure traces associated with each volume clamping, and the *numbered points* correspond to the minimum pressure during each occlusion. This minimum pressure is assumed to be the value at end-relaxation. **d** shows the P-V relation determined by eliminating time as a parameter from **b** and **c**. Note particularly the relative ease with which the volume at zero transmural pressure (V_o) can be determined from the data of **c**. V_{ES}, ventricular end-systolic volume the cessation of relaxation. Finally, panel d shows the result of removing time as a parameter and plotting the complete diastolic pressure-volume relation for this run.

Fig. 3 shows the diastolic pressure-volume relations for the 8 dogs, determined by the method of Fig. 2. Because V_o, should be invariant in the course of a study, the different runs were superimposed with a common V_o and all volumes are relative to the end-systolic volume.

To solve Eqs. 1 and 3, we determined absolute volume from left ventricular weight with the method described in detail elsewhere [6]. The measured left ventricular weights and the volumes calculated from the regressions for the 8 dogs of this study are presented in Table 1. These values were used to determine the stiffness parameters.

Estimation of the Stiffness Constants, S_p and S_n

All the data points above V_o in each dog were pooled and used with the estimated absolute volumes (Table 1) in a regression analysis of the log-linearized Eq. 1 With this approach, S_p is the slope of the regression line. The results for each dog are shown in Table 1. An analysis of covariance revealed no significant difference among the values of S_p. An ensemble mean S_p was also calculated from a single regression analysis by pooling all the data points from all the runs to give a mean of $S_p = 14.6$ mmHg ($r = 0.917$; $n = 318$). This value is identical to the mean value of Table 1 ($S_p = 14.5 \pm 1.5$ mmHg; $n = 8$) and reflects the viability of the volume scaling approach and the low variance in the data ($CV = 10\%$).

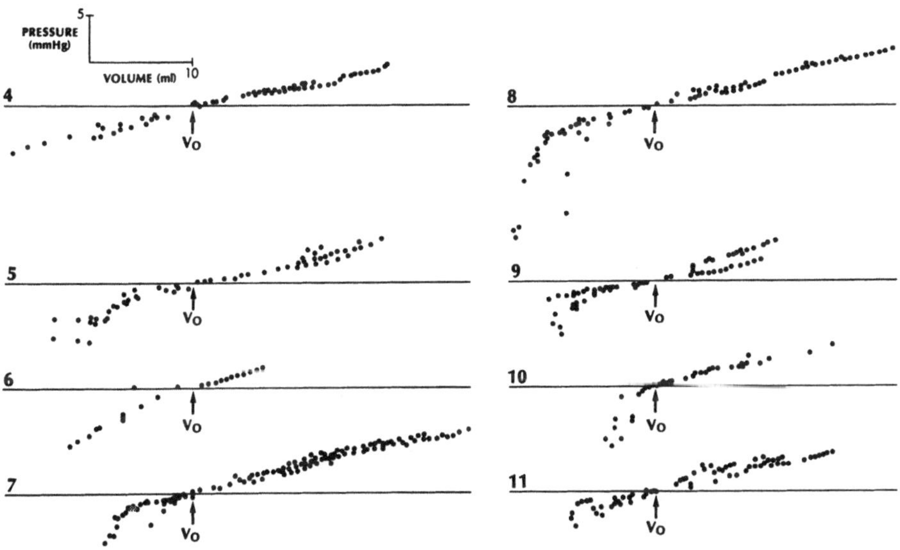

Fig. 3. Passive pressure-volume coordinates in the individual dogs (*4–11*) obtained from volume clamped and from normal cycles. Different runs are pooled by matching them at the volume at zero transmural pressure (V_o), determined in each run; hence, all volumes are relative to the end-systolic volume

Table 1. Diastolic properties of the left ventricle

Dog	BW (kg)	LVW (g)	LVW/BW (g/kg)	V_o* (ml)	Vo*/LVW (ml/100 g)	V_o (ml)	V_m (ml)	V_d (ml)	V_m-V_o (ml)	V_o-V_d (ml)	S_p (mmHg)	S_n (mmHg)	$dP_p/dV_{(o)}$ (mmHg/ml)	$dP_n/dV_{(o)}$ (mmHg/ml)
4	27.2	157	5.77	13.3 ± 7.5	8.5	44.7	139.7	15.7	95.0	29.0	13.3	4.6	.140	.138
5	25.6	124	4.84	12.7 ± 1.2	10.2	35.3	110.4	12.4	75.1	22.9	12.7	7.1	.169	.274
6	25.0	93	3.72	14.2 ± 1.8	15.3	26.5	82.8	9.3	56.3	17.2	14.1	2.9	.250	.218
7	25.3	126	4.98	8.9 ± 2.1	7.1	35.9	112.1	12.6	76.2	23.3	14.5	8.5	.190	.183
8	25.8	145	5.62	14.0 ± 2.8	9.7	41.3	129.1	14.5	87.8	26.8	16.3	12.3	.186	.259
9	27.3	133	4.87	13.9 ± 2.8	10.5	37.8	118.4	13.3	80.6	24.5	13.3	1.8	.165	.073
10	26.7	120	4.49	1.4 ± 4.8	1.2	34.1	106.8	12.0	72.7	22.1	16.8	26.5	.231	.359
11	27.8	146	5.25	7.4 ± 2.9	5.1	41.6	129.9	14.6	88.3	27.0	15.1	6.0	.171	.170
Mean	26.3	131	4.94	9.2	8.5	37.2	116.2	13.1	79.0	24.1	14.5	8.7	.188	.209
± SD	1.0	20	.65	5.8	4.2	5.6	17.6	2.0	12.0	3.7	1.5	7.9	.036	.089
Pooled											14.6	5.1		

BW, body weight; LVW, left ventricle weight; V_o*, equilibrium volume relative to end systolic volume (mean ± SD for all runs); V_o, absolute value of equilibrium volume determined from LVW (see text); V_m, maximal volume; V_d, minimal volume; V_m-V_o, operating range in positive phase; V_o-V_d, operating range in negative phase; S_p, normalized coefficient of stiffness for positive phase; S_n, normalized coefficient of stiffness for negative phase (pooled value for each dog); $dP_p/dV_{(o)}$, dP/dV at V_o determined from positive P-V relation; $dP_n/dV_{(o)}$, dP/dV at V_o determined from negative P-V relation at low inotropic state (see text and Table 2)

Assuming that all the pressure-volume coordinates below the equilibrium volume could be pooled in each dog, we calculated S_n from Eq. 4 using the data of Table 1. The results for each dog are also shown in Table 1. There were no significant differences among the values of S_n. The ensemble mean S_n was calculated by pooling all the points from each dog to give a mean of $S_n = 5.1$ mmHg $r = 0.632$; $n = 175$). The large difference between the ensemble mean S_n and the means from Table 1 ($S_n = 8.7 \pm 7.9$ mmHg; $n = 8$) is due partly to the variance within and between dogs, which in turn may be due to the more *dynamic* nature of the negative phase of the P-V relation, as compared to the more *passive* positive phase. The large variation may also be due to the influence of inotropic state on diastolic suction [5] in contrast to the known lack of such influence on the positive portion of the P-V relation as described later.

To test the hypothesis that the negative portion of the diastolic P-V relation was influenced by inotropic state [5], we selected one run at high — and one run at low-inotropic state (Table 2). Six of the 7 dogs showed significant differences in S_n due to inotropic state; the pooled values at each inotropic state for all 7 dogs were significantly different (Table 2). As expected, S_p was not influenced by inotropic state (Table 2).

Effect of Ventricular Weight

S_p did not correlate with left ventricular (LV) weight (r, 0.015). Fig. 4 illustrates the entire diastolic P-V relation in 2 dogs of different LV weights. In particular, note that although the LV weights and functional operating ranges of these two ventricles differ widely, and that the chamber compliances, dP/dV, are different at either equal volumes or pressures, the stiffness parameters, S_p, are equal to each other

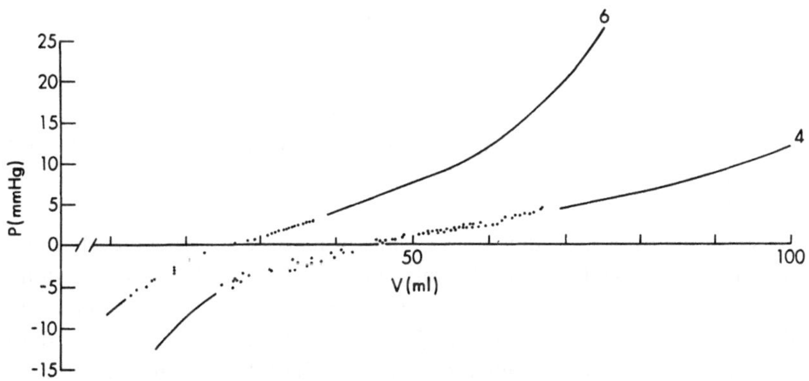

Fig. 4. Pressure-volume (*P-V*) data points from two dogs (4 and 6, Table 1) with different left ventricular weights (LVW), illustrating the influence of LVW on the chamber properties. The stiffness coefficient, S_p, is independent of LVW. Because the operating volume range, V_m–V_o, increases with LVW, the chamber stiffness, dP/dV, at any given pressure is inversely proportional to the LVW. Similarly, at any given volume, dP/dV is smaller in large ventricles. *Dots* are the raw data points, *solid curves* are the extrapolations from the logarithmic fit. See Table 1 for individual data

(Table 1). There was a strong negative correlation between the value of dP/dV at V_o and LV weight ($r = 0.856$; $p < 0.01$; $n = 8$). The widely scattered points in the negative pressure range preclude drawing similar conclusions for the sector.

Discussion

We have demonstrated here and elsewhere that the left ventricle can relax to a negative pressure in the absence of filling [3, 5, 6]. We have also shown that end-systolic volume clamping need not necessarily lead to negative pressures, but it does lead to a pressure that is lower than the pressure which would have existed with filling [3, 6].

Contraction below the equilibrium volume is analogous to contraction below slack length in an isolated muscle. It is well-known that an unloaded papillary muscle or myocyte will contract and relengthen in the absence of any external forces [10]. There must, then, be internal restoring forces in the ventricular chamber when it contracts to below the slack length of the sarcomeres within its wall. There may also be restoring forces due to a ventricular shape change that occurs after sarcomere shortening has ended. The pressure-flow relations across the aortic valve have long been known to reflect an inertial system. It is not unreasonable to conclude that, because of momentum in aortic outflow [11], blood may continue to leave the ventricle after the sarcomeres have ceased shortening, thereby leading to a deformation that can store elastic energy even in the absence of contraction below equilibrium volume. The more vigorous the outflow, the greater the likelihood of a shape change. Perhaps this is why the negative portion of the P-V curve was influenced by the inotropic state (increased S_n, see Table 2) and showed so much variability.

We define diastolic suction as property of the left ventricle that relies upon internal restoring forces to establish the pressure gradient for filling. Since storage of energy requires an expenditure of at least an equal amount of energy, it appears that the functional role of diastolic suction is to maintain the integrity of the lungs by keeping the filling pressure low. Although it is not the scope of this paper to determine the source of restoring forces, it is worth noting a recent study relating the extracellular matrix of the myocardium and its mechanical parameters [12].

We used the data from Table 2 along with the ensemble mean values for S_p and S_n to create the 'average' dog in this study. $P_p = -14.6 \ln[(116-V)/79]$ and $P_n = \ln[(V-13)/24]$.

Given these estimates of absolute volumes and properties, we calculate that our 'average' dog functioned with an end-diastolic volume of 52.3 ml, and end-systolic volume of 28.1 ml, and an ejection fraction of 46%. We can further define an end-systolic volume reserve as $(V_o - V_{es})/(V_o - V_d)$ and determine that our dogs used 38% of that reserve. Similarly, the end-diastolic volume reserve is $(V_{ed} - V_o)/(V_m - V_o)$, and our dogs used only 19% of the reserve. Thus, the Frank-Starling mechansim is readily available. Finally, at 0.95 V_m the ventricular pressure is 40 mmHg; this is consistent with the maximum end-diastolic pressures found by others [9, 13]. It is also consistent with the known tendency for the cardiovascular system to regulate pressure as volumes vary.

There are no data available with which to compare our value of stiffness in the negative phase (S_n). In a previous study [5], we reported the intresting observation

Table 2. Effect of inotropic state on S_n and S_p

Dog	Inotropic state	dP/dt_{max} (mmHg/s)	S_n (mmHg)	($_n$)	$P<$	S_p (mmHg)	($_n$)	P
4	L	1760	4.0	(6)		13.2	(7)	
					.001			NS
	H	2200	4.9	(6)		14.2	(5)	
5	L	2420	6.3	(7)		12.1	(10)	
					.05			NS
	H	2860	9.5	(8)		11.0	(8)	
6	L	1650	3.7	(7)		13.9	(6)	
					NS			NS
	H	1980	3.6	(6)		14.1	(6)	
7	L	1540	4.2	(10)		14.5	(11)	
					.001			NS
	H	3080	9.6	(5)		14.1	(12)	
8	L	2970	7.0	(4)		16.8	(8)	
					.002			NS
	H	3740	14.9	(5)		16.4	(9)	
9	L	1870	1.8	(6)		13.1	(4)	
					/			NS
	H	/	/	/		13.3	(11)	
10	L	1760	7.9	(4)		14.0	(6)	
					.025			NS
	H	2420	24.6	(4)		15.0	(8)	
11	L	1540	4.6	(7)		15.8	(8)	
					.001			NS
	H	30	19.7	(5)		16.3	(6)	
	L	1950 ± 540		(7)				
					.05			
	H	2810 ± 650		(7)				
Mean ± SD	L		4.9 ± 2.0	(7)		14.1 ± 1.4	(8)	
					.001			NS
	H		12.4 ± 7.2	(7)		14.4 ± 1.6	(8)	
Pooled	L		4.6	(51)		14.4	(60)	
					.05			NS
	H		6.1	(39)		14.5	(65)	

S_p and S_n, stiffness constants; ($_n$), negative portion of the P-V relation; L, low inotropic state; H, high inotropic state

that ventricular pressure fell to a greater negative value at higher levels of contractility and similar end-systolic volumes. The tendency for S_n to be greater at higher inotropic states (Table 2) supports this result and suggests that diastolic suction may be due to a combination of elastic recoil and one or more mechanisms related to the ventricular systole.

Ideally, if Eqs. 1 and 3 are to accurately characterize the entire pressure — volume relation, they should have the same slope at their common point, the equilibrium volume. In the 'average' dog, dP/dV at V_o is 0.185 and 0.123 mmHg/ml for the positive and negative phases of the P-V relation, respectively. Alternatively, a paired t-test between the values of dP/dV in Table 1 showed no significant difference. Note also that the ratio, $S_p/S_n = 14.6/4.6 = 3.17$, is very similar to the ratio of operating volumes for the positive and negative phases, $79/24.1 = 3.28$, indicating again the equality of dP/dV at V_o and supporting our scaling approach to the description of the passive P-V relation.

Conclusion

We have described a unique method of separating the effects of relaxation from the effects of filling so that we may determine the passive pressure-volume relation of the left ventricle above and below the equilibrium volume. The technique of end-systolic volume clamping has also allowed us to determine the negative portion of the P-V relation and, thus, to elucidate the mechanism and importance of diastolic suction. Finally, we have described the P-V relation with a logrithmic approach that is physically more realistic than the traditional exponential. We look forward to the application of these concepts to characterize the changes in the pathological heart.

References

1. Brecher GA (1958) Critical review of recent work on ventricular diastolic suction. Circ Res 6: 554–566
2. Tyberg JV, Keon WJ, Sonnenblick EH, Urshel CW (1970) Mechanics of ventricular diastole. Cardiovasc res 4: 423–428
3. Yellin EY, Hori M, Yoran C, Sonnenblick EH, Gabbay S, Frater RWM (1986) Left ventricular relaxation in the filling and nonfilling intact canine heart. Am J Physiol 250: H620–H629
4. Suga H, Goto Y, Igarashi Y, Yamada O, Nozawa T, Yasamura Y (1986) Ventricular suction under zero source pressure for filling. Am J Physiol 251: H47–H55
5. Hori M, Yellin EL, Sonnenblick EH (1982) Left ventricular suction as a mechanism of ventricular filling. Jpn Circ J 46: 124–129
6. Nikolic S, Yellin EL, Tamura K, Vetter H, Tamura T, Meisner, JS, Frater RWM (1988) Passive properties of canine left ventricle: diastolic stiffness and restoring forces. Cric Res 62: 1210–1222
7. Abe H, Nakamura T, Motomiya M, Konno K, Arai S (1978) Stresses in left ventricular wall and biaxial stress-strain relation of the cardiac muscle fiber for the potassium-arrested heart. J Biomech Eng 100: 116–121
8. Hardy HH, Collins RE (1982) On the pressure-volume relationship in circulatory elements. Med Biol Eng Comput 20: 565–570
9. Nakamura T, Abe H, Arai S, Kimura T, Kushibiki H, Motomiya M, Konno K, Suzuki N (1982) The stress-strain relationship of the diastolic cardiac muscle and left ventricular compliance in the pressure-overload canine heart. Jpn Circ J 46: 76–83
10. Krueger JW, Forletti D, Wittenberg BA (1980) Uniform sarcomere shortening behavior in isolated cardiac muscle cell. J Gen Physiol 76: 587–607
11. Noble MIM (1968) The contribution of blood momentum to left ventricular ejection in the dog. Circ Res 23: 663–670
12. Robinson TF, Cohen-Gould L, Remily RM, Capasso JM, Factor SM (1985) Extracellular structures in heart muscle. In: Harris P, Poole-Wilson PA (eds) Advances in myocardiology, vol 5 Plenum, New York
13. Glantz SA, Kernoff RS (1975) Stiffness determined from canine left pressure-volume curve. Circ Res 37: 787

Discrepancy Between Slow Relaxation and Increased Myocardial Stiffness

Takashi Serizawa, Osami Kohmoto, Masahiko Iizuka, Tetsuo Ohya, Shin-ich Momomura, and Tsuneaki Sugimoto[1]

Summary. The time constant of left ventricular pressure fall (T_B, $P = P_0 + A.e^{-t/T_B}$) is a good indicator of myocardial relaxation, but the physiological nature of the asymptote (P_0) is not yet clear. The effects of hypoxia and the combined effects of hypoxia and DPI 201-106 (DPI, 10^{-6} *M* and 3×10^{-6} *M*), a new cardiotonic agent which increases the sensitivity of contractile elements to Ca^{2+}, on diastolic myocardial function in isolated isovolumic rabbit hearts were investigated. There were close correlations between the change in left ventricular end-diastolic pressure (⊿ LVEDP) and ⊿ P_0 ($r = 0.98$, $P < 0.01$) and between ⊿ T_B and ⊿ LVEDP ($r = 0.73$, $P < 0.01$) following hypoxia. The combination of hypoxia and DPI induced further increments in LVEDP (25 ± 8 mmHg, $P < 0.05$ vs hypoxia) and P_0 (15 ± 7 mmHg, $P < 0.05$), and still there was a close linear correlation between ⊿LVEDP and ⊿ P_0 ($y = 1.08x + 1.5$, $r = 0.90$, $P < 0.05$). However, T_B(52 ± 12 ms) was not altered and no relationship was observed between ⊿ T_B and ⊿ LVEDP.

P_0 reflected LVEDP in isolated isovolumic heart. The discrepancy between the changes in T_B and P_0 induced by DPI suggest different mechanisms underlying the changes in left ventricular diastolic stiffness and relaxation.

Introduction

The mechanisms of elevated left ventricular diastolic pressure relative to volume induced by ischemia have been highly controversial [1–4]. Recent studies have suggested an important role for abnormal left ventricular diastolic properties other than extrinsic compression in diastolic left ventricle by pericardium or right ventricle [5-9]. Furthermore, the definition of abnormal diastolic properties is itself controversial. We previously proposed that diastolic abnormalities be separated into two parts with each considered a potentially independent aspect of the process: (1) slow relaxation reflected by the increase in the time constant of left ventricular isovolumic pressure decline (T), and (2) incomplete relaxation due to failure of systolic tension to be completely dissipated by the end-diastole [10]. However, others have indicated that slow relaxation and elevated diastolic pressure are part of a sequential process, by showing elevated left ventricular end-diastolic pressure (LVEDP) exclusively from incomplete relaxation [11].

[1] The Second Department of Internal Medicine, Faculty of Medicine, University of Tokyo, 7-3-1 Hongo, Bunkyo-ku, Tokyo, 113 Japan

This review shows our recent study which investigated the relationship between myocardial diastolic distensibility and relaxation in isolated isovolumic rabbit hearts through the intervention of hypoxia and DPI 201-106 (4-[3-(4-diphenyl-methyl-1-piperazinyl)-2-hydroxypropoxy]-1H-indole-2-carbonitrile, DPI), a new cardiotonic agent not mediated by cyclic AMP, and briefly surveys our current understanding of the physiological meaning of the time constant T in relation to left ventricular diastolic distensibility, i.e., myocardial diastolic stiffness.

Effects of Hypoxia at Constant Coronary Flow

In isolated rabbit heart, hypoxia in which the partial pressure of oxygen in the perfusate was reduced from 524 ± 17 (equilibrated by 5% CO_2 and 95% O_2) to 166 ± 10 mmHg (equilibrated by 5% CO_2, 20% O_2 and 75% N_2) induced an abrupt fall in peak systolic pressure, LVdP/dt, and coronary perfusion pressure. Left ventricular end-diastolic pressure rose gradually during the hypoxia. However, all changes were stabilized at 5 min after the onset of hypoxia and sustained for another 10 min [11]. At the end of 15 min of hypoxia, significant reductions occurred in peak systolic pressure (101 ± 8 to 67 ± 7 mmHg, $P < 0.01$), LVdP/dt (1631 ± 285 to 999 ± 204 mmHg/s, $P < 0.01$) and LV(−) dP/dt (1218 ± 167 mmHg/s, $P < 0.01$). Hypoxia led to increase in LVEDP 10 ± 3 to 21 ± 7 mmHg, $P < 0.05$), time constant of left ventricular pressure decay (T_B) [12] (41 ± 8 to 54 ± 10 ms, $P < 0.01$), and asymptote (P_o) (−1 ± 6 to 12 ± 5 mmHg, $P < 0.01$).

Effects of Hypoxia and DPI

DPI 201-106 (DPI), a new cardiotonic agent that increases the sensitivity of contractile elements to Ca^{2+}, is quite different in its inotropic action from other inotropic agents mediated by increased intracellular cyclic AMP or Na^+-K^+ exchange inhibition [12]. In our preliminary study using the same experimental preparation, 10^{-6} *M* DPI increased LV dP/dt/P by 14%, and decreased coronary vascular resistance and left ventricular end-diastolic pressure by 10% and 23%, respectively. T_B was not altered at this dosage.

When 10^{-6} *M* DPI was infused for 10 min in hypoxic hearts, further increments were observed in LVEDP (from 21 ± 7 to 25 ± 8 mmHg, $P<0.05$), and P_0 (12 ± 5 to 15 ± 7 mmHg, $P < 0.05$). Peak systolic pressure, LV(−) dP/dt, and T_B (54 ± 10 to 52 ± 12 ms) were not altered.

Relationship Between Myocardial Relaxation and Diastolic Stiffness

A close linear correlation between ΔP_0 and ΔLVEDP was observed following hypoxia ($y = 0.99x - 1.3$, $r = 0.98$, $P < 0.01$), indicating that P_0 reflected myocaridal diastolic stiffness because the heart contracted isovolumically. The same relationship was found in the effect of 3×10^{-6} *M* DPI on hypoxia ($y = 1.08x + 1.5$, $r = 0.90$, $P < 0.05$), suggesting that common mechnisms are involved in increased myocardial diastolic stiffness by myocardial oxygen deficit and DPI (Fig. 1).

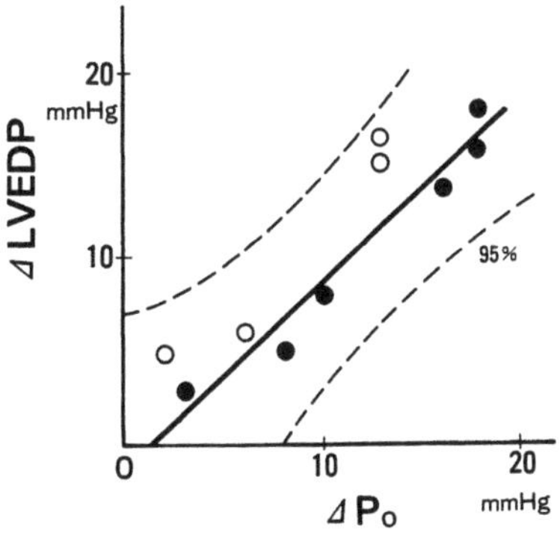

Fig. 1. Relationships between the change in left ventricular end diastolic pressure (*⊿LVEDP*) and asymptote (*⊿P_0*) induced by hypoxia and DPI 201-106 (DPI). A close linear correlation was observed between ⊿LVEDP and ⊿P_0 following hypoxia ($y = 0.99\ x - 1.3$, $r = 0.98$, $P<0.01$). The same relationship was observed in the effect of 3×10^{-6} *M* DPI on hypoxia ($y = 1.08x + 1.5$, $r = 0.93$). All changed values were within the 95% confidence limit of the regression line. ●, ○, indicate changes induced by hypoxia and hypoxia with 3×10^{-6} *M* DPI, respectively

A close linear relationship was observed between ⊿LVEDP and ⊿T_B following hypoxia, but no such relationship was observed when DPI was added to hypoxia (Fig. 2).

Although the actual relationship between the biochemical process and the time constant of left ventricular isovolumic pressure fall (T) has not yet been clarified, evidence suggests that T reflects the rate of Ca^{2+} uptake by sarcoplasmic reticulum (SR) [13].

Initially, T was calculated on the assumption that the left ventricular Isovolumic pressure fall curve fit a monoexponential curve with a 0 mmHg asymptote (T_W, $P = A.e^{-t/T_W}$) [14]. However, there is no theoretical framework leading to the assumption that the asymptote is a constant 0 mmHg, and if the end point of relaxation is defined by the extent of left ventricular distensibility, i.e., diastolic myocardial stiffness, T_W may contain erroneous information, especially if diastolic pressure is changed through intervention. Thus, it is preferable to calculate the time constant T_B and asymptote P_0 simultaneously by the best fit method (T_B, $P = P_0 + A.e^{-t/T_B}$) [15].

In previous clinical studies [15, 16], P_0 increased following ischemia, but one study showed the lowest P_0 in the severest ischemic patients at rest [15]. T_W is known to be affected by loading conditions, i.e., left ventricular volume [14, 17], which is determined by the extent of systolic shortening, but considered to represent the stored potential energy for elastic recoil. T_B and P_0 are also affected by left ventricular volume. One reason our results differed from those of others is that we used an isovolumic heart model. Results highly similar to ours were reported in the isovolumic rat heart [9] and suggested the usefulness of P_0 and T_B in discussions of relaxation and stiffness. The close linear correlation between ⊿ LVEDP and ⊿ P_0 induced by hypoxia, as demonstrated by our previous study [18] indicates that P_0 can be used as an index of mycordial diastolic stiffness.

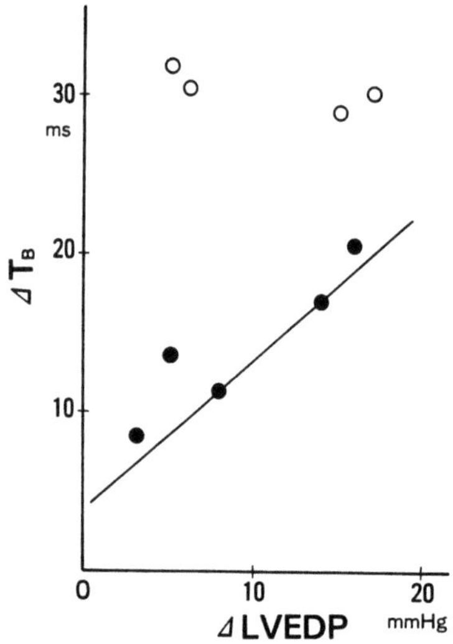

Fig. 2. Relationship between the change in time constant of left ventricular decay (ΔT_B) and left ventricular end-diastolic pressure (*ΔLVEDP*). A close linear correlation was observed between ΔT_B and ΔLVEDP ($y = 0.97x + 3.5$, $r = 0.73$) following hypoxia. However, no definite correlation was observed in the changes induced by hypoxia with DPI 201-106 (DPI). ●, ○ indicate changes following hypoxia and hypoxia with DPI, respectively

DPI acted differently in hypoxic and aerobic conditions. LVEDP and P_0, which were already increased secondary to hypoxia, were elevated further. These facts suggest that the increased Ca^{2+} affinity to contractile elements caused by DPI resulted in an increased number of cross-bridges not only in systole, but also in diastole by disturbing Ca^{2+} sequestration from contractile proteins, an energy-requiring process that is affected by insufficient O_2 supply. As a result, some Ca^{2+} remained on the contractile proteins and tension was maintained throughout the period. The same linear relationships between ΔLVEDP and ΔP_0 induced by hypoxia and hypoxia with DPI in the present study suggest a common mechanism to explain increased diastolic stiffness following sole hypoxia and hypoxia with DPI.

We analyzed our previous study [18] further and found a close linear correlation between ΔLVEDP and ΔT_B following hypoxia. The result does not necessarily contradict the hypothesis that elevated LVEDP induced by ischemia is due not to elevated myocardial stiffness but to prolonged myocardial relaxation [11]. However, this hypothesis cannot explain the exclusive increase in LVEDP with unaltered T_B in conditions like DPI under hypoxia. Because the detachment of Ca^{2+} from contractile proteins and Ca^{2+} sequestration from cytoplasma are energy-requiring processes, it was not surprising to see simultaneous deterioration in myocardial relaxation and stiffness following myocardial oxygen deficit. On the other hand, if SR function is not altered and only detachment of Ca^{2+} from contractile proteins is impaired, increased myocardial diastolic stiffness without any changes in myocardial relaxation ensues, as shown in the present study. Myocardial diastolic abnormalities might be governed by separate mechanisms.

References

1. Glantz SA, Parmley WW (1978) Factors which affect the diastolic pressure-volume curve. Circ Res 42: 171–180
2. Grossman W, Barry WH (1980) Diastolic pressure-volume relations in the diseased heart. Fed Proc 39: 148–155
3. Grossman W (1985) Why is left ventricular diastolic pressure increased during angina pectoris? J Am Coll Cardiol 5: 607–608
4. Ross J Jr (1979) Acute displacement of the diastolic pressure-volume curve of the left ventricle: role of the pericardium and the right ventricle. Circulation 59: 32–37
5. Serizawa T, Carabello BA, Grossman W (1980) Effects of pacing-induced ischemia on left ventricular diastolic pressure-volume relations in dogs with coronary stenoses. Circ Res 46: 430–439
6. Paulus WJ, Serizawa T, Grossman W (1982) Altered left ventricular diastolic properties during pacing-induced ischemia in dogs with coronary stenoses: potentiation by caffeine. Circ Res 50: 218–227
7. Bourdillon PD, Lorell BH, Mirsky I, Paulus WJ, Wynne J, Grossman W (1983) Increased regional myocardial stiffness of the left ventricle during pacing-induced angina in man. Circulation 67: 316–323
8. Sasayama S, Nonogi H, Miyazaki S, Sakurai T, Kawai C, Eiho S, Kuwahara M (1985) Changes in diastolic properties of the regional myocardium during pacing-induced ischemia in human subjects. J Am Coll Cardiol 5: 599–606
9. Lorell BH, Wexler LF, Momomura S, Weinberg E, Apstein CS (1986) The influence of pressure overload left ventricular hypertrophy on diastolic properties during hypoxia in isovolumically contracting rat hearts. Circ Res 58: 653–663
10. Serizawa T, Vogel WM, Apstein CS, Grossman W (1981) Comparison of acute alterations in left ventricular relaxation and diastolic chamber stiffness induced by hypoxia and ischemia: role of myocardial oxygen supply-demand imbalance. J Clin Invest 64: 91–102
11. Weisfeldt ML, Frederiksen JW, Yin FCP, Weiss JL (1978) Evidence of incomplete left ventricular relaxation in the dog. J Clin Invest 62: 1296–1302
12. Scholtysik G, Salzmann R, Berthold R, Herzig JW, Quast U, Markstein R (1985) DPI 201-106, a novel cardioactive agent. Combination of cAMP-independent positive inotropic, negative chronotropic, action potential prolonging and coronary dilatory properties. Arch Pharm 329: 316–325
13. Parmley WW, Sonnenblick EH (1969) Relation between mechanics of contraction and relaxation in mammalian cardiac muscle. Am J Physiol 216: 1084–1091
14. Weiss JL, Frederiksen JW, Weisfeldt ML (1976) Hemodynamic determinats of the time course of fall in canine ventricular pressure. J Clin Invest 58: 751–776
15. Thomspon DS, Waldron CB, Juul SM, Naqvi N, Swanton RH, Coltart DJ, Jenkins BS, Webb-Peploe MM (1982) Analysis of left ventricular pressure during isovolumic relaxation in coronary artery disease. Circulation 65: 690–697
16. Carroll JD, Hess OM, Hirzel HO, Krayenbuehl HP (1983) Exercise-induced ischemia: The influence of altered relaxation on early diastolic pressures. Circulation 67: 521–527
17. Raff GL, Glantz SA (1981) Volume loading slows left ventricular isovolumic relaxation rate: evidence of load-dependent relaxation in the intact dog heart. Circ Res 48: 813–824
18. Serizawa T, Momomura S, Kohmoto O, Ohya T, Sato H, Takahashi T, Mochizuki T, Iizuka M, Sugimoto T (1987) Mechanisms of abnormal relaxation induced by ischemia: comparison of low flow ischemia and hypoxia in isolated rabbit heart. Jpn Circulation J 51: 90–97

Coronary Circulation and Cardiac Function

Yukio Maruyama, Shoichi Satoh, Kazuhira Maehara, and Tamotsu Takishima[1]

Summary. In this article, the close link between coronary circulation and cardiac function is reviewed from several points of view. Cardiac function is greatly determined by the loading condition, and on the contrary, the latter definitely regulates coronary blood supply through myocardial oxygen demand. Since many other variables also affect both coronary circulation and cardiac function simultaneously, it is difficult to establish in the in situ heart how the relation between the two factors is modified by loading change alone. From the clinical point of view, this problem is serious in respect to how to change loading in order to manage myocardial ischemic injury with coronary stenosis. In that matter, our experimental model using an excised, perfused heart preparation has given some valuable insight. Although many questions remain unanswered and controversies remain, our results support the concept that preload elevation leads to aggravation of ischemic injury while alleviating it in afterload elevation. Effects of loading on coronary circulation are, in particular, discussed. We stress the important role of the pericardium and also of the surrounding heart pressure on determining Pf = 0.

Introduction

Cardiac function is intimately related to coronary circulation and vice versa, because coronary driving pressure is largely determined by left ventricular function. Although there are many interesting problems concerning the link between coronary circulation and cardiac function, only the following four effects will be discussed.

Effects of Coronary Flow on Ventricular Diastolic Function

In 1974 Barry et al [1]. reported that the left ventricular diastolic pressure-volume relationship varies during angina pectoris. Since then, various interventions to induce myocardial ischemia in coronary artery disease, such as pacing, have been known to alter diastolic ventricular properties. These results suggest that severe coronary flow reduction may reduce diastolic ventricular compliance.

In contrast to these expectations, left ventricular global ischemia did not experimentally produce any significant changes in the left ventricular pressure-

[1] The First Department of Internal Medicine, Tohoku University School of Medicine, 1-1 Seiryo-machi, Sendai, 980 Japan

circumference relationship [2], although there are large species-related differences concerning this point. Moreover, regional myocardial ischemia due to 50% coronary flow reduction or complete coronary occlusion shifted the diastolic left ventricular pressure-volume or pressure-segmental length curve to the right [3, 4]. In this phenomenon, there was an increased, unstressed segmental length component [5]. Although it has been reported that the leftward shift of diastolic pressure-length relationship of the ischemic segment occurs, it is not due to acute coronary flow reduction, but due to rigor bond formation [6] or scar formation [7] in the chronic stage. Thus, acute regional myocardial ischemia itself may increase diastolic ventricular compliance, because the segmental pressure-length relationship following acute ischemia did not show any leftward or upward shift indicating reduced diastolic segmental compliance. However, in reports also showing a rightward shift of the pressure-length relationship, the slope of the relationship was steeper than that prior to coronary occlusion, suggesting reduced compliance [8]. Thus, understanding of the data mentioned may require further verification, because the elastic stiffness of the ischemic myocardium is more precisely defined by the stress-strain relation than by the pressure-length relation.

In contrast to previously reported rightward displacement or absence of significant shift in the pressure-volume (length) relationship, Serizawa et al. [4], by increasing heart rate after reducing coronary flow to the left ventricle by 50% and then, after stopping atrial pacing, found a similar upward shift of the left ventricular diastolic pressure-volume relationship observed clinically in anginal attack. Although the post-pacing ischemic state may differ from that during exercise-induced ischemia in neurohumoral and hemodynamic responses, the latter pressure-length relationship also showed an upward shift from that with ischemia alone (Fig. 1) [5]. Paulus et al. [9] observed that an upward shift is exaggerated by caffeine and reduced by Ca antagonist. Accordingly, it is likely that some further intervention to alter intracellular Ca^{2+} distribution in addition to coronary flow reduction decreases diastolic ventricular compliance.

Effects of Coronary Flow on Systolic Ventricular Pump Function

It is well known that the capacity of the left ventricle to develop tension at any given resting tension is directly correlated with O_2 delivery [10]. However, there was little data until 1975 concerning the relations among global left ventricular function, ischemic segmental wall motion, and coronary blood supply in one major coronary arterial branch stenosis [11].

According to previous experimental data, coronary flow reduction by approximately 50% led to the onset of both regional lactate production and segmental mechanical dysfunction, and thereafter caused a progressive decrease in both variables. On the other hand, no significant changes in global left ventricular performance were noted at the onset of regional ischemia. Moreover, even coronary flow reduction to a more extended perfusion area of the left main coronary artery by 30% showed only minimal impairment of left ventricular function [12]. Thus, it is likely that until near-critical coronary stenosis there is no dramatic change in overall ventricular function irrespective of the extent of the affected ventricular mass.

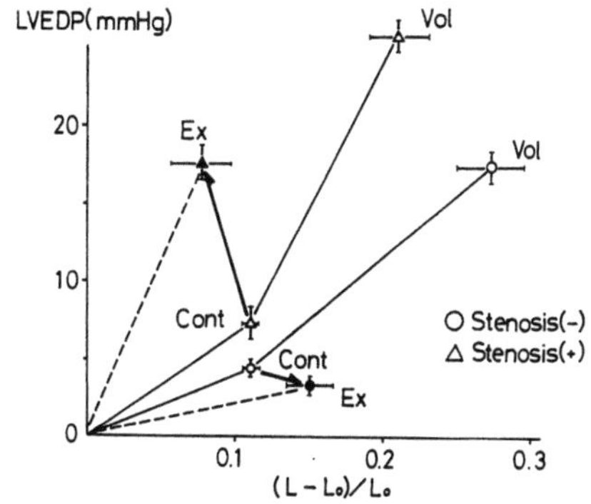

Fig. 1. Left ventricular end-diastolic pressure (*LVEDP*)-Lagrangian strain relationship during exercise (*Ex*) and volume loading (*Vol*) with and without left anterior descending (LAD) and left circumflex coronary artery stenoses at the 30% coronary blood flow reduction level in 7 anesthetized dogs. Exercise was performed by rhythmic contraction of the lower legs induced by electrical stimulation. Segment length (*L*) was measured by ultrasonic crystals in the LAD area. Compared to the relationship between LVEDP and $L\text{-}L_o/L_o$ without coronary stenosis (○ Vol), that of ischemia (△ Vol) is shifted to the left, and the further leftward shift was observed during exercise with coronary stenosis (▲ Ex), suggesting lesser diastolic compliance in the ischemic segment. ●, exercise without coronary stenosis; ———, directional change following Vol; -------, directional change following Ex; *Cont*, control state prior to Ex or Vol

Effects of Preload and Afterload Changes on Ischemic Myocardium and Coronary Circulation

A decrease in afterload pressure increases cardiac output while varying coronary blood flow through a decrease of coronary driving pressure and also coronary adjustment to ventricular unloading. On the contrary, in afterload elevation, cardiac output basically decreases, and coronary circulation is also modified by this loading change. Thus, in spite of loading changes coronary blood supply and myocardial oxygen demand are matched to each other in a physiological loading condition. However, it has been clarified [13] that close coupling between the two variables is not maintained when aortic pressure is reduced to the extent of causing a deficit of coronary flow relative to metabolic need.

The mean aortic pressure which increased cardiac output maximally was 65 mmHg, which was closely associated with the lower limit of the autoregulation of coronary blood flow, and, with less than this value, cardiac function began to decrease due to an imbalance between myocardial oxygen demand and coronary blood supply [13]. Thus, when coronary autoregulation has expired due, for example, to the presence of coronary stenosis, how are myocardial ischemia and coronary circulation modified by loading changes? When considering this problem, it is necessary to take into account whether myocardial ischemia is global or regional. This is because in global ischemia it is possible to assess further ischemic changes during the intervention by assessing total cardiac function alone, whereas in regional ischemia, ischemic regional and global functions should be tested simultaneously [14].

With varied afterload resistances, relations between global ventricular function and left coronary flow have been investigated at constant heart rate and preload, using the isolated, perfused canine heart preparation, in which the coronary perfusion pressure was 30 mmHg less than mean aortic pressure, mimicking 90% coronary stenosis. The mean aortic pressure showing maximal cardiac output was 72 mmHg. If coronary stenosis is more severe, it is quite probable that this value of aortic pressure showing maximal cardiac output will increase [13].

Although DaLuz et al. [15] reported a preferential effect of sodium nitroprusside on both global and ischemic regional functions, there have been several reports in which afterload reduction in the presence of an experimental coronary occlusion or acute myocardial infarction may lead to further functional deterioration of the ischemic area. In our previous study using coronary stenosis of 50% flow reduction in the control loading condition, the predominat effect observed in the ischemic area during afterload reduction was due to a reduced perfusion pressure rather than from reduced oxygen consumption [14]. However, it should be noted that global ventricular function does not clearly appear to be depressed during the intervention.

Though the reason why the effects of afterload reduction on the ischemic region lead to different results in different studies is unclear, it should be noted in our study [14] that pure afterload reduction without varying other variables was performed, using ultrasonic crystals instead of inaccurate mercury-in-silastic rubber length guages. As a result, it is likely that in pure afterload reduction with coronary arterial stenosis the function of the ischemic myocardium might be further aggravated, unless afterload reduction is done carefully.

We are thus led to ask, does afterload elevation improve regional function of the ischemic area? This question, however, has not been fully elucidated, not only in a heart with an partial occlusion, but also in a heart with a complete occlusion [16]. According to our study [17] (in which left circumflex coronary flow was reduced to about one-third of the pre-stenotic value, and thereafter, afterload elevation with constant preload was induced as shown in Fig. 2), coronary blood flow increased and ischemic injury was ameliorated, i.e., intramyocardial CO_2 ($PmCO_2$), 60 vs 47 mmHg (mean, $P < 0.01$), aortic pressure 73 vs 117 mmHg ($P < 0.01$), and left circumflex coronary flow, 15 vs 29 ml/min ($P < 0.01$); whereas cardiac output decreased from 1349 to 844 ml/min ($P < 0.01$). On the contrary, afterload elevation combined with preload elevation did not improve ischemic injury. From these results, it is suggested that, although coronary flow increased after both preload and afterload elevation, the preload elevation counterbalanced the beneficial effects of afterload elevation on salvaging ischemic injury.

It has been reported that preload elevation causes a decrease in myocardial blood flow, especially in the subendocardial layer. However, in the in situ condition, preload elevation leads to afterload pressure elevation, therefore it is difficult to observe the effect of preload change alone. As shown by the example in Fig. 3, left ventricular end-diastolic pressure elevation alone from 7.1 to 16.3 mmHg ($n = 18$) aggravated ischemic injury assessed by ECG ST elevation and $PmCO_2$ increase [18]. One possible reason is that left ventricular dimension facing the contraction, that is "afterload stress" might be increased despite a lack of change in coronary perfusion pressure. On the other hand, a beneficial effect of preload elevation on the non-ischemic area as a result of the Frank-Starling mechanism has been reported [19]. There are several methodological differences between their and our prior

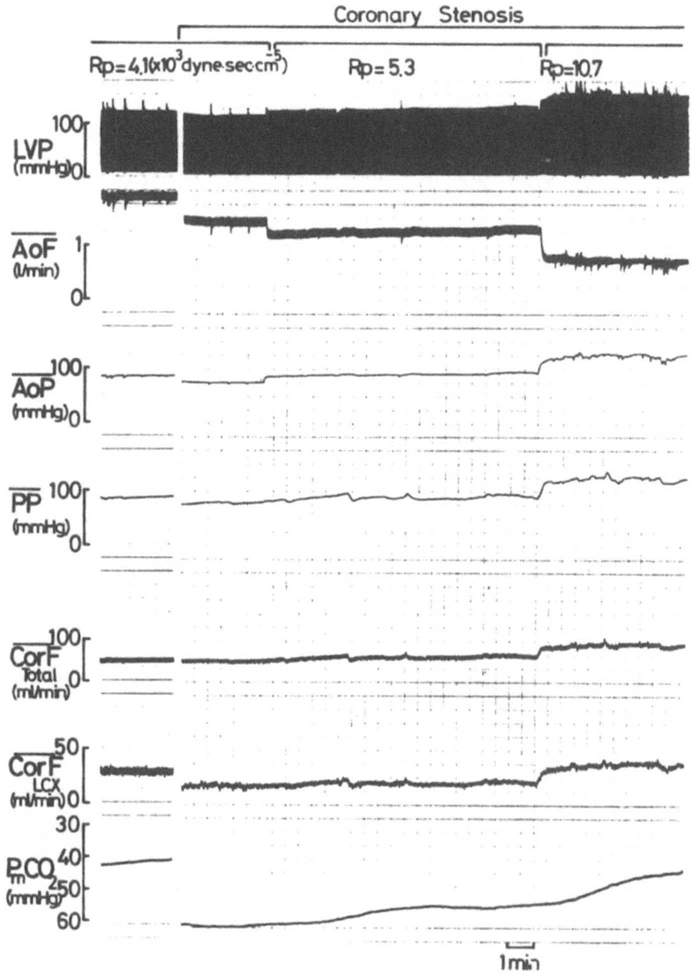

Fig. 2. The isolated heart is loaded with a hydraulic system simulating the aortic input impedance of the dog. A large variable-height reservoir is connected to the left atrium through the pulmonary vein. The heart is perfused with the arterial blood from the support dog by a servo-controlled peristaltic pump. The mean perfusion pressure was maintained equal to the mean aortic pressure during measurements. Shown are representative tracings of pressures, flows, and myocaridal CO_2 tension (P_mCO_2) in afterload elevation with constant preload. The *left panel* shows the control state before stenosis of the left circumflex branch (LCx) at 4.1×10^3 dyne.sec.cm^{-5} of peripheral resistance (*Rp*). Ten min after the LCx stenosis, pressures, flows, and the $PmCO_2$ reached steady state. Thereafter, Rp was elevated sequentially at 10-min intervals under constant left ventricular end-diastolic pressure. When aortic pressure increased duë to Rp increments, P_mCO_2 in the ischemic segment decreased (*upward deflection*). *LVP*, left ventricular pressure; $\overline{AoF}$, mean aortic flow; $\overline{AoP}$, mean aortic pressure; $\overline{PP}$, mean coronary perfusion pressure; $CorF_{Total}$, mean blood flow of the left common coronary artery; $Cor\ F_{LCX}$, mean blood flow of the left circumflex branch

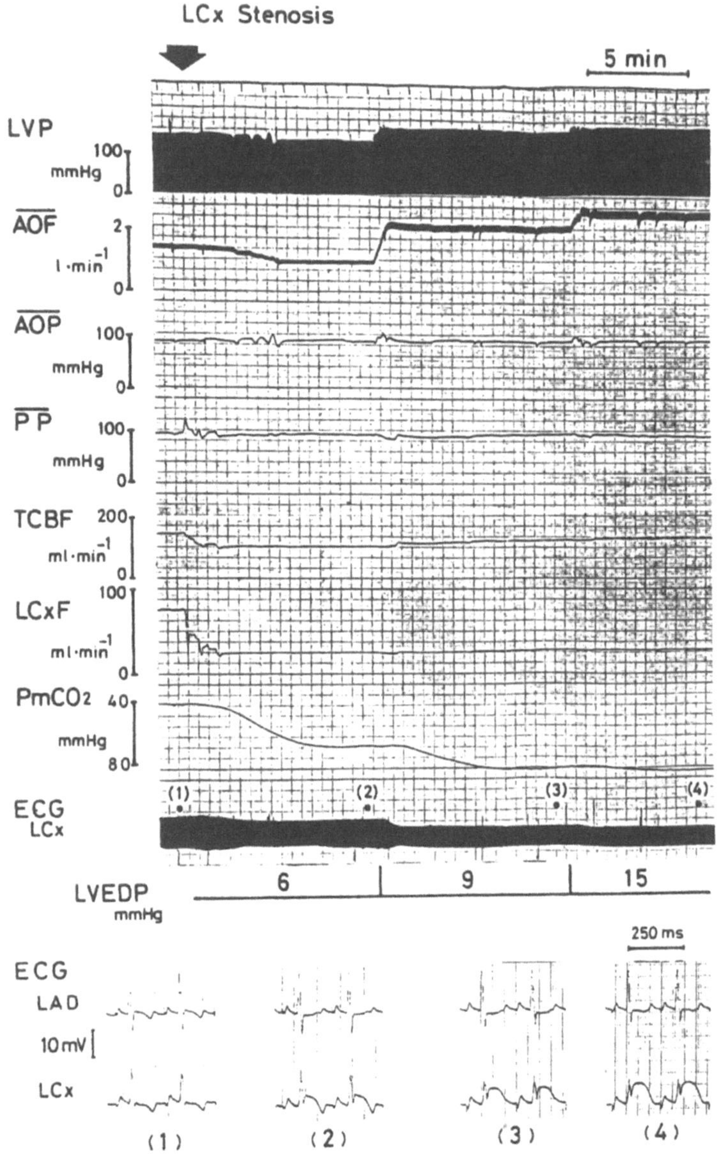

Fig. 3. Representative traces of the hemodynamic parameters, P_mCO_2, and epicardial ECGs in the left circumflex branch (*LCx*) area, when left ventricular end-diastolic pressure (*LVEDP*) was increased in a stepwise fashion following LCx stenosis with constant mean aortic pressure (mean perfusion pressure). LCx stenosis induced (*arrow*) the marked increase in P_mCO_2 (*downward deflection*) and ST elevation in the LCx area. The slow ECG recordings in the *upper panel* are more clearly shown in the *lower panel* with a high-speed recording, in the order (*1*), (*2*), (*3*), and (*4*). The elevation of LVEDP caused a further increase in both P_mCO_2 and the ECG-ST level of the LCx area. *LVP*, left ventricular pressure; $\overline{AOF}$, mean aortic flow; $\overline{AOP}$, mean aortic pressure; $\overline{PP}$, mean coronary perfusion pressure; *TCBF*, mean total left coronary blood flow; *LCxF*, mean left circumflex flow; P_mCO_2, myocardial carbon dioxide tension in the LCx area; *ECG*, epicardial ECG of each area; *LAD*, left anterior descending artery

studies, e.g., coronary artery, occlusion vs stenosis, and heart rate not fixed vs fixed. Thus, direct comparison may not be possible. It should be noted that despite the appearance of the Frank-Starling effect in the ischemic area, in contrast to their results, ischemic injury proceeded depending on preload increase; consequently, it is dangerous to draw a conclusion only from mechanical advantages in cardiac function.

Effects of Diastolic and Systolic Ventricular Pressures on Coronary Pressure-Flow Relationship

Coronary circulation has been reported to be regulated by many factors. One of the features which most distinguishes coronary circulation from circulation in other organs is the vascular system which is embedded in the myocardium (excepting epicardial vessels). With the intramyocardial tissue, pressure changes during one cardiac cycle, therefore, coronary circulation is profoundly affected. That is, coronary flow is supplied in diastole, whereas in systole it is impeded (Fig. 4a). Accordingly, when diastolic intramyocardial pressure increases, coronary blood supply is greatly disturbed, especially in the subendocardial layer [20]. This has also been shown by an altered diastolic coronary pressure-flow relationship. In other words, following preload elevation zero-flow pressure ($Pf = 0$) in the diastolic coronary pressure-flow relationship elevated significantly, indicating reduced coronary flow through an increase of effective back pressure. The presence of the pericadium [21] or an increase in surrounding heart pressure [22] enhanced this effect. Rather, the surrounding heart pressure determined $Pf = 0$ much more definitively than did intracardiac pressure elevation. These data supported a previous hypothesis that the subepicardial muscle layer has the lowest value of $Pf = 0$, the pressure of the last part of the myocardium to be perfused.

When an increase in preload augments systolic developed pressure, the beating effect on the phasic coronary flow pattern in systole is more inhibitory than that prior to preload elevation. On the contrary, by blood squeezing out through myocardial contraction, the so-called massaging action, diastolic flow is promoted (Fig. 4b). Also, by the autoregulatory vasodilating mechanism in the case still having vasodilating action, coronary inflow does not fall greatly despite an increased preload as shown in Fig. 3, though myocardial blood flow distribution might be changed. Thus, interaction between systole and diastole has been pointed out experimentally and theoretically [22]. In this phenomenon of systolic and diastolic interaction, vascular compliance plays an important role [23], and accordingly, the diastolic coronary pressure-flow relationship must be analyzed carefully [20, 22]. Finally, since the previous results have mainly been obtained in the left coronary artery, direct application to right coronary circulation must be made carefully. This is because right and left coronary circulation has been pointed out to differ in various respects [24].

In conclusion, the interaction between systole and diastole on phasic coronary flow is not fully clarified yet. Thus, it is unclear whether or not the magnitude of the interaction is the same in different myocardial layers. Further experiments are required, including flow measurement in various sizes of vessels.

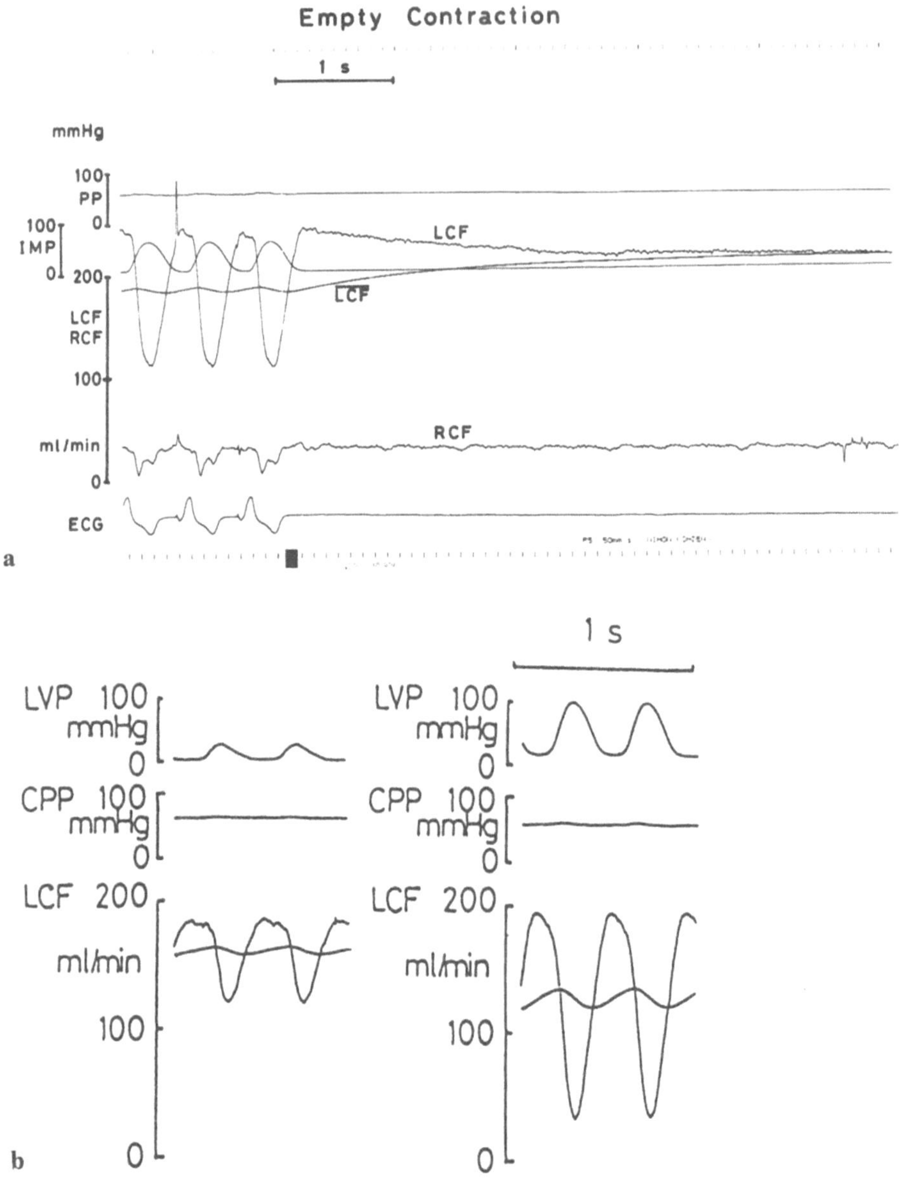

Fig. 4a. Phasic and mean left coronary flow (*LCF*, $\overline{LCF}$) in maximally vasodilated (due to adenosine infusion) state of beating, but empty, canine heart, in which, although systolic shortening is greater, coronary flow inhibition in systole is not larger than isovolumetrically beating (i.e., minimal shortening, if any) (not shown), suggesting that factor of primary importance in inhibiting systolic coronary flow is not shortening but intra-myocardial pressure itself. *RCF*, right coronary flow; *IMP*, intramyocardial pressure; *PP*, perfusion pressure. **b** Phasic and mean left coronary flow (*LCF*) in low (*left panel*) and high (*right panel*) preload in maximally vasodilated state when coronary perfusion pressure (*CPP*) and heart rate are maintained constant. As shown in *right panel*, diastolic LCF is higher, while systolic minimal flow is smaller, when left ventricular end-diastolic pressure is increased. *LVP*, left ventricular pressure

References

1. Barry WH, Brooker JZ, Alderman EL, Harrison DC (1974) Changes in diastolic stiffness and tone of the left ventricle during angina pectoris. Circulation 49: 225-263
2. Palacios I, Johnson RA, Newell JB, Powell WJ (1976) Left ventricular end-diastolic pressure-volume relationships with experimental acute global ischemia. Circulation 53: 428-436
3. Theroux P, Ross J Jr, Franklin D, Covell JW, Bloor CM, Sasayama S (1977) Regional myocardial function and dimensions early and late after myocardial infarction in the unanesthetized dog. Circ Res 40: 158-165
4. Serizawa T, Carabello BA, Grossman W (1980) Effect of pacing-induced ischemia on left ventricular diastolic pressure-volume relations in dogs with coronary stenoses. Circ Res 46: 430-439
5. Maehara K, Maruyama Y, Yoshikata S, Takahashi T, Takishima T (1987) Effects of stress on left ventricular diastolic stiffness and relaxation with and without coronary stenosis. Automedica 9: 143
6. Wexler LF, Weinberg EO, Ingwall JS, Apstein CS (1986) Acute alterations in diastolic left ventricular chamber distensibility: Mechanistic differences between hypoxemia and ischemia in isolated perfused rabbit and rat hearts. Circ Res 59:515-528
7. Hood WB Jr, Bianco JA, Kumar R, Whiting RB (1970) Experimental myocardial infarction. IV. Reduction of left ventricular compliance in the healing phase. J Clin Invest 49: 1316-1323
8. Theroux P, Franklin D, Ross J Jr, Kemper WS (1974) Regional myocardial function during acute coronary artery occlusion and its modification by pharmacologic agents in the dog. Circ Res 35: 896-908
9. Paulus WJ, Serizawa T, Grossman W (1982) Altered left ventricular diastolic properties during pacing-induced ischemia in dogs with coronary stenoses. Potentiation by caffeine. Circ Res 50: 218-227
10. Bacaner MB, Lioy F, Visscher MB (1971) Coronary blood flow, oxygen delivery rate and cardiac performance. J Physiol (London) 216: 111-127
11. Wyatt HL, Forrester JS, Tyberg JV, Goldner S, Logan SE, Parmley WW, Swan HJC (1975) Effect of graded reductions in regional coronary perfusion on regional and total cardiac function. Am J Cardiol 36: 185-192
12. Vatner SF, McRitchie RJ, Maroko PR, Patrick TA, Braunwald E (1974) Effects of catecholamines, exercise and nitroglycerin on the normal and ischemic myocardium in conscious dogs. J Clin Invest 54: 563-575
13. Isoyama S, Maruyama Y, Koiwa Y, Ishide N, Kitaoka S, Tamaki K, Satoh S, Shimizu Y, Ino-Oka E, Takishima T (1981) Experimental study of afterload reducing therapy. The effect of the reduction of systemic vascular resistance on cardiac output, aortic pressure and coronary circulation in isolated, ejecting canine hearts. Circulation 64: 490-499
14. Maruyama Y, Ashikawa K, Isoyama S, Satoh S, Suzuki H, Watanabe J, Shimizu Y, Ino-Oka E, Takishima T (1984) Pressure-length loop in the ischemic segment during left circumflex coronary artery stenosis and its modification by afterload reducing in excised perfused canine hearts. Basic Res Cardiol 79: 155-163
15. DaLuz PL, Forrester JS, Wyatt HL, Tyberg JV, Chargrasuli SR, Parmley WW, Swan HJC (1975) Hemodynamic and metabolic effects of sodium nitroprusside on the performance and metabolism of regional ischemic myocardium. Circulation 52: 400-407
16. DaLuz PL, Forrester JS, Wyatt HL, Tyberg JV, Chargrasuli SR, Parmley WW, Swan HJC (1975) Hemodynamic and metabolic effects of sodium nitroprusside on the performance and metabolism of regional ischemic myocardium. Circulation 52: 400-407
17. Maruyama Y, Isoyama S, Ashikawa K, Satoh S, Suzuki H, Nishioka O, Watanabe J, Takishima T (1989) Effects of afterload elevation on the ischemic myocardium in isolated, paced canine heart with partial coronary stenosis. Am J Cardiol 63: 40E-44E
18. Satoh S, Maruyama Y, Ashikawa K, Isoyama S, Suzuki H, Watanabe J, Nishioka O, Ino-Oka E, Takishima T (1986) Effects of preload alteration on the degree of ischemia

and function of ischemic myocardium under constant mean aortic pressure, coronary perfusion pressure and heart rate in isolated perfused canine heart. Jpn Circ J 50: 1100–1112
19. Lew WYW, Ban-Hayashi E (1985) Mechanisms of improving regional and global ventricular function by preload alterations during acute ischemia in the canine left ventricle. Circulation 72: 1125–1134
20. Hoffman JIE (1987) Transmural myocardial perfusion. Prog Cardiovasc Dis 29: 429–464
21. Watanabe J, Maruyama Y, Satoh S, Keitoku M, Takishima T (1987) Effects of the pericardium on the diastolic left coronary pressure-flow relationship in the isolated dog heart. Circulation 75: 670–675
22. Satoh S, Watanabe J, Keitoku M, Itoh N, Maruyama Y, Takishima T (1988) Influences of pressure surrounding the heart and intracardiac pressure on the diastolic coronary pressure-flow relationship in excised canine heart. Circ Res 63: 788–797
23. Spaan JAE, Breuls NPW, Laird JD (1981) Diastolic-systolic coronary flow differences are caused by intramyocardial pump action in the anesthetized dog. Circ Res 49: 584-593
24. Maruyama Y, Satoh S, Watanabe J, Keitoku M, Hangai K, Takishima T (1987) Diastolic filling pressure of the left and right ventricles alters coronary circulation of each perfusion bed in a different manner. Automedica 9: 154

Pressure-Volume Relation of the Left Atrium in Man

Michio Arakawa, Tsutomu Tanaka, and Senri Hirakawa[1]

Summary. The reservoir function of the left atrium is important in producing an appropriate driving pressure and in providing an appropriate amount of blood for the following ventricular rapid filling.

To quantify this reservoir function in terms of the compliance, we simultaneously measured left atrial volume with biplane cineangiography and left atrial pressure, and reconstructed the pressure-volume relation in 10 patients with normal cardiac function. We tested the hypothesis that the pressure-volume relation of the left atrium during the ascending limb of the v-loop is largely a passive state. For this purpose, we observed whether the pressure-volume relation could be described as an exponential function ($p = b \cdot e^{aV}$: p, pressure; a and b, constants; V, Volume) which is widely accepted to describe the passive filling characteristics of the left ventricle in diastole.

The correlation coefficient of the regression equation was 0.96 ± 0.04 (mean ± SD), indicating that the strength of the relationship is great between the pressure and volume and that the slope of this regression line is gentle at low pressure and is steeper at high pressure. The left atrial compliance at the pressure of the peak v-wave (13.5 ± 3.6 mmHg) was 1.03 ± 0.26 $ml \cdot mmHg^{-1} \cdot m^{-2}$.

We conclude that the pressure-volume relation is curvilinear during the ascending limb of the v-loop and that the reservoir function of the left atrium appears to be relatively small.

Introduction

Although there are no valves at the junctions of the pulmonary veins and the left atrium, sphincter-like structures of the atrial muscle around the veins and extended myocardial fibers to the veins exert a valve action pressure and prevent the reflux of blood from the atrium into the pulmonary veins on atrial contraction [1].

The atrial muscle operates under the Frank-Starling mechanism, and so increased resting muscle length results in a more forceful contraction [2–4].

The left atrium fills passively and the intraatrial pressure increases during the ventricular systole. The left atrium favors the subsequent left ventricular rapid filling by releasing this accumulated blood with the aid of the elevated driving pressure. This reservoir function of the left atrium is a function of the diastolic compliance and can

[1] The Second Department of Internal Medicine, Gifu University School of Medicine, 40 Tsukasa-machi, Gifu, 500 Japan

be quantified from the pressure-volume relation in the passive state. However, the diastolic compliance of the left atrium has seldom been studied in man because of methodological limitations.

There are two periods in which the left atrium fills passively: (1) the atrial "rapid"-filling period during the ventricular systole, from the x-trough to the peak of the v-wave and (2) the atrial "slow"-filling period during the ventricular diastole, from the y-trough to the d-point.

The atrial rapid-filling period occurs during the ventricular systole. Initially, the volume increases but the pressure decreases, forming the x-trough. Then, the volume and pressure increase concomitantly, forming the ascending limb of the v-loop. The magnitude of the volume change and pressure change during the ascending limb of the v-loop is wide enough for one to reliably construct the pressure-volume relation. During the atrial rapid-filling period, the left atrium becomes an isolated chamber with the mitral valve to the left ventricle closed (isolated chamber phase).

The atrial slow-filling period occurs during the ventricular diastole. During the ventricular rapid-filling period, the atrial volume and pressure decrease concomitantly, forming the descending limb of the v-loop and the y-trough. During the subsequent ventricular slow-filling period, the atrial volume and pressure increase concomitantly, forming a short segment of the pressure-volume relation (from the y-trough to the d-point). The magnitude of the volume change and pressure change during this period is not wide enough to permit reliable construction of the pressure-volume relation. During the atrial slow-filling period, the left atrium assumes a common chamber with the left ventricle through the opened mitral valve (common chamber phase).

We tested the hypotheses, A and B:

A. The pressure-volume relation of the left atrium during the ascending limb of the v-loop (the atrial rapid-filling period) reflects the diastolic properties of the left atrium. If this hypothesis is correct, the pressure-volume relation should be basically the same as the pressure-volume relation of the left ventricle in diastole, and should be exponential. To partly validate this hypothesis, we fitted the pressure-volume relation to the monoexponential equation [5] with the method of least squares.

B. The pressure-volume relation is basically the same between the atrial rapid-filling period and the artrial slow-filling period. If this hypothesis is correct, the volume at a certain pressure during the artrial rapid-filling period should be equal to the volume corresponding to that pressure during the atrial slow-filling period. Therefore, we observed the agreement of volumes between the atrial rapid-filling period and slow-filling period at the pressure at the end of the slow-filling period, the d-point.

Finally, the diastolic chamber compliance of the left atrium at the peak of the v-wave was determined, which determines incremental blood volume and internal pressure during the left atrial rapid-filling period.

We found that the pressure-volume relation is curvilinear, and the diastolic compliance is relatively small in the left atrium.

Subjects and Methods

Subjects

A total of ten patients, six with mild coronary artery disease and four without, underwent diagnostic cardiac catheterization. Informed consent was obtained. The mean age was 53 years, ranging from 50 to 71 years. The mean left atrial pressure was less than 12 mmHg, and the left ventricular ejection fraction was more than 50%. All patients were in normal sinus rhythm and mean heart rate was 70 beats/min, ranging from 52 to 92 beats/min. Mitral regurgitation was not present in any patients. The left atrial branch of the coronary artery was intact in all patients.

Methods

The left atrial pressure was directly obtained through a Sones catheter-transducer (Statham P23 Db) system by the Brockenbrough technique. The frequency characteristics of this fluid-filled catheter-transducer system were flat up to 12 Hz. This frequency response is fairly acceptable for recording the phasic pressure change of the left atrium [6, 7]. This system has a transmission delay of about 35 ms. In one patient, a catheter-tip transducer was used (Fig. 1). The zero pressure was assigned to the midchest level. The left atrium was visualized by injecting contrast material (Isopaque 370, 77%) into the pulmonary artery trunk at a rate of 20 ml/s during the lightly-held expiration. The left atrial cineangiography was done in biplanar projec-

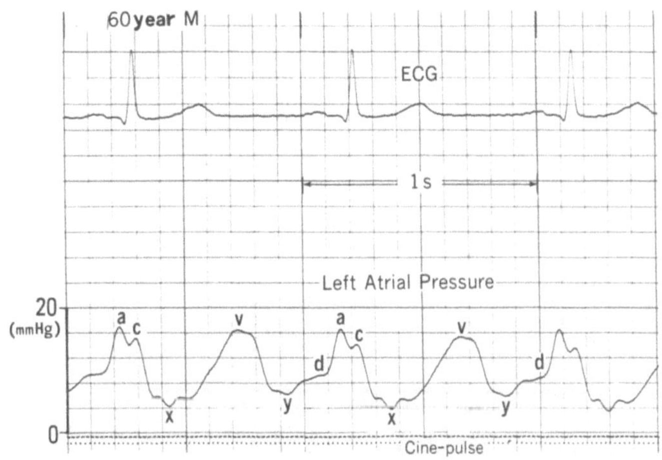

Fig. 1. Left atrial pressure recorded by the catheter-tip transducer system. The *a-wave* results from atrial contraction. The *c-wave* occurs at the beginning of the ventricular systole as the mitral valve bulges into the atrium. The *c-wave* is followed by the *x-trough*, which is considered to result from the combined effects of the descent of the mitral valve during the ventricular contraction and the continued atrial relaxation. The *v-wave* results from the passive filling of the atrium, followed by the *y-trough*, which results from the opening of the mitral valve and from the blood release into the left ventricle. The *d-point* is the end point of the slow filling of the ventricle as well as the atrium

tions (posteroanterior and left lateral projections) at a speed of 50 frames/s (Toshiba Medical). The left atrial pressure, electrocardiogram, and signals of cineexposure pulses were recorded at a paper speed of 100 mm/s (VR-12, Electronics for Medicine, Honeywell).

The left atrial volume was measured at intervals of 40 ms throughout one cardiac cycle, using an area-length method [8] with the help of a film motion analyzer (Cardias 300A, Nac). Each left atrial volume was identified from the coordinated cineframe counter. To compensate for the pressure transmission delay of 35 ms, the pressure was moved forward by two cineframes (40 ms) from the encountered cineframe number.

The pressure-volume relation of the left atrium is illustrated in Fig. 2.

The diastolic pressure-volume relation of the left ventricle is an exponential function and curvilinear, being a large increase in volume with a small increase in pressure at a low pressure, and a small increase in volume with a large increase in pressure at a high pressure.

The pressure-volume relation of the left atrium during the ascending limb of the v-loop was fitted to the exponential function $p = b \cdot e^{aV}$, where p is pressure, a and b are constants (a, volume elastic constant; b, extrapolated y-intercept), e is the base of natural logarithm, and V is volume. If the pressure-volume relation is the exponential curve as in the left ventricle during diastole, we can partly verify that the pres-

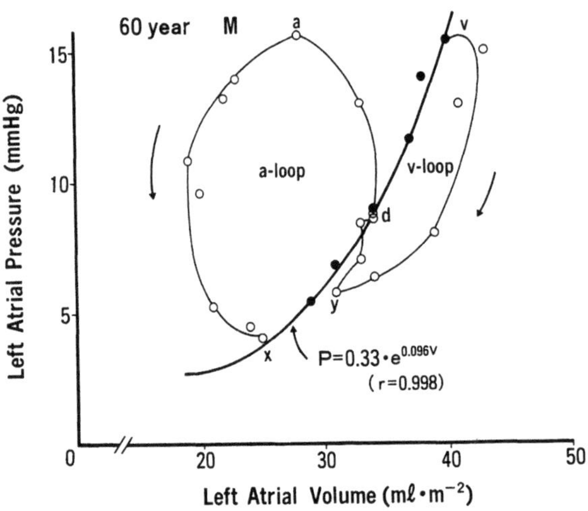

Fig. 2. Pressure-volume relation of the left atrium in one cardiac cycle. As the atrium contracts, the pressure increases but the volume deceases, forming the *a-loop*. As the ventricle contracts, the atrial volume increases, but the pressure decreases. Then the volume and the pressure increases concomitantly and rapidly, forming the *ascending limb* of the *v-loop* (the atrial rapid-filling period). As the ventricle relaxes, the atrial volume and pressure rapidly decrease, forming the *descending limb* of the *v-loop*. For the following ventricular slow filling period, beginning at the *y-trough* and ending at the *d-point*, both the left atrium and the left ventricle fill passively, and the volume and pressure increase concomitantly and slowly (the atrial slow-filling period)

sure-volume relation of the left atrium during the ascending limb of the v-loop results from the passive filling. The exponential equation was converted to the natural logarithmic function, $\ln p = \ln b + aV$, to obtain the best straight line with the method of least squares. The correlation coefficient was used to quantify the strength of the relationship between the two variables, pressure and volume.

The left atrial volume at the end of the atrial slow-filling period (d-point) was compared with the volume at the corresponding pressure on the ascending limb of the v-loop during the atrial rapid-filling period. If the volumes in the two periods closely agree, we can partly verify that, in the left atrium, the pressure-volume relation is equal between the atrial rapid- and slow-filling periods.

The exponential equation was converted to the differential function, $1/(dp/dV) = 1/ap$, to estimate the instantaneous chamber compliance (dV/dp) of the left atrium (C_{LA}).

The values are given as means ± SD. Student's *t*-test was used for the paired data. $P < 0.05$ was considered to be statistically significant.

Results

The main hemodynamic variables are listed in Table 1. The correlation coefficient of the regression equation was 0.96 ± 0.04, indicating the close and strong relationship between pressure and volume; and the regression line is curvilinear, as that of the left ventricle in diastole.

The volumes at the end of the atrial slow-filling period (the d-point) and at the corresponding pressure during the ascending limb of the v-loop were 28.6 ± 6.5 $ml \cdot m^{-2}$ and 29.7 ± 6.6 $ml \cdot m^{-2}$, respectively. Although the difference was statistically significant ($P < 0.05$), this close agreement of the values would indicate that the pressure-volume relation during the atrial rapid-filling period fairly reflects the pressure-volume relation during the left atrial slow-filling period.

The diastolic compliance at the peak of the v-wave (13.5 ± 3.6 mmHg) was 1.03 ± 0.26 $ml \cdot mmHg^{-1} \cdot m^{-2}$.

Table 1. Main variables collected from 10 patients mostly with mild coronary artery disease

LAP (m) mmHg	LAP (v) mmHg	LAV (v) $ml.m^{-2}$	LAP (d) mmHg	LAV (d) $ml.m^{-2}$	LAV (d)* $ml.m^{-2}$	a_{LA} $ml^{-1}.m^{-2}$	$C_{LA\,(v)}$ $ml.mmHg.m^{-2}$
9.0 ± 2.0	13.5 ± 3.6	36 ± 7	8.6 ± 1.9	28.6 ± 6.5	29.7 ± 6.6	0.08 ± 0.01	1.03 ± 0.26

LAP (m), mean left atrial pressure; LAP (v), left atrial pressure at the v-pressure (top of the v-loop); LAV (v), left atrial volume at the v-pressure; LAP (d), left atrial pressure at the d-point (the end of the slow filling period); LAV (d), left atrial volume at the pressure of the d-point; LAV (d)*, left atrial volume during the ascending limb of the v-loop at the corresponding pressure to the d-point, calculated from the regression equation of the pressure-volume relation fitted to the exponential equation; $C_{LA\,(v)}$, compliance of the left atrium at the v-pressure

Discussion

Only a few papers are available on the compliance of the left atrium in man.

Baxley et al. [9] estimated the average left atrial compliance, defined as the ratio of the total atrial volume change to the total atrial pressure, to be 5.14 ± 4.31 (mean $\pm$ SD) $ml \cdot mmHg^{-1} \cdot m^{-2}$ in patients with rheumatic mitral regurgitation and 1.61 ± 1.52 $ml \cdot mmHg^{-1} \cdot m^{-2}$ in patients with ruptured chordae tendineae. Their method of estimating the average left atrial compliance includes the pressure changes as well as volume changes during atrial systole and uses a chord rather than a tangent to the curve. Therefore, this compliance represents only one point somewhere on the diastolic chamber compliance line between early and late diastole. Consequently, their compliance is necessarily larger than the compliance at late diastole.

The average distensibility of the postmortem human left atrium, estimated by us from Liu et al.'s data [10] as the ratio of the fractional change in volume (ΔV) to the fractional change in pressure (Δp), were 4.4 $ml \cdot mmHg^{-1}$ between 0 and 5 mmHg, 2.3 $ml \cdot mmHg^{-1}$ between 5 and 10 mmHg, and 1 $ml \cdot mmHg^{-1}$ between 10 and 20 mmHg.

Nakajima et al. [11] estimated the compliance of the left atrium in man, applying the monoexponential equation to the ascending limb of the v-loop. The volume elastic constant was $4.04 \pm 0.55 \cdot 10^{-2} \cdot ml^{-1}$ in six patients with neurocirculatory asthenia, indicating that the compliance of the left atrium is about 2 times less compliant compared with our current data.

The factors which likely affect the pressure-volume relation of the left atrium besides the intrinsic properties of the myocardium are as follows:

1. *Vascular engorgement of the left atrial wall.* Coronary flow to the left atrium occurs during both ventricular systole and diastole except for the atrial contraction [12]. Because of this effect, the left atrial wall will engorge progressively and the pressure-volume relation will shift to the left from the original intrinsic state. Atrial fibrillation shifts the pressure-volume relation to the less compliant state due to vascular engorgement resulting from an increased blood flow to the left atrium [13].
2. *Left ventricular contraction.* A powerful contraction of the left ventricle pulls down the atrial floor. Also, acute volume reduction of the left ventricle will decrease the intrapericardial volume. These factors appear to enhance filling of the left atrium extrinsically [14]. In this study, the effect of the left ventricular contraction appeared to be constant, because all of our patients presented almost normal ejection fraction.
3. *Left atrial contraction.* The loss of atrial contraction greatly decreases the pulmonary venous flow during ventricular systole [14–16]. Thus, the atrial rapid-filling rate is decreased in atrial fibrillation. It is generally accepted that peak negative dp/dt of the ventricle has a direct linear relationship to myocardial contractility [17]. It seems reasonable to speculate that this relationship in the ventricle is also present in the atrium. The left atrial systolic function, defined as the ratio of ejection volume to the volume at the beginning of atrial contraction (preload of the left atrium), appears to alter the filling rate of the left atrium [18]. However, we do not know how this alters the pressure-volume relation during the ascending limb of the v-loop.

4. *Pericardial constraint*. The pericardium is by far less compliant [19] and may restrain the filling of the left atrium; hence, it may shift the pressure-volume relation of the left atrium to the left. The pressure-volume relation during the atrial rapid-filling period is used for the analysis of compliance, which coincides with the period in mid-to late- ventricular systole. The pericardial pressure seems to remain unchanged during the latter half of the ventricular systole [19, 20]. This would indicate that the intraatrial pressure reflects the transmural pressure. On the other hand, the pressure-volume relation during the atrial slow filling period coincides with the period of slow filling in ventricular diastole. The pericardial pressure seems to progressively rise during the ventricular diastole [19, 20]. This would indicate that the intraatrial pressure does not reflect the transmural pressure. However it seems inadequate to conclude what the effect of pericardial constraint on the pressure-volume relation is without further information.
5. *Other factors*. There occurs a temporal change in the pressure-volume relation when the pressure overload or volume overload is imposed upon the left atrium. The left atrial pressure is exceedingly elevated because of unchanged pressure-volume relation in the acute stage but tends to return to the normal pressure because of the rightward and downward shift of the pressure-volume relation in the chronic stage. An increased heart rate shortens the ventricular systolic period, which, in turn, shortens the diastolic period of the left atrium and appears to increase the filling rate of the left atrium. Atrio-ventricular interaction and interatrial interaction may influence the pressure-volume relation of the left atrium. These factors remain to be determined.

Briefly, we have discussed the importance of the left atrial compliance as a determinant of the left ventricular filling during the slow filling period. During the slow filling period, the pressure gradient across the mitral valve is zero or fluctuates closely around zero, and both the left atrium and left ventricle become a common chamber through the opened mitral valve. Therefore, the volume in both the left atrium and left ventricle increases concomitantly [21] and assumedly in proportion to their respective compliances, since the pressure equalizes in this common chamber. The compliances of the left atrium and the left ventricle (determined from the pressure-volume relation during the ventricular slow-filling period) in the same patient population, at the pressure at the end of the ventricular slow-filling period, were 1.60 ± 0.41 (means $\pm$ SD) $ml \cdot mmHg^{-1} \cdot m^{-2}$ and 4.22 ± 1.12 $ml \cdot mmHg^{-1} \cdot m^{-2}$, respectively [22]. These results indicate that the left atrium is two to three times less compliant than the left ventricle. We suggest that during the slow filling period, the interaction between the left atrial and left ventricular diastolic compliances provides the preferential delivery of blood to the left ventricle and enhances the preload of the left ventricle.

References

1. Nathan H, Eliakim M (1966) The junction between the left atrium and the pulmonary veins: An anatomic study of human hearts. Circulation 34: 412–422
2. Williams JF Jr, Sonnenblick EH, Braunwald E (1965) Determinants of atrial contractile force in the intact heart. Am J Physiol 209: 1061–1068

3. Matsuda Y, Toma Y, Ogawa H, Matsuzaki M, Katayama K, Fujii T, Yoshino F, Moritani K, Kumada T, Kusukawa R (1983) Importance of left atrial function in patients with myocardial infarction. Circulation 67: 566–571
4. Yamaguchi M, Arakawa M, Tanaka T, Takaya T, Nagano T, Hirakawa S (1987) Study on left atrial contractile performance: Participation of Frank-Starling mechanism. Jpn Circ J 51: 1001–1009
5. Gaasch WH, Cole JS, Quinones MA, Alexander JK (1975) Dynamic determinants of left ventricular diastolic pressure-volume relations in man. Circulation 51: 317–323
6. Sutterer WF, Swan HJC, Wood EH (1957) Minimal dynamic response requirements of manometer systems for recording left atrial and ventricular pressures in man. Physiologist 1: 81
7. Krovetz LJ, Goldbloom SD (1974) Frequency content of intravascular and intracardiac pressures and their time derivatives. IEEE Trans Biomed Eng 21: 489–501
8. Sauter HJ, Dodge HT, Johnston RR, Graham TP (1964) The relationship of left atrial pressure and volume in patients with heart disease. Am Heart J 67: 635–642
9. Baxley WA, Kennedy JW, Feild B, Dodge HT (1973) Hemodynamics in ruptured chordae tendineae and chronic rheumatic mitral regurgitation. Circulation 48: 1288–1294
10. Liu CK, Piccirillo RT, Ellestad M (1964) Distensibility of the postmortem human left atrium in nonrheumatic and rheumatic heart disease. Am J Cardiol 7: 232–238
11. Nakajima K, Iizuka M, Natsume T, Sakaguchi A, Shibata H, Shimomura K, Ikeda M (1982) Assessment of left atrial stiffness in man. Circulation (Suppl II) 66: 327
12. Kajiya F, Wada Y, Ogasahara Y, Hiramatsu O, Mito K, Kagiyama M, Matsuoka S, Tsujioka K, Goto M (1988) Measurement of blood flow in the atrial artery and vein using laser Doppler velocimeter. (abstract) Jpn J Med Electr Biol Eng 26 (Suppl): 390 (in Japanese)
13. White CW, Kerber RE, Weiss HR, Marcus ML (1982) The effects of atrial fibrillation on atrial pressure-volume and flow relationships. Circ Res 51: 205–215
14. Kalmanson D, Veyrat C, Chiche P (1971) Atrial versus ventricular contribution in determining systolic venous return: A new approach to an old riddle. Cardiovasc Res 5: 293–302
15. Keren G, Sherez J, Megidish R, Levitt B, Laniado S (1985) Pulmonary venous flow pattern—its relationship to cardiac dynamics: A pulsed Doppler echocardiographic study. Circulation 71: 1105–1112
16. Keren G, Bier A, Sherez J, Miura D, Keefe D, LeJemtel T (1986) Atrial contraction is an important determinant of pulmonary venous flow. J Am Coll Cardiol 7: 693–695
17. Weisfeldt ML, Scully HE, Frederiksen J, Rubenstein JJ, Pohost GM, Beierholm E, Bello AG, Daggett WM (1974) Hemodynamic determinants of maximum negative dP/dt and periods of diastole. Am J Physiol 227: 613–621
18. Toma Y, Matsuda Y, Moritani K, Ogawa H, Matsuzaki M, Kusukawa R (1987) Left atrial filling in normal human subjects: Relation between left atrial contraction and left atrial early filling. Cardiovasc Res 21: 255–259
19. Holt JP, Rohde EA, Kines H (1960) Pericardial and ventricular pressure. Circ Res 8: 1171–1181
20. Smiseth OA, Frais MA, Kingma I, White AM, Knudtson ML, Cohen JM, Manyari DE, Smith ER, Tyberg JV (1986) Assessment of pericardial constraint: The relation between right ventricular filling pressure and pericardial pressure measured after pericardiocentesis. J Am Coll Cardiol 7: 307–314
21. David D, Michelson EL, Naito M, Chen CC, Schaffenburg M, Dreifus LS (1983) Diastolic "locking" of the mitral valve: the importance of atrial systole and intraventricular volume. Circulation 67: 640–645
22. Arakawa M, Nagano T, Takaya T, Noda T, Miwa H, Kagawa K, Tanaka T, Yamaguchi M, Hirakawa S (1987) Diastolic compliance of the left atrium in man: A challenging concept as a determinant of preload of the left ventricle. Automedica 9: 120

Chapter C.
Ventricular Function and Interaction Between Heart and Vessels

JAN BAAN

While each of the papers in this section can stand alone as an interesting description and particular view of left ventricular pump function, whether it be diastolic, systolic, or a combination of both, the common denominator is the aspect of interaction between ventricular function and other parts or aspects of the circulation. These aspects are: myocardial oxygen consumption (Suga et al., Rankin et al., and Sunagawa et al.), matching between ventricular power or work and arterial load (Westerhof et al., Sunagawa et al., and Piene), effect of arterial impedance on inotropism (Baan and van der Velde), left atrial and mitral valve dynamics (Yellin) and right ventricular and pericardial pressures (Tyberg and Smith). The effects of ischemia on ventricular function are covered by the contributions of Rankin et al. and Tyberg and Smith. All papers represent animal studies, and search for basic mechanisms in the interpretation of results, but most authors also discuss clinical implications. Controversial results as well as different viewpoints become manifest to the careful reader; some of these will be highlighted here, since they may form the starting points of new investigations.

Myocardial oxygen consumption can easily be translated into the energy equivalent supplied to the heart to maintain basic metabolic rate, excitation-contraction coupling, and external work. However, which measurable parameter(s) representing energy demand should energy supply be related to? While part of the problem resides in the fact that heat released during contraction cannot be measured in the intact heart, an important difference in viewpoints concerns the definition of "total mechanical energy." Both Suga's concept of "pressure-volume-area," representing external stroke work plus potential energy at end-systole, and the concept of Rankin et al. which represents external stroke work plus energy required to maintain systolic pressure at a given end-diastolic volume, contain attractive components and certainly go beyond earlier notions based on external work (times frequency) or tension-time index alone. Suga et al. describe experiments of quick release in which the PVA concept clearly falls short of explaining the results. More specifically designed studies of this type will be needed to prove or disqualify theories about energy demand. One general problem with the current views based on the end-systolic pressure-volume relation (ESPVR) as held particularly by Suga is the finding that this relation, especially in vivo, is curved, making extrapolation to zero volume invalid. Even more importantly, the phenomenon reported by our group that the slope and volume intercept of the ESPVR depends on changes in aortic input impedance (Baan and van der Velde), perhaps accompanied by changes in myocardial perfusion, puts into question whether a simple relationship between oxygen consumption and ESPVR-based mechanical energy exists.

The papers by Westerhof et al., Sunagawa et al. (both left ventricle), and Piene (right ventricle) all address the question of optimization of work or power performed by the ventricle faced with its arterial load. The common view is that the ventricle operates close to the point of optimal power, which, according to Sunagawa et al. is more or less maintained during maximal exercise in dogs. These conclusions are based on studies employing linear ESPVR and the concept of arterial elastance, E_a. In contrast, using the cat as a model and employing mean-ventricular pressure vs flow relations, Westerhof et al. find that changes in arterial resistance by a vasodilator and in inotropic state by noradrenalin result in sizable shifts of the optimum power point. Piene, also using the cat model and pressure-flow relations, discusses some of the differences in species comparing his results to dog data of others. While the variance in findings must be reconciled, most pressing appears to be the need to improve our understanding of the control mechanisms involved in bringing about optimization of matching, or the lack of it, during hemodynamic or inotropic interventions. The implications of these studies for oxygen utilization and efficiency, touched upon by Sunagawa et al., are obviously of great importance.

A comprehensive review of filling dynamics and its implications for relaxation disturbances is given by Yellin. The studies of the significance of prevention of filling, also originating primarily from Yellin's group, are discussed briefly in the contribution of Rankin et al. Intriguing aspects of diastolic and systolic interaction between right and left ventricle are described by Tyberg and Smith. Especially interesting are his observations on left ventricular performance and failure as a consequence of right ventricular (RV) overload, caused by septal shifts and explained in terms of stroke work vs effective transmural diastolic pressure. The latter is co-determined both by RV and pericardial pressures, quantities which also play a crucial role in the maintenance of RV pump function despite RV infarction.

The short-and long-term effects of local ischemia upon ventricular pressure-segment relations are shown and discussed by Rankin et al., who advocate the use of "preload recruitable stroke work" rather than the ESPVR to assess contractility. In view of the afterload dependency of the latter relation this suggestion deserves more consideration. Rankin et al. also show how pressure-volume (P-V) loops may be obtained in patients by digital subtraction angiography during transient vena cava occlusion, but applying the Leiden conductance volume catheter for this purpose in patients would seem to be preferable.

Most important, however, is that the study of dynamic P-V relations, diastolic as well as systolic during hemodynamic, surgical, or pharmacologic interventions in man, is now within our reach. This creates the possibility to test most of the mechanisms, hypothetical or real, discovered in animal or isolated heart preparations, in patients undergoing cardiac catheterization or cardiac surgery. Eventually, such studies should lead to more refined diagnosis and better treatment of cardiac disease.

Energetics of the Heart

Hiroyuki Suga, Shiho Futaki, and Yoichi Goto[1]

Summary. A new measure of the total mechanical energy (TME) generated by ventricular contraction was proposed to be quantified by a specific area in the pressure-volume (P-V) diagram that is bounded by the end-systolic and end-diastolic P-V relations and the systolic P-V trajectory. This area is called the *systolic P-V area (PVA)*, interchangeably used with TME. We found in the left ventricle of the excised, cross-circulated dog heart preparation that PVA correlated linearly with myocardial O_2 consumption (V_{O_2}) regardless of loading conditions in a stable contractile state. The load-independent linear V_{O_2}-PVA relation was elevated in a parallel manner by positive inotropic interventions, primarily due to an increased V_{O_2} for the augmented excitation-contraction coupling. The slope of the V_{O_2}-PVA relation can be considered to reflect inversely the efficiency of energy conversion to PVA from the excess V_{O_2} above unloaded V_{O_2}. This efficiency is primarily the product of the oxidative phosphorylation efficiency from V_{O_2} to ATP and the contractile machinery efficiency from ATP to PVA. Thus, the V_{O_2}-PVA relation seems a promising tool for the anlayses of cardiac energetics and mechanoenergetic coupling.

Introduction

Mechanical parameters that correlate best with cardiac O_2 consumption (V_{O_2}) have been extensively researched over 7 decades [1]. Among many proposed parameters, myocardial tension or its time integral and contractility have been widely accepted as the two most important determinants of V_{O_2} per beat, and heart rate is another important determinant for V_{O_2} per minute [2]. However, over the last decade, many pieces of evidence have been accumulated in our laboratory that the total mechanical energy (TME) generated by contraction as measured by a specific area (PVA) in the pressure-volume diagram correlates best with V_{O_2} per beat in a stable contractile state [3], and a change in contractile state is characteristically correlated with a change in V_{O_2} fraction for the excitation-contraction (EC) coupling [4]. Our findings have been corroborated by similar findings in different heart preparations in other laboratories [5, 6]. The key findings on the relation between V_{O_2} and PVA are reviewed here.

[1] Department of Cardiovascular Dynamics, National Cardiovascular Center Research Institute, 5-7-1 Fujishirodai, Suita, Osaka, 565 Japan

Total Mechanical Energy (TME)

A new method was proposed to quantify TME generated by a ventricular contraction about a decade ago [7]. TME consists of external work (EW) and mechanical potential energy [7]. TME is different from the sum of EW and the conventional internal work that was assumed to be performed on the series elastance by the contractile element in the Hill model of myocardium [8]. Since there was evidence that the ventricle could release EW more than the conventional internal work when the afterload was appropriately reduced [8], we used a different measure of TME, as follows.

Our method is based on the time-varying elastance model of the ventricle [9]. This model can simulate the systolic and diastolic pressure-volume (P-V) relationships of a contracting left ventricle (LV) as a counterclockwise rotating P-V relation line from a diastolic compliant level to an end-systolic stiff level with a pivot on a small positive volume (V_{O_2}) on the volume axis [9].

In this mechanical model, a time-varying increment in elastance and hence a development of ventricular pressure are theoretically accompanied by generation of mechanical energy in the time-varying elastic wall of the ventricle [7]. This mechanical energy corresponds to the *triangular area* between the end-systolic and end-diastolic P-V relation curves on the *left* or *origin side* of the *area* for EW [7], as schematically shown in Fig. 1a. This mechanical energy is a mechanical or elastic potential energy (PE) like that in a Hookean spring when it has been stretched [7]. TME at end systole consists of the sum of the PE at end systole and EW performed during systole [7] and corresponds to the *area* (PVA) circumscribed by the *thick lines* and *curves* shown in Fig. 1a. PVA and TME are equivalent and used interchangeably. PE is assumed to be irreversibly converted into heat during relaxation and diastole unless it is converted to EW during relaxation by appropriate load reduction [7]. PVA is an abbreviation of the *systolic P-V area* [7]. Although PVA has dimensions of mmHg·ml/beat, this can be converted to J/beat with a physical conversion factor of 1 mmHg·ml = 1.33×10^{-4} J.

Oxygen Consumption (V_{O_2})

Since myocardial V_{O_2} is equivalent to the total energy (or enthalpy) input to myocardium (1 ml O_2 = 20 J) under aerobic conditions [1], we examined how V_{O_2} would correlate with PVA under a variety of loading and contractile conditions in the excised, cross-circulated dog LV [3, 4].

First, we studied the correlation between V_{O_2} and PVA of variably loaded contractions in a stable contractile state [3]. We used V_{O_2} per beat instead of V_{O_2} per min because PVA to be correlated with V_{O_2} is TME generated by each contraction. (If V_{O_2} per min is used, PVA times heart rate per min must be correlated with it.) We obtained a linear regression equation of $V_{O_2} = A \cdot PVA + D$ in each contractile state in a given LV as schematically shown in Fig. 1b [3]. A is a slope coefficient and D is a constant. When both V_{O_2} and PVA were expressed in (J/beat) per 100 g, A was about 2.5 (dimensionless) and D was about 0.5 (J/beat) per 100 g in the control contractile state without any intentional inotropic interventions on LV [3]. The correlation coefficient (r) between V_{O_2} and PVA was above 0.95 in most cases [3]. The

a,b

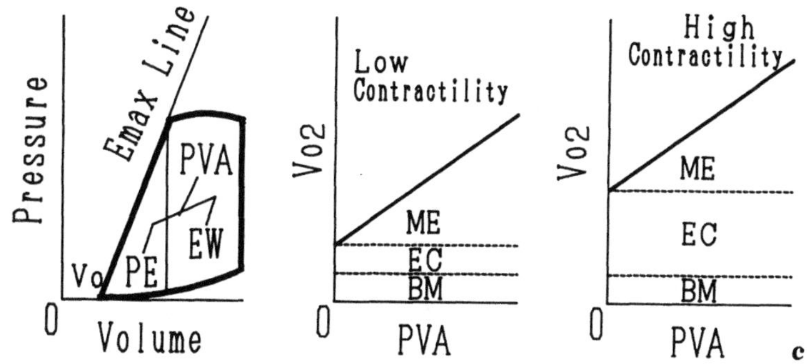

c

Fig. 1a. Pressure-volume diagram and systolic pressure-volume area (*PVA*), external mechanical work (*EW*), and elastic or mechanical potential energy (*PE*) of a ventricle. **b** Schematic diagram of the *V_{O2}-PVA* relation (*diagonal line*) in a low contractile state. The constant fraction of V_{O2} labeled *BM* at the *bottom* is V_{O2} for basal metabolism (BM). The constant fraction of V_{O2} labeled *EC* in the *middle* is V_{O2} for excitation-contraction (*EC*) coupling. The *triangular* fraction of V_{O2} labeled *ME* is V_{O2} for mechanical contraction to generate total mechanical energy (TME) in terms of *PVA*. **c** Schematic diagram of the *V_{O2}-PVA* relation in a high contractile state. As for *BM*, *EC*, and *MS*, see the legend for **b**. V_{O2}, myocardial oxygen consumption

coefficient of determination (r^2) was therefore usually above 0.90, indicating that as much as 90% of the variation of V_{O_2} can be ascribed to the variation of PVA and only 10% of the variance of V_{O_2} should be ascribed to some other factors which remain yet unknown [3].

Secondly, the load-independent V_{O_2}-PVA relation was confirmed in a specific protocol. When PVA was maintained unchanged while end-diastolic and stroke volumes were increased from isovolumic contraction, V_{O_2} also remained virtually unchanged despite gradual decreases in peak pressure, peak wall tension, and their time integrals [10, 11]. This result indicates conclusively that none of these conventional indices which are believed to be major determinants of V_{O_2} can consistently correlate with V_{O_2}.

Thirdly, the course of the diastolic P-V trajectory was shifted to change the relative magnitudes of EW and PE without changing PVA [12]. However, as long as PVA remained unchanged, V_{O_2} also remained unchanged [12]. This result indicates that PVA as a whole, but not the individual fractions of EW and PE, is uniquely correlated with V_{O_2}.

One intriguing exception is the result of a quick release of ventricular pressure and volume in the middle or at the end of systole [13]. At the time of quick release, generation of TME was forced to stop in the middle of systole, or alternatively the same TME had been generated whether quick release occurred or not at the end of systole. In either case, the V_{O_2}-PVA relation of quick-release contractions had a gentler slope than control [13]. The slope tended to be gentler as the quick-release speed increased [13]. In other words, V_{O_2} for a given PVA decreased in quick-release contractions and as the quick-release speed increased. The decrease in V_{O_2} by quick release contradicts the straightforward induction of the time-varying elastance model [7]. A similar but greater decrease in V_{O_2} in quick-release contractions has been observed in ferret papillary muscle experiments in which the linear V_{O_2}-PVA relation was confirmed in isometric and shortening contractions [5].

Contractile State (E_{max})

When the contractile state was enhanced by epinephrine or calcium, E_{max} (a global index of ventricular contractility represented by the slope of the end-systolic P-V relation or ventricular maximum elastance) increased [4, 9]. Simultaneously, the load-independent V_{O_2}-PVA relation was elevated in a parallel manner as schematically shown in Fig. 1c [4]. When contractile state was depressed by propranolol or verapamil, E_{max} decreased and the V_{O_2}-PVA relation was lowered in a parallel manner [14]. There was a statistically significant r between E_{max} and the elevation of the V_{O_2}-PVA relation for each given inotropic agent [15]. Therefore, $V_{O_2} = A \cdot PVA + D$ could be expanded to $V_{O_2} = A \cdot PVA + B \cdot E_{max} + C$, where C corresponds to basal metabolism per beat and B indicates the sensitivity of the elevation of the V_{O_2}-PVA relation to E_{max}. A remained constant at about 2.5 (dimensionless). Sensitivity coefficient B was about 0.0013 ml O_2 per (mmHg/ml) per $(100\ g)^2$ for calcium and 0.0023 ml O_2 per (mmHg/ml) per $(100\ g)^2$ for epinephrine [15]. These B values are corrected from the previously reported values because the right ventricular unloaded V_{O_2} was not subtracted previously [15].

An intriguing positive inotropic intervention was myocardial cooling from 36°C to 29°C. Although the cooling increased E_{max} by 1.4 times, the V_{O_2}-PVA relation was elevated only slightly [16]. B was not significantly greater than zero. So far, cooling seems to be the only positive inotropism that requires minimal increment in V_{O_2} for a given increment in E_{max} [16].

Efficiency

Efficiency is the ratio of the useful energy output to the total energy input. If one looks at the V_{O_2}-PVA relation, total V_{O_2} is the total energy input to myocardium, and TME (or PVA) can be considered to be the useful mechanical energy output. Note that TME does not mean the useful work output of the heart as a pump. Then, the reciprocal of the slope of the line connecting the origin and a working V_{O_2}-PVA point indicates an overall efficiency from the total V_{O_2} to PVA, as schematically shown in Fig. 2a. This efficiency varies with PVA and E_{max} [4]. Note that this efficiency is different from the conventional mechanical efficiency of the heart, which is the ratio of EW to total energy input or total V_{O_2} [1, 2].

If the unloaded V_{O_2} (or V_{O_2}-axis intercept, or the V_{O_2} fraction for both basal metabolism and EC coupling, or PVA-independent V_{O_2}, or non-mechanical V_{O_2}) is subtracted from the total V_{O_2} to obtain excess V_{O_2}, another efficiency can be defined from the excess V_{O_2} to PVA [3, 4]. This efficiency is represented by the reciprocal (1/A) of the slope coefficient (A) of the V_{O_2}-PVA relation, as shown in Fig. 2b. Since A was 2.5 (dimensionless) on average, this efficiency is 40% ($100 \times 1/2.5$) on average (range: 30%–50%). This efficiency is considered to be the efficiency of the contractile machinery from excess V_{O_2} to PVA. We have not yet identified the factors which caused the interindividual variation of this efficiency. Intraindividual variation of the efficiency with inotropic interventions was relatively small.

Since myocardial cooling decreases the myosin ATPase activity and hence cross-bridge (CB) cycling rate ($Q_{10} = 2–3$, where Q_{10} is the decrease in rate for a drop of

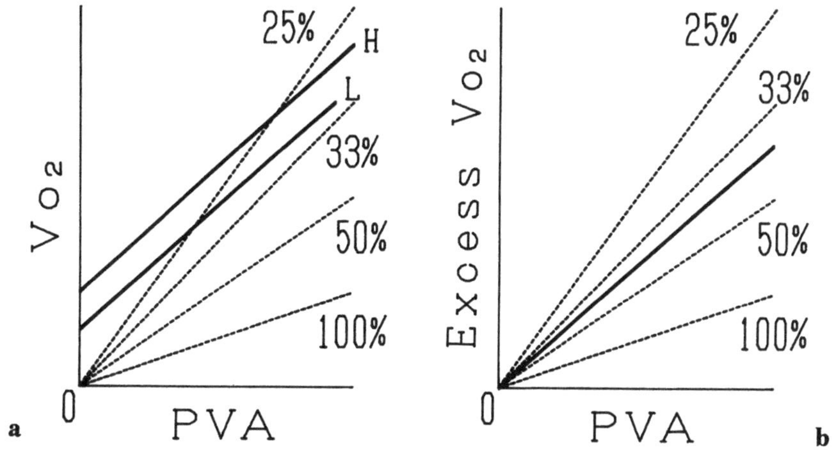

Fig. 2a. The relation between total V_{O2} and total mechanical energy (TME = *PVA*) and a family of isoefficiency lines (25%, 33%, 50%, and 100%). Two V_{O2}-*PVA* relations are shown: one for a low contractility (*L*) and the other for a high contractility (*H*). The isoefficiency is a constant efficiency from the total V_{O2} to *PVA*. **b** The relation between excess V_{O2} and *PVA* (= TME) and a family of isoefficiency lines. Two excess V_{O2}-*PVA* relations are superimposable. The excess V_{O2} is equal to the total V_{O2} minus unloaded V_{O2} or V_{O2}-axis intercept. The isoefficiency is a constant efficiency from the excess V_{O2} to *PVA*. V_{O2}, myocardial oxygen consumption, *PVA*, pressure-volume area

10°C) increasing the heat economy of force generation [17], we expected that cooling would also increase the efficiency of the contractile machinery and hence decrease the slope of the V_{O_2}-PVA relation. However, the 7°C cooling did not change the slope despite almost 50% prolongation of the duration of contraction and hence, presumably, 50% slowing of CB cycling rate [16].

The efficiency of the contractile machinery is the product of the efficiency from the enthalpy of V_{O_2} to the enthalpy of ATP synthesized in the oxidative phosphorylation and the efficiency from the enthalpy of ATP hydrolysis to PVA generated by CB cycling in the contractile machinery. The former is known to be 65% (60%-70%) [18]. Then, the latter is calculated to be 65% (60%-70%) regardless of loading and contractile conditions because their product must be 40% (30%-50%) on average, as shown in the energy flow diagram (Fig. 3). Therefore, the relatively constant slope of the V_{O_2}-PVA relation regardless of changes in loading conditions in a given contractile state and changes in contractile state suggests a relatively constant efficiency (or stoichiometry) of the energy conversion from the enthalpy change of ATP hydrolysis to PVA in the contractile machinery regardless of ventricular loading and contractile conditions.

Conclusion

The introduction of PVA as a measure of TME and the load- and contractility-independent slope of the V_{O_2}-PVA relation have revealed a fundamental law dominating

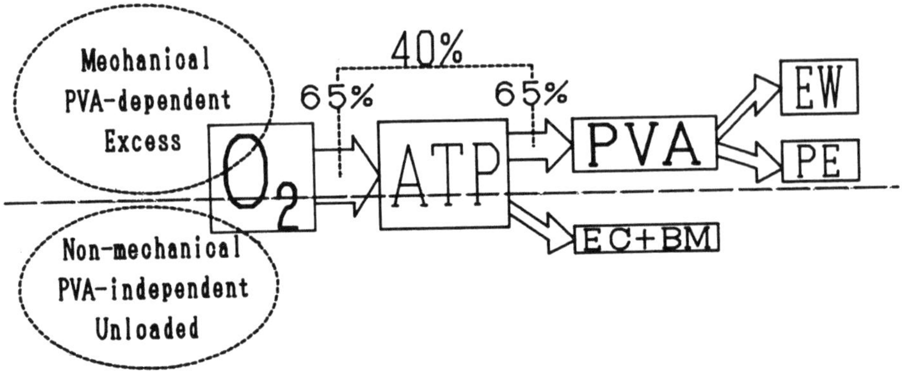

Fig. 3. Energy flow diagram from total V_{O2} to *EW* via ATP and *PVA* (= total mechanical energy TME). The total V_{O2} consists of two fractions: one for mechanical contraction and called *PVA* (or TME)-dependent V_{O2} or excess V_{O2} , and the other for nonmechanical activity such as basal metabolism (*BM*) and excitation-contraction coupling (*EC*). The latter is called *PVA* (or TME)-independent V_{O2} or unloaded V_{O2} . The *horizontal dashed line* separates the two fractions of V_{O2} as well as ATP synthesized in the oxidative phosphorylation. A fraction of ATP is utilized for *BM* and *EC*, and the rest is utilized for generation of *PVA* (= TME) and mechanical contraction. *PVA* (= TME) is used for external mechanical work (*EW*) and the rest remains as potential energy (*PE*). When *PE* is not used, it is dissipated as heat. V_{O2}, myocardial oxygen consumption; *EW*, external work; *PVA*, pressure-volume area

the chemomechanical energy transduction in the myocardial contractile machinery. This inherent law was discovered thanks to the PVA concept proposed by Suga a decade ago [7]. The generality of this law, found in canine LV, has been gradually confirmed in other animal species and different myocardial preparations by other investigators [5, 6, 19]. Since the skeletal muscle does not have a similar law [19], we would suppose that the contractile machinery of myocardium is provided with an intrinsic autoregulatory mechanism for the stoichiometry from ATP hydrolysis to the generation of mechanical energy in agreement with the supposition made by Gibbs Chapman [18]. Recently, Yasumura and Suga [20] were successful in accounting for the concept of PVA and its correlation with V_{O_2} or the number of ATP split, with a constitutive CB model using fundamental assumptions of CB behavior.

Acknowledgments. We gratefully thank all co-workers whose names appear in the reference list for their energetic help throughout our studies over the last decade. Without their help, we could not have carried out these studies efficiently.

References

1. Gibbs CL (1978) Cardiac energetics. Physiol Rev 58: 174–254
2. Braunwald E, Ross J, Sonnenblick EH (1976) Mechanisms of contraction of the normal and failing heart, 2nd edn. Little Brown, Boston, pp 166–199
3. Suga H, Hayashi T, Shirahata M (1981) Ventricular systolic pressure-volume area as predictor of cardiac oxygen consumption. Am J Physiol 240: H39–H44

4. Suga H, Hisano R, Goto Y, Yamada O, Igarashi Y (1983) Effect of positive inotropic agents on the relation between oxygen consumption and systolic pressure-volume area in canine left ventricle. Circ Res 53: 306–318
5. Hisano R, Cooper G (1987) Correlation of force-length area with oxygen consumption in ferret papillary muscle. Circ Res 61: 318–328
6. Goto Y, Slinker BK, LeWinter MM (1988) Similar normalized Emax and O_2 consumption-pressure volume area relation in rabbit and dog. Am J Physiol 255: H366-H374
7. Suga H (1979) Total mechanical energy of a ventricular model and cardiac oxygen consumption. Am J Physiol 236: H498–H505
8. Suga H (1980) Relaxing ventricle performs more external work than quickly released elastic energy. Eur Heart J 1(Suppl A): 131–137
9. Suga H, Sagawa K, Shoukas AA (1973) Load independence of the instantaneous pressure-volume ratio of the canine left ventricle and effects of epinephrine and heart rate on the ratio. Circ Res 32: 314–322
10. Suga H, Hayashi T, Suehiro S, Hisano R, Shirahata M, Ninomiya I (1981) Equal oxygen consumption rates of isovolumic and ejecting contractions with equal systolic pressure-volume area in canine left ventricle. Circ Res 49: 1082–1091
11. Suga H, Goto Y, Nozawa T, Yasumura Y, Futaki S, Tanaka N (1987) Force-time integral decreases with ejection despite constant oxygen consumption and pressure-volume area in dog left ventricle. Circ Res 60: 797–803
12. Suga H, Goto Y, Yamada O, Igarashi Y (1984) Independence of myocardial oxygen consumption from pressure-volume trajectory during diastole in canine left ventricle. Circ Res 55: 734–739
13. Yasumura Y, Nozawa T, Futaki S, Tanaka N, Suga H (1989) Time-invariant oxygen cost of mechanical energy in dog left ventricle: Consistency and inconsistency of time-varying elastance model with myocardial energetics. Circ Res 64: 764–778
14. Suga H, Goto Y, Yasumura Y, Nozawa T, Futaki S, Tanaka N, Uenishi M (1988) O_2 consumption of dog heart under decreased coronary perfusion and propranolol. Am J Physiol 254: H292–H303
15. Suga H, Igarashi Y, Yamada O, Goto Y (1985) Mechanical efficiency of the left ventricle as a function of preload, afterload and contractility. Heart Vessels 1: 3–8
16. Suga H, Goto Y, Igarashi Y, Yasumura Y, Nozawa T, Futaki S, Tanaka N (1988) Cardiac cooling increases Emax without affecting relation between O_2 consumption and systolic pressure-volume area in dog left ventricle. Circ Res 63: 61–71
17. Alpert NR, Mulieri LA (1986) Determinants of energy utilization in the activated myocardium. Fed Proc 45: 2697–2600
18. Gibbs CL, Chapman JB (1985) Cardiac mechanics and energetics: chemomechanical transduction in cardiac muscle. Am J Physiol 249: H199–H206
19. Gibbs CL (1987) Cardiac mechanics and the Fenn effect. Basic Res Cardiol 82 (Suppl 2): 61–68
20. Yasumura Y, Suga H (1988) Crossbridge model compatible with the linear relation between left ventricular oxygen consumption and pressure-volume area. Jpn Heart J 29: 335–347

An Energetic Analysis of Myocardial Performance*

J. Scott Rankin, Joseph R. Elbeery, John C. Lucke, William Gaynor, David H. Harpole, Michael P. Feneley, Srdjan Nikolić, George W. Maier, George S. Tyson, Jr., Craig O. Olsen, and Donald D. Glower[1]

Summary. Cardiovascular dynamics is one of the oldest lines of medical research, having its origins in the work of William Harvey in the seventeenth century. Yet despite its long history, conceptual understanding of cardiac performance is advancing more rapidly than ever, and many different scientifc approaches are currently yielding exciting new insights. This chapter reviews 15 years of work from our laboratories at Duke University on the quantitative assessment of diastolic and systolic ventricular function. Our approach to the analysis of chamber geometry, ventricular interaction, and diastolic mechanical properties is described, leading to the observation of a fundamentally linear relationship between myocardial energy production (net external work) and end diastolic fiber length. This relationship is further validated and expanded to provide a useful estimate of myocardial inotropism that is applicable to pathophysiologic analysis of myocardial ischemia and hypertrophy. Finally, recent extensions of this technique to human studies have proven useful to the understanding of cardio-pulmonary interactions and valvular heart disease. As knowledge of myocardial adaptive mechanisms improves, enhanced diagnostic and therapeutic capabilities could translate into significant advances in patient care.

"When I first tried animal experimentation for the purpose of discovering the motions and functions of the heart, I found it so truly difficult that I almost believed that the motion of the heart was to be understood by God alone." *William Harvey (1628)*

A Brief History of Cardiac Physiology

Speculation about circulatory function began in the dawn of history, and existing theories were catalogued by Galen in the second century A.D. [1]. During the middle ages, most of Galen's work was lost in the West, but his concepts of cardiac function were sustained in the Islamic world through the writings of Avicenna. With the reawakening of European scientific thought in the fifteenth and sixteenth centuries, Servetus and also Columbus in Padua described the passage of blood through the lungs, and it was during William Harvey's medical studies in Padua that he

* Supported by NIH Grants HL-09315, HL-29436, and HL-17670.

[1] The Department of Surgery, Duke University Medical Center, Durham, NC 27710, USA

assimilated Columbus' theories, learned of the directionality of blood flow from Fabricius, and was influenced by Galileo's use of the scientific method [2, 3]. Upon returning to London, Harvey began a series of direct experimental observations of circulatory dynamics, both in animals and human cadavers, and his theories were published in 1628 in *De Motu Cordis* [4]. Most would agree that Harvey's full description of the circulation in this book was the single most revolutionary and important advance in the history of medical science, and the majority of Harvey's principles are still relevant today, three and one half centuries later. Every physician working in the area of circulatory function probably should be familiar with this work [4].

Subsequent discoveries by Richard Lower and Stephen Hales further advanced the emerging discipline of hemodynamics, but it was the development of the Kymograph pressure recording device by Carl Ludwig in 1849 that firmly established experimental cardiac physiology [3, 5]. This rudimentary technology evolved over the next 50 years into highly accurate optical recording systems that, in fact, would rival today's methodology in precision and fidelity. Using these techniques, Ludwig and his students Adolph Fick, Julius Cohnheim, Henry Bowditch, and Otto Frank established most of the important principles of contemporary cardiac physiology, including the length-dependency of myocardial function, before the turn of the twentieth century.

Simultaneous with the work of the Ludwig school, Francois-Frank and Marey in France [6], and Roy, Martin, Howell, Donaldson, and Henderson in Britain and the United States made significant contributions, including the development of a cardiometer to measure changes in ventricular volume indirectly. It was on this background that Ernest Starling began his experiments at University College, London in the early 1900s [7]. Working with an isolated mammalian heart-lung preparation, he showed that the heart adapts to increasing arterial pressure by a compensatory rise in filling pressure, which then tends to maintain cardiac output. In his Linacre lecture, he went on to state: "The law of the heart is thus the same as the law of muscular tissue generally, that *the energy of contraction, however measured, is a function of the length of the muscle fibre.*" He continued: "Many lines of argument converge similarly in the direction of the view that the chemical changes, occurring in a muscle as a result of excitation, affect surfaces arranged longitudinally in the muscle. Now the effect of lengthening the muscle will be to increase the extent of active surface, so that we may restate our general proposition in the following way. *Any increase in the extent of active surface increases the energy of change.*" This prophetic statement antedated by decades the sliding filament/myosin cross-bridge theory of myocardial contraction and formed the basis of all subsequent research in this area.

While many have contributed importantly to this subject in recent years, it was perhaps Sarnoff and his co-investigators [8, 9, 10], working at the National Institutes of Health in the 1950s and 1960s, who rekindled a general interest in cardiac function. Sarnoff showed that the external work of the heart (manifested as cardiac output and arterial blood pressure) rose steeply as filling pressure increased, and then at a certain level of filling, the work output was maximized and plateaued. In 1956, Sarnoff also suggested that the relationship between stroke work and end diastolic volume might be linear, although technical limitations precluded direct volume measurements at that time. The experimental work of this group, together with later studies of afterload reduction and pulmonary artery balloon catheters, formed the basis of clinical cardiac care to the present day. More recent classic contributions by Suga, Sagawa, and associates and many other contemporary studies of ventricular pressure-volume relationships have stimulated renewed interest in this area and are nicely reviewed elsewhere [11]. With this background, the remainder of this chapter will outline a particular viewpoint of basic cardiac function, formulated over the past 15 years by our research group at Duke University. Physiologic principles will be illustrated primarily by publications from our laboratory, although within each paper, full reference lists are available.

An Energetic Model of Cardiac Function

Ventricular Geometry

Before a comprehensive analysis of myocardial performance can be undertaken, it is necessary to understand the geometric characteristics of the heart under a variety of physiologic conditions. Of course, the left ventricle is the major pumping chamber of the heart, ejecting blood into the aorta during systole and filling during diastole (Fig. 1). Diastolic filling occurs in 3 phases: early rapid-filling, during which the velocity and extent of filling are greatest; diastasis in mid-diastole; and atrial systole in late diastole. Ninety percent of the change in ventricular volume with filling is

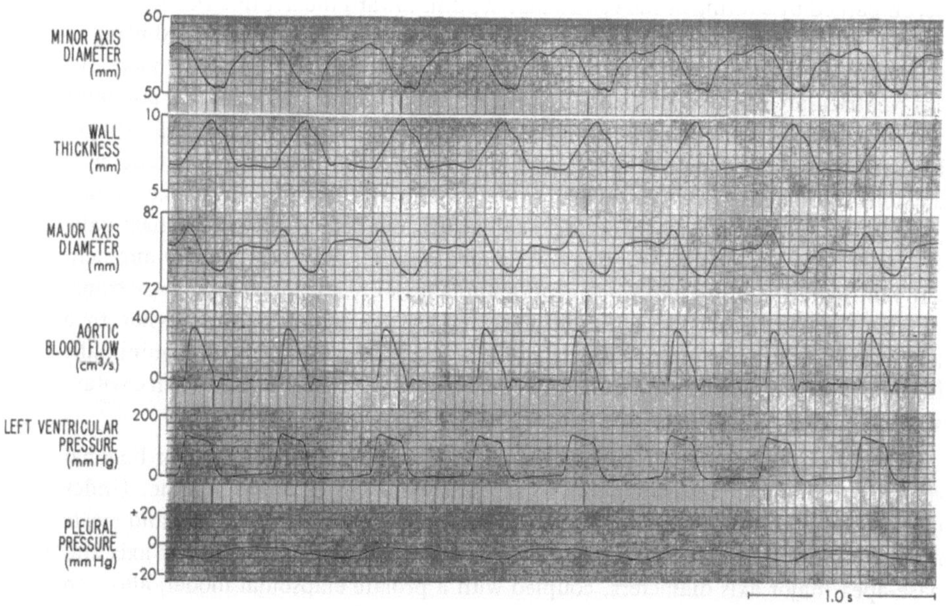

Fig. 1. Typical measurements of left ventricular dimensions, flow, and pressure observed in a conscious dog

produced by expansion of the minor axis circumference, and 10% relates to major axis elongation [12–14]. The consequent change in chamber geometry produces a more spherical ventricular shape at end-diastole [15, 16]. This concept can be illustrated quantitatively by relating chamber shape or eccentricity to ventricular volume, and with such an analysis, eccentricity has been shown to be an inverse linear function of diastolic volume (Fig. 2). Ventricular geometry during systole seems to be different, in that dimensional changes during ejection occur in a more concentric fashion (Fig. 2). It was hypothesized that ejection-phase geometry is related almost solely to average fiber orientation, which is more evenly distributed in all directions. In contrast, diastolic anisotropic mechanical properties, and consequently shape during filling, are potentially determined by many other factors, such as myocardial collagen distribution, papillary-mitral continuity, etc. [17]. An alternative theory predicts this type of behavior based on a "basket-weave" fibre angle distribution [18]. As a practical consequence of differing diastolic and systolic geometry, the ventricle undergoes significant shape changes during the transitional periods of isovolumic contraction and relaxation. As would be surmised from Fig. 2, these isovolumic rearrangements in geometry become more prominent at lower ventricular volumes [12].

Another important geometric consideration is diastolic ventricular interaction [19]. Because a common septum is shared by both ventricles, it might be anticipated that events in one would affect the geometry of the other. During the cardiac cycle, complex transmural forces are operational across the interventricular septum, and its diastolic position relative to the remainder of the heart is influenced by relative left and right ventricular pressures, i.e., transseptal pressure. For example, pulmonary artery occlusion increases right ventricular pressure relative to left, and is accompanied by shifting of the ventricular septum toward the left free wall with little change in the other diameters (Fig. 3). The physiologic effect of septal shifting is related primarily to altered diastolic compliance. Leftward septal shifting increases left ventricular diastolic pressure relative to volume and produces an apparent decrease in diastolic chamber compliance. The change in left ventricular shape, however, has little effect on left ventricular systolic function at a constant end-diastolic volume, as evidenced by maintained pressure generation and stroke volume [19]. In fact, pulmonary artery occlusion of a sufficient degree to produce maximally tolerated right ventricular hypertension, if anything, tends to very slightly augment left ventricular pressure and work generation (Fig. 4), via transseptal energy transmission [20]. In the reverse situation, maximal left ventricular decompression by a left atrial-aortic assist pump affects right ventricular systolic function minimally [21]; thus, direct systolic ventricular interaction appears to be minimal when cavitary volume is used as the index of preload.

An assumption inherent in each of these studies is the validity of using orthogonal ventricular diameter measurements to assess changes in chamber volume. Under normal conditions, the minor axis circumference of the left ventricle fills and ejects fairly concentrically. Measurements of epicardial anterior-posterior minor axis and base-apex major axis diameters, coupled with a prolate ellipsoidal model, allow an accurate and highly linear estimate of changes in absolute ventricular volume [19]. The limitations of this method include significant under-estimation of absolute volume change [12], and perhaps slightly different linear relationships between

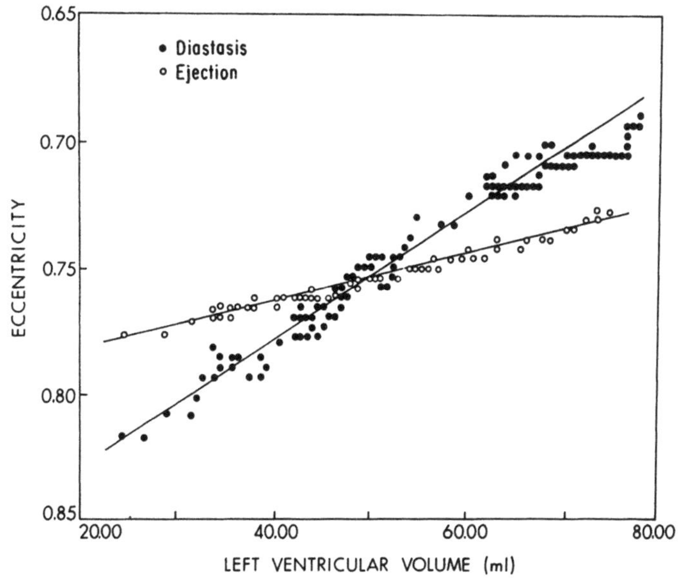

Fig. 2. Relationship between left ventricular shape (eccentricity) and volume for diastole (diastasis) and ejection in a conscious dog

calculated and absolute cavitary volume for systole and diastole (unpublished data). Another potential problem is the need to measure ventricular wall volume in order to subtract it from epicardial shell volume in the calculation of cavitary volume. In our original studies, an additional pair of ultrasonic transducers was used to measure dynamic wall thickness [12], but the results were variable and subject to error [22]. Later, a measured postmortem wall volume was used for this purpose [23], and more recently, echocardiographically determined wall volume [24]. This last technique is simple and accurate, and allows serial assessment of changing wall volume, for example during the development of ventricular hypertrophy [25]. The ability to estimate dynamic ventricular volume in the intact heart has been a significant advance in cardiac physiology.

If acute or chronic septal shifting is anticipated, measurement of left ventricular septal-free wall diameter and utilization of a general ellipsoidal model are required [19]. Addition of a right ventricular free wall transducer and measurement of right ventricular septal-free wall diameter allows accurate simultaneous assessment of dynamic cavitary volume in both ventricles using a shell subtraction model [26]. Thus, the understanding of biventricular chamber geometry has improved significantly over the past 15 years, and this information now will be used to further develop physiologic concepts of diastolic and systolic cardiac function.

Diastolic Mechanical Properties

For the purpose of this discussion, diastole will be defined as that portion of the cardiac cycle in which all influences of systolic contraction are absent [27]. In effect,

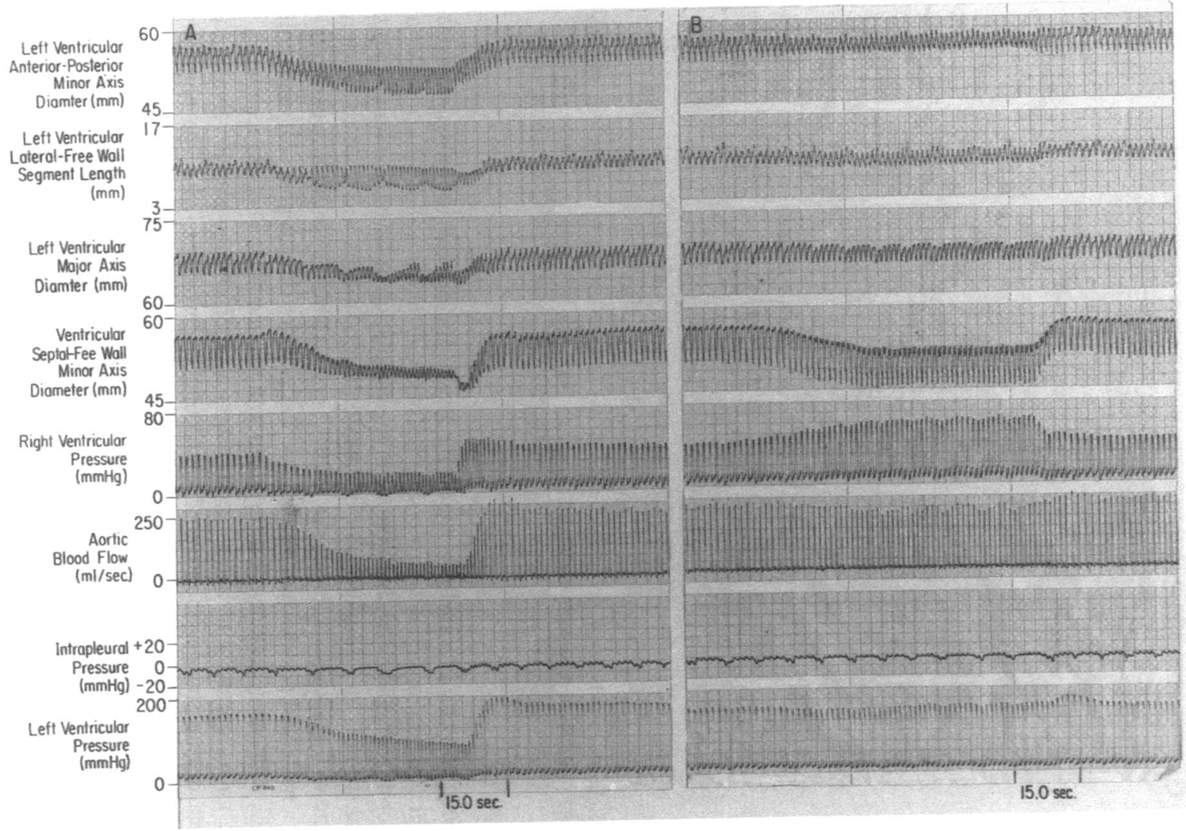

Fig. 3A, B. Left ventricular dimension characteristics with **A** vena caval occlusion and **B** marked septal shifting induced by pulmonary artery occlusion

Fig. 4. With pulmonary artery occlusion (*PAO*) and septal shifting, left ventricular stroke work increases slightly as compared to control vena caval occlusion (*VCO*). The two panels represent the maximal variability observed in 8 dogs

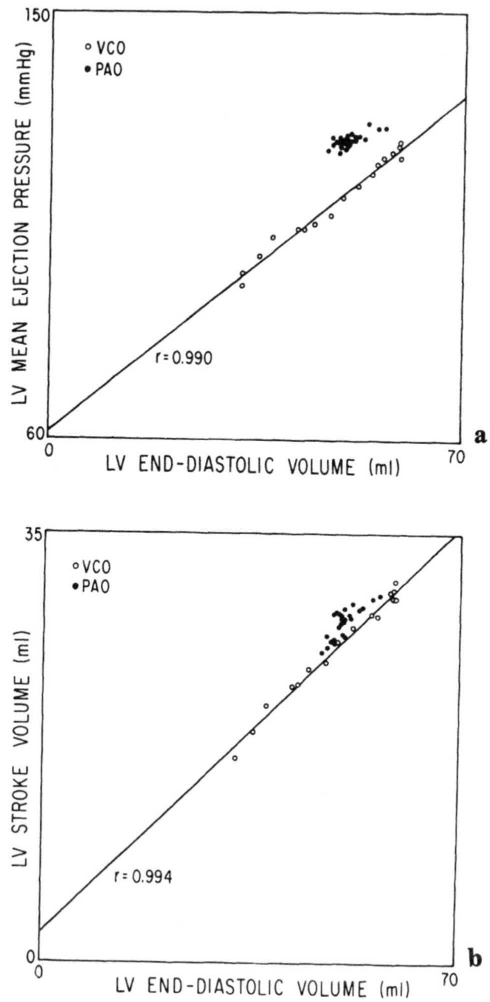

this definition excludes relaxation characteristics, which are covered in another section. Thus, diastolic properties are the purely passive mechanical characteristics of the fully relaxed myocardium and have been shown to be relatively independent of acute pharmacologic alterations in systolic function [28].

As with most biologic tissues, the relationship between diastolic ventricular pressure (P) and volume (V) or between myocardial force and length is exponential, and a finite x-intercept also exists. An exponential model predicting zero pressure at a finite volume must be used to quantify the relationship [22, 29], and a modification of the Glantz equation is ideal for this purpose:

$$P = \alpha \quad (e^{\beta\left[\frac{V - V_0}{V_0}\right]} - 1) \tag{1}$$

where α and β are nonlinear elastic coefficients, and V_0 is the x-intercept or the unstressed diastolic volume [29].

During rapid diastolic filling or atrial systole, measured ventricular pressure is always higher than would be predicted by a simple exponentially elastic relationship, probably reflecting viscous properties. These viscous forces are proportional to the rate of lengthening in the intact heart [22] and need to be taken into consideration. The simplest technique involves omitting dynamic data from the analysis and fitting only diastatic measurements (dV/dt < 5% per second) to Eq. 1.

The final factor to be considered is proper normalization of force and dimension data. When dealing with different hearts or with ventricles that are undergoing geometric changes over time, pressure and dimension measurements must be converted to wall stress (σ) and strain (ϵ). Our group generally uses an ellipsoidal modification of LaPlace's law [30] to calculate stress:

$$\sigma = \frac{\mathrm{Pr}_b}{h}\left(1 - \left[\frac{r_b^3}{r_a^2(2r_b + h)}\right]\right) \tag{2}$$

where r_b is minor axis midwall radius, r_a is major axis radius, and h is wall thickness. The Lagrangian strain formula is used to normalize dimensions:

$$\epsilon = \frac{\ell - \ell_0}{\ell_0} \tag{3}$$

where ℓ is the instantaneous midwall minor axis circumference, and ℓ_0 is the unstressed dimension at 0 mmHg diastolic transmural pressure (obtained at maximal vena caval occlusion). After conversion of pressures and dimensions to stress and strain, normalized myocardial elasticity can be calculated from Eq. 1, and the mechanical properties of each unit of myocardium can be estimated, even in hearts with dynamically changing ventricular geometry (Fig. 5) [22]. Specific analytic requirements, such as obtaining an adequate data range by vena caval occlusion, measuring intrapleural pressure in intact subjects, etc. are discussed elsewhere [27]. With this technique, directional changes or differences in diastolic mechanical properties can be quantified using standard covariant analyses of: (1) the unstressed volumes or lengths, and (2) the nonlinear elastic coefficients, α and β [17, 19].

Systolic Myocardial Performance

According to the most accepted theory, the fundamental unit of muscular contraction is the myosin cross-bridge [31]. With electrical depolarization of the myocardial cell membrane, ionized calcium fluxes into the cytoplasm causing the myosin molecule to hydrolyze adenosine triphosphate (ATP) into adenosine diphosphate and inorganic phosphate. When ATP is split, a considerable amount of chemical energy is released from the ATP molecule and transferred into a conformational change in the myosin cross-bridge. This chemo-mechanical alteration in the cross-bridge produces sliding of myosin filaments relative to actin, and shortening of the sarcomere. Over the physiologic range of sarcomere lengths (1.6–2.0 μ), the surface area of available cross-bridge interactions and, therefore, the metabolic energy transferred into mechanical energy during sarcomere contraction are linearly proportional to end-diastolic sarcomere length. This length-dependency of cross-bridge interaction at the sarcomere level constitutes the histochemical basis of the Frank-Starling relationship. As the final steps in the process, calcium is

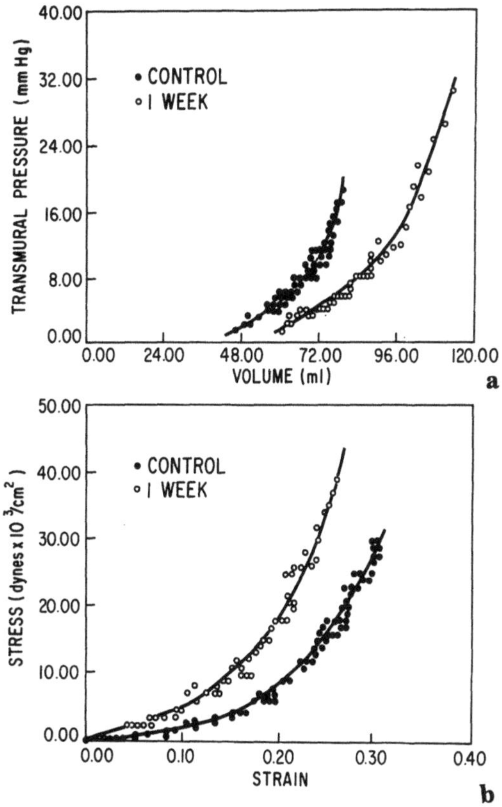

Fig. 5a, b. After 1 week of an aortocaval shunt in a conscious dog, the diastolic pressure volume curve of the left ventricle shifts to the right (**a**). However, normalization for differences in chamber geometry and mass (**b**) produces a leftward shift of the myocardial stress and strain curve

removed from the cell by active transport of the cytoplasmic reticulum, high-energy phosphate bonds are regenerated at the mitochondrial level by aerobic metabolism of oxygen and substrates, and this chemical energy is transmitted back to the cross-bridge region via the creatine phosphate shuttle.

Conceptual understanding of global ventricular performance is enhanced by considering the heart as a pump that imparts energy (or work) to the circulation. Basic variables for the analysis include left ventricular pressure and volume, and as illustrated in Fig. 6a, each cardiac cycle can be characterized by a distinct pressure-volume loop. The area of each loop represents net external mechanical energy (or stroke work) expended for that cycle, while end-diastolic volume reflects the mean end-diastolic fiber length or preload. The relationship between stroke work (SW) and end-diastolic volume (EDV) has been found to be quite linear under all conditions thus far tested (Fig. 6b) and has been termed the preload recruitable stroke work relationship [23]. Thus, as suggested by both Starling and Sarnoff [see 7, 8], energy output of the heart seems to be a linear function of EDV, and with this

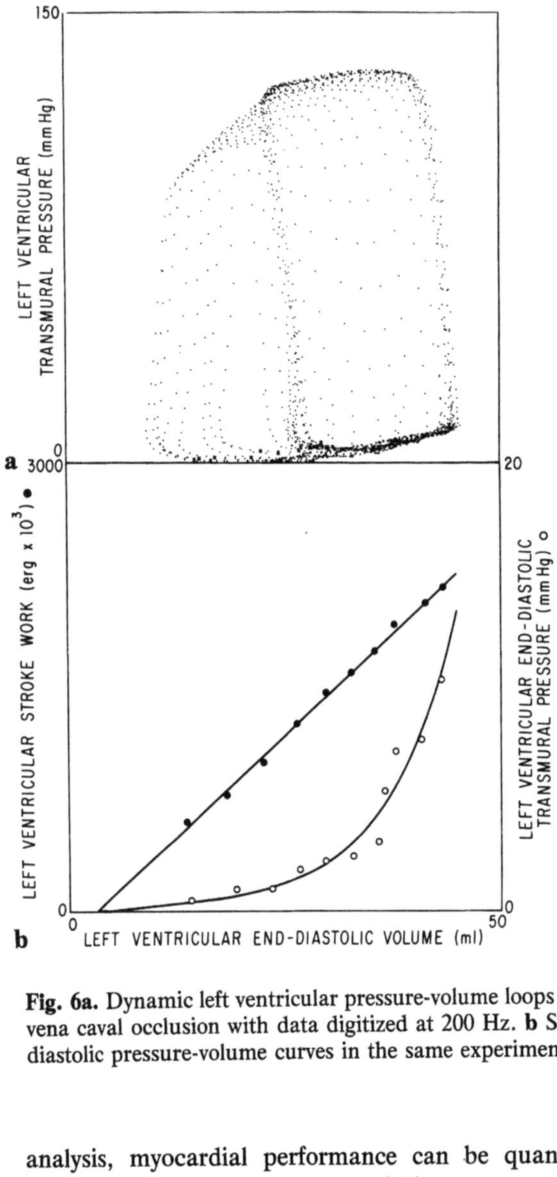

Fig. 6a. Dynamic left ventricular pressure-volume loops obtained in the conscious dog during vena caval occlusion with data digitized at 200 Hz. **b** Stroke work-end diastolic volume and diastolic pressure-volume curves in the same experiment as **a**

analysis, myocardial performance can be quantified simply by a linear slope constant (M_w) and an x-intercept (V_0):

$$SW = M_w\,(EDV - V_0). \tag{4}$$

In the intact heart, the slope constant (or work capacity at a given EDV) is insensitive to physiologic variations in afterload but is altered significantly by inotropic interventions (Fig. 7). This relationship, therefore, is a conceptually clear representation of myocardial inotropism, and as such, provides a useful framework for analysis and understanding of ventricular function. Variations in heart rate, ventricular geometry, and myocardial mass can be taken into account by converting pressure volume data to stress and strain using Eqs. 2 and 3 (Fig. 8). Then, heart

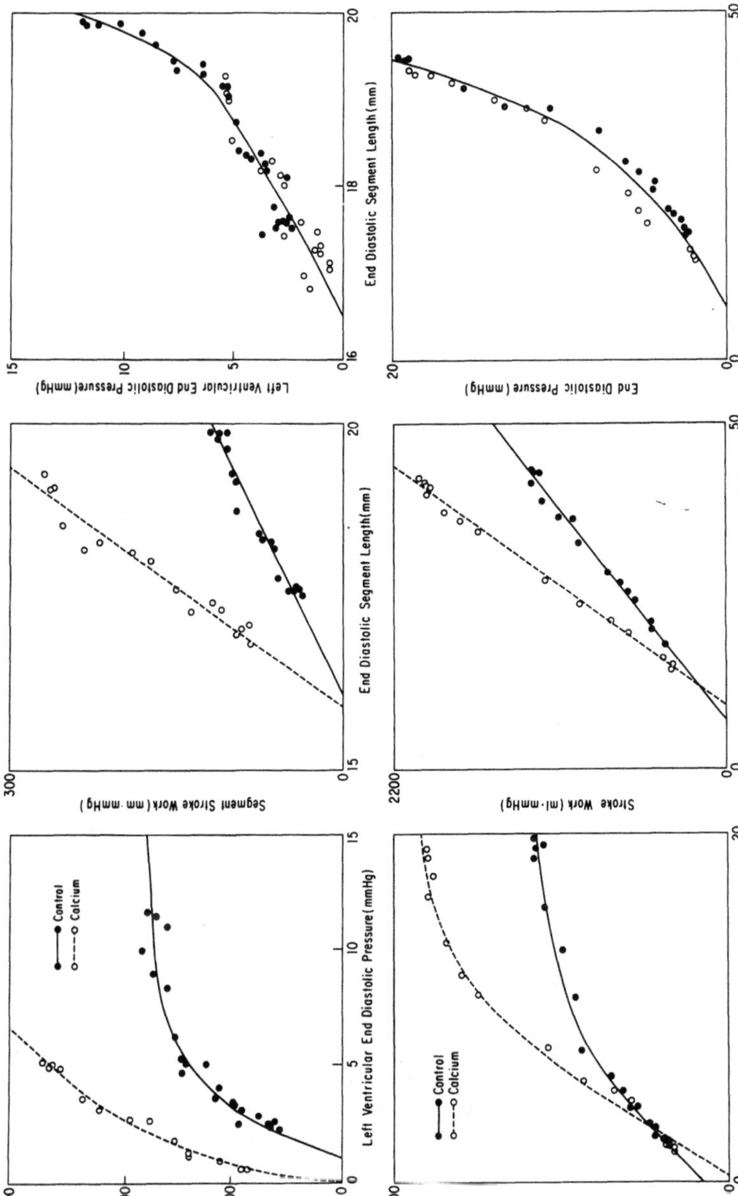

Fig. 7. Typical effects of calcium infusion on stroke work vs end-diastolic presssure relationships *(left panels)*, stroke work vs end-diastolic dimension relationships *(middle panels)*, and end-diastolic pressure vs dimension relationships *(right panels)*. *Top panels* show typical data from one regional study, and *bottom panels* depict a representative global study

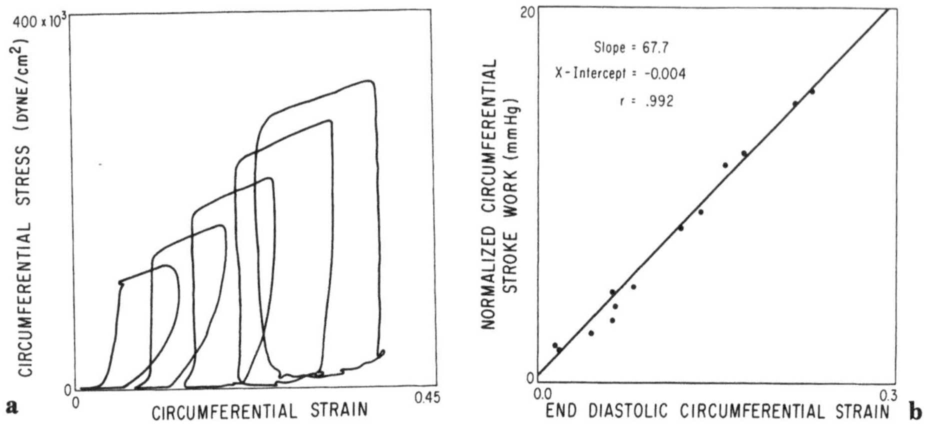

Fig. 8a. Circumferential stress-strain loops in one study with the resultant relationship between end-diastolic circumferential strain and normalized circumferential stroke work (**b**) calculated as the area enclosed by each circumferential stress-strain loop

rate is considered by converting stress-strain stroke work to myocardial power output (normalized work per unit time) and plotting that variable versus end-diastolic strain (Fig. 9). The slope of this relationship (M_{MP}) reflects the external mechanical energy per unit time expendable per unit diastolic deformation for one cubic centimeter of myocardium at the midwall minor axis hoop circumference. Thus, this one number (M_{MP}) meets all the requirements for a quantitative measure of myocardial inotropic state in the intact heart [23].

In acute experiments, where myocardial volume, geometry, and heart rate are not changing, the slope and x-intercept of the stroke work-end diastolic volume relationship can be used to assess directional changes in ventricular function employing analysis of covariance (Hotelling's t^2-test). This method has proven to be simple, practical, and highly reproducible. When geometry, heart rate, and myocardial mass are different (e.g., when studying different patients with valvular heart disease), the more complicated, but still practical, method of calculating M_{mp} is required. This single coefficient hypothetically can be used to define differences in myocardial inotropism, independent of loading, geometry, wall mass, and heart rate [23].

Relaxation Characteristics

Systole can be divided into 3 phases: isovolumic contraction, ejection, and isovolumic relaxation. During relaxation, the myocardium makes a transition from systolic energy generation to a passive resting diastolic state. The detailed physiology of relaxation has been nicely reviewed elsewhere [32], but for the purpose of this paper, several relaxation phenomena should be discussed. Relaxation is not fully complete in the normal heart until the first diastolic minimum of ventricular pressure, and interestingly, half of rapid filling is accomplished before this point

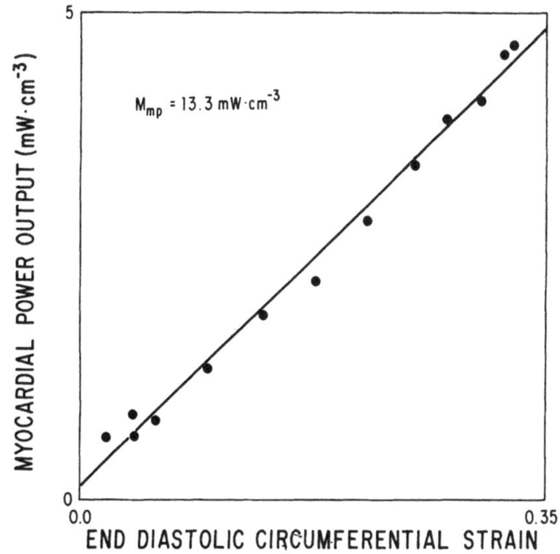

Fig. 9. Typical linear relationship between myocardial power output and end-diastolic circumferential strain with intrinsic myocardial function presented as M_{mp}. Data points represent successive cardiac cycles during transient vena caval occclusion, during which the spontaneous heart rate changed by less than 10%

[12, 13, 22]. Evidence now exists that active restorative forces associated with relaxation assist rapid filling during this period [33]. Because these restorative forces are usually manifested by early lengthening of the muscle, they are very difficult to discern in the normal heart. By implanting a prosthetic valve in the mitral position, Yellin's group has been able to study these forces with occlusion of the disc during systole and elimination of filling for the next diastole. When filling and myocardial lengthening are prevented, relaxation-phase restorative forces then become manifested as negative diastolic pressure or a fall in diastolic pressure below the passive pressure-volume curve (Fig. 10) [34].

While the interpretation of this phenomenon is somewhat controversial, the primary author believes the most likely hypothesis relates to "relaxation restorative energy" released from the contractile apparatus during the transition from systole to diastole. Conceivably, a certain population of cross-bridges may be left in an attached and deformed condition as active state reaches its concluding stages. The potential energy stored within these deformed cross-bridges then could be released in the form of either muscle lengthening or negative cavitary pressure during late relaxation.

Two lines of evidence support this theory. First, in collaborative studies between our group and Yellin's, interruption of normal cross-bridge dynamics by clamping volume at end-systole prevents relaxation-phase lengthening and generates a negative pressure (Fig. 10) [34]. The ensuing systole, however, is markedly perturbed, producing considerably less work or energy at a given end-diastolic volume than for control conditions [35]. This finding suggests an integral link

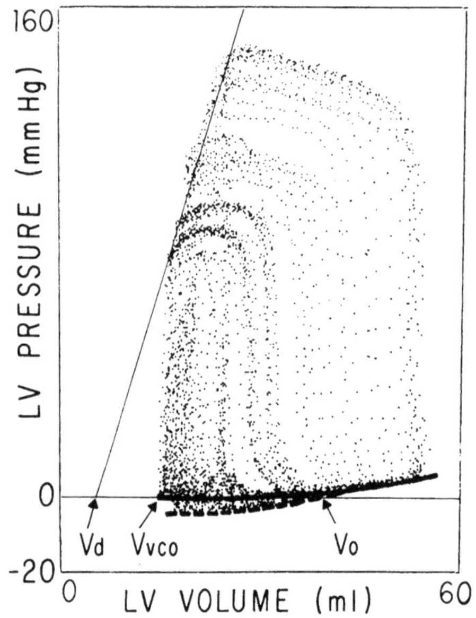

Fig. 10. Left ventricular pressure volume data during vena caval occlusion in an open-chest dog. Every 3–4 beats, the mitral disc was occluded, preventing filling and causing ventricular pressure to fall below the passive diastolic curve. V_d, volume intercept of end-systolic pressure-volume relation; V_{vco}, volume at maximal vena caval occlusion; V_o, diastolic volume at which occlusion curve crosses abscissa

between relaxation restorative energy and the machinery of systolic contraction, perhaps at the cross-bridge level. Furthermore, drug interventions that either increase or decrease the force of contraction tend to produce parallel changes in restorative forces at any given end-diastolic volume. Although this hypothesis requires further experimental testing, these findings suggest that relaxation-phase restorative forces are a property of the energy-generating component of the sarcomere, rather than reflecting simply a passive diastolic spring in compression below its slack length [33].

Myocardial Energetics

Cardiac metabolic activity over time can be estimated by myocardial oxygen consumption. Each milliliter of oxygen utilized by the heart provides 2.02 joules of energy to the contractile apparatus via aerobic pathways and high energy phosphate metabolism. Thus, oxygen consumption can be used to quantify myocardial *energy utilization*. Myocardial *energy expenditure* has two components: (1) external energy, which is stroke work or the integral of the ventricular pressure (P) volume (V) loop, and (2) internal energy, which is the thermodynamic cost of maintaining systolic ventricular pressure at a given volume [36, 37]. Internal energy expenditure is estimated as the product of left ventricular mean ejection pressure (P_{me}) and end diastolic volume (V_{ed}), and constitutes the sole mechanical energy produced during isovolumic contraction. Thus, total mechanical energy expenditure (TME) for each cardiac cycle can be calculated as:

$$TME = \int PdV + (P_{me} \cdot V_{ed}). \tag{5}$$

Because 1 mmHg·ml is equivalent to 1.33×10^{-4} joules, total mechanical energy expenditure can be compared to metabolic energy utilization to obtain a measure of metabolic to mechanical energy transfer (Fig. 11). Oxygen consumption seems to be linearly related to TME both on a steady state [37] and beat-to-beat basis [38]. Moreover, the relationship remains constant in the intact heart with variations in heart rate, during inotropic stimulation by calcium infusion, and with afterload interventions (Fig. 11). Alterations in myocardial energetics may have a significant pathophysiologic role in myocardial failure and can be quantified by the regression coefficients of this relationship.

Pathophysiology

Myocardial Ischemia

Because of high metabolic demand and tight coupling between energy utilization and expenditure, acute coronary occlusion produces an almost immediate decrement in myocardial segment shortening and work. Even when full reperfusion is accomplished after a 15-min period of reversible ischemia, dysfunction can be prolonged, requiring up to 24–48 h for complete recovery. Ischemic myocardial dysfunction in this setting is characterized by a diminished slope of the stroke work-end diastolic segment length relationship (Fig. 12), together with a rightward shifting of the x-intercept (ℓ_0), termed *diastolic creep* [39, 40]. With reperfusion, the slope recovers rapidly, but ℓ_0 remains over-stretched, diminishing the work capacity at any given preload. Subsequent recovery of systolic function occurs slowly and is associated with a reversal of creep [39]. Catecholamine infusion, pressure afterloading, or a second ischemic event during early reperfusion can worsen dysfunction and prolong recovery [41–43].

Postischemic systolic dysfunction and diastolic creep can be correlated at an ultrastructural level with sarcomere over-stretching, increased z-band separation, and widening of the I-bands [39]. It seems that the sarcomeres become stretched beyond their normal working range, and disengagement of actin filaments from the M-band may reduce the number of possible cross-bridge interactions and potentially account for the mechanical findings. Moreover, diastolic properties during ischemia become sensitive to standard pharmacologic interventions, suggesting persistent cross-bridge attachments during diastole and/or disarray of normal cross-bridge cycling. This entire process undoubtedly will provide useful areas of investigation at the histochemical level in coming years and will yield further insights into ischemic dysfunction.

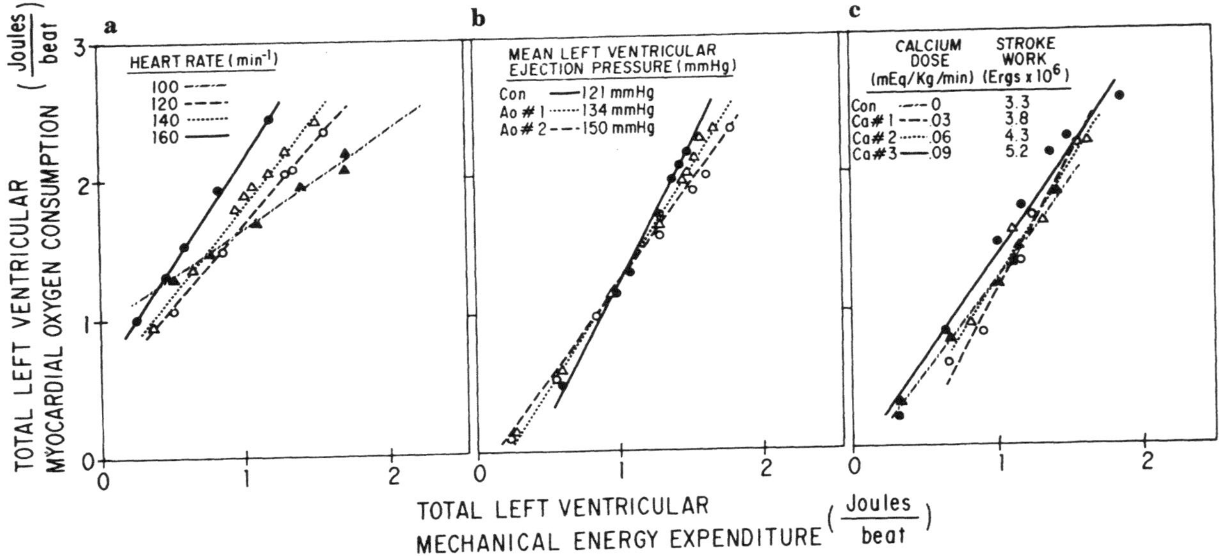

Fig. 11a-c. Typical relationships observed in conscious dogs between myocardial oxygen consumption and total mechanical energy with variations **a** in heart rate, **b** afterload, and **c** inotropism. *Con*, control; *Ao*, aorta

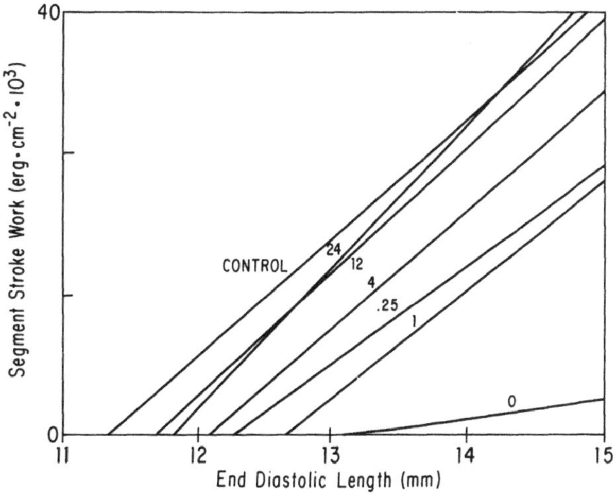

Fig. 12. Segment stroke work-end diastolic segment length relationships for a representative dog under control conditions, after 15 min of coronary occlusion (*O*), and after 0.25, 1, 4, 12, and 24 h of coronary reperfusion

Myocardial Hypertrophy

Cardiac valvular insufficiency reduces ventricular pumping efficiency and produces an excessive volume overload [44]. In a chronic animal model of aortic regurgitation, end-diastolic pressure and volume increase, and total stroke volume is augmented in proportion to the regurgitant volume [24]. Forward cardiac output is maintained by a combination of Frank-Starling reserves (increasing end diastolic volume) and inotropic reserves recruited by sympathetic ganglionic reflexes [45]. The chronically elevated diastolic pressure, however, produces a time-dependent increase in diastolic ventricular dimensions at 0 mmHg transmural pressure, which also can be termed creep. As opposed to ischemic creep, dimensional lengthening in volume overload is associated with an increased number of sarcomeres in series and parallel and a more normal sarcomere morphology [46]. Hypertrophy probably is induced by an increasing systolic wall stress secondary to enlarging chamber radius, and requires around 2 weeks to become evident (Fig. 13) [47]. Over the next several weeks, hypertrophy produces wall thickening in proportion to circumferential creep, and wall stress is brought back toward control values. Hypertrophy also tends to normalize diastolic properties (Fig. 14) and to bring V_0 back toward control [48]. With the chronically increased work load, however, myocardial inotropism progressively declines (Fig. 15), and for some unknown reason, myocardial failure ensues [24]. Preliminary results from our laboratory suggest a possible link between a reduced energy utilization capacity and the heart failure process (Fig. 16), although much more work will be required to define fully the physiologic causes of myocardial decompensation.

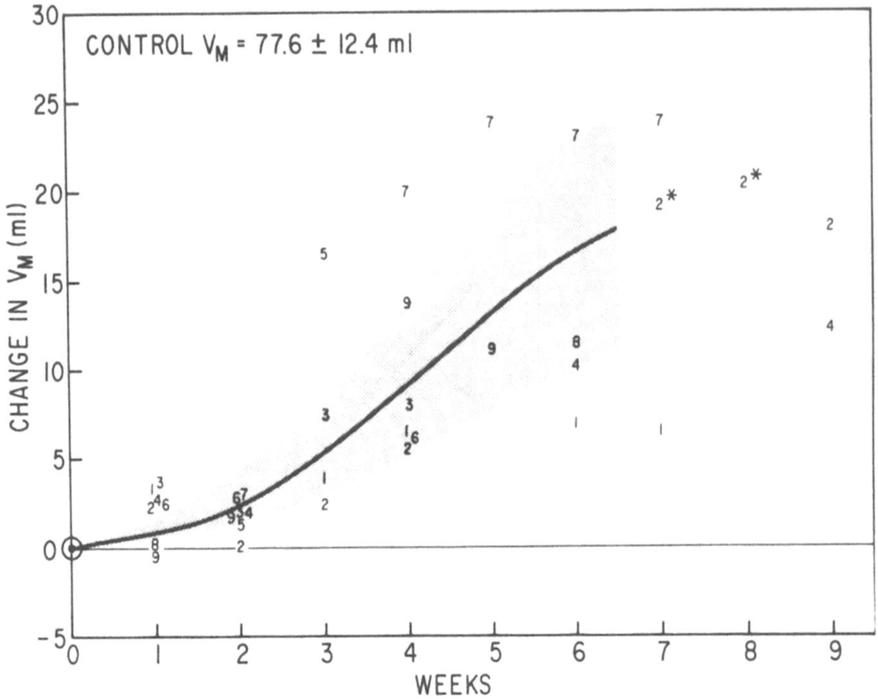

Fig. 13. Time course of hypertrophy after creating aortic insufficiency in conscious dogs. Each number (1–9) represents a given animal, and successive time points are shown. V_m, myocardial volume

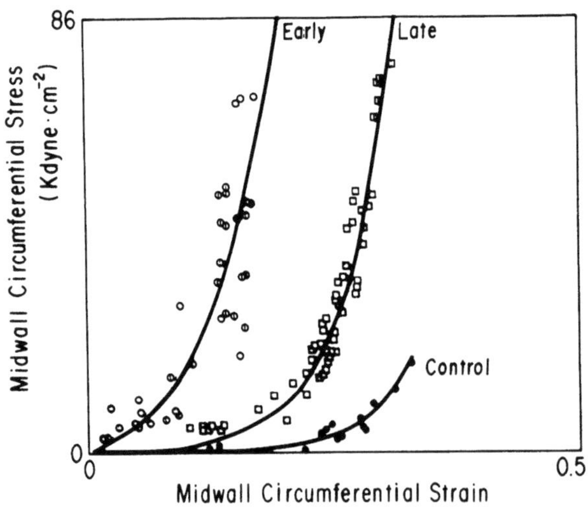

Fig. 14. Early after creation of aortic insufficiency, diastolic creep causes myocardial overstretching and a diminished compliance. With hypertrophy, compliance returns toward control

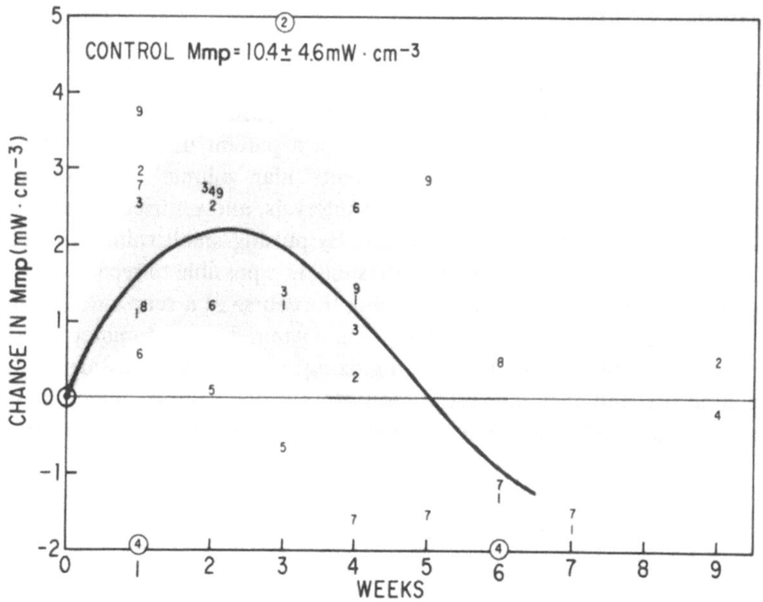

Fig. 15. Early after creation of aortic insufficiency, myocardial power is augmented, probably by reflex autonomic control. Thereafter, inotropism progressively deteriorates. M_{mp}, slope of plot of normalized work per unit time vs end-diastolic strain

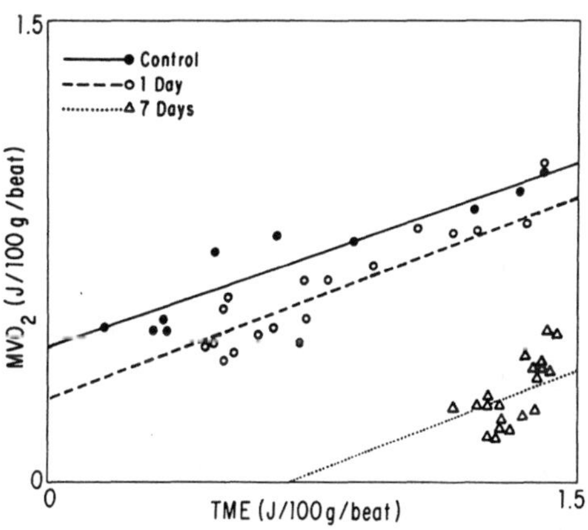

Fig. 16. Relationship between myocardial oxygen consumption (MVO_2) and total mechanical energy (*TME*) production for the left ventricle of a conscious dog in the control state and after 1 day and 7 days of massive aortic valve insufficiency. With progression of myocardial failure, oxygen utilization diminishes

Human Studies

It is now possible to perform detailed measurements of myocardial function in man on a routine basis [49]. Left ventricular pressure volume loops obtainable during transient inferior vena caval occlusion in a patient undergoing cardiac catheterization are shown in Fig. 17. Left ventricular volume is measured with digital subtraction ventriculography at 30-ms intervals, and ventricular pressure is obtained with a high-fidelity micromanometer. By pulsing small volumes of dilute contrast into the left ventricle during each diastole, it is possible to record continuous ventriculographic data on digital disc during the course of a vena caval occlusion without perturbing ventricular function. Employing semi-automated edge definition programs and simultaneously digitizing ventricular pressure reduce analysis complexity and facilitate estimation of pressure volume curves in man. Human stroke work-end diastolic volume characteristics appear highly linear and seem

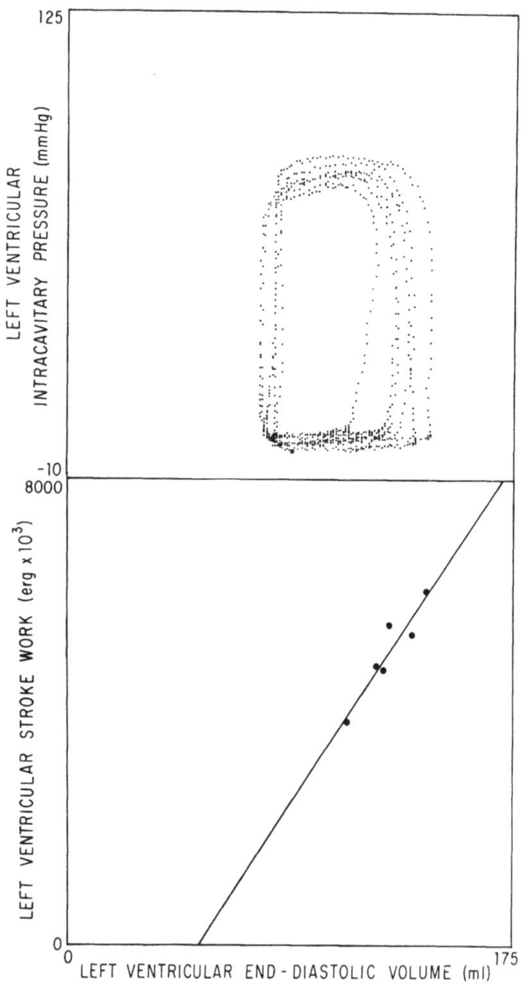

a

b

Fig. 17a. Left ventricular pressure volume loops and **b** stroke work-end diastolic volume relationship with inferior vena caval occlusion in man

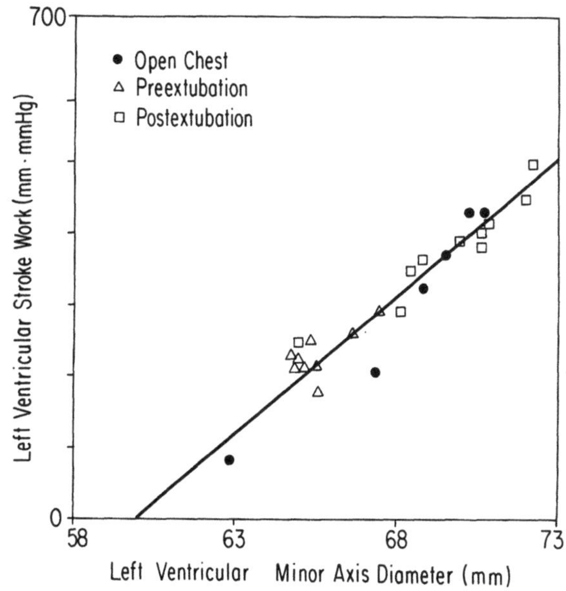

Fig. 18. Improved stroke work after endotracheal extubation seemed to be caused simply by improved left ventricular filling, since stroke work-end diastolic volume characteristics did not deviate from intraoperative curve

similar to those from dog studies, although it is impossible to obtain as much of the curve for reasons of patient safety [49]. Similar data measured in postoperative patients (Fig. 18) allowed our group to show that application or withdrawal of airway pressure affects cardiac function almost purely by limiting or enhancing ventricular preload [50].

Recent studies of ventricular function also have been performed in valvular heart disease patients before and after prosthetic valve replacement [51]. Utilizing radionuclide ventriculography and micromanometry, ventricular pressure volume curves and stroke work were measured at multiple end diastolic volumes in patients with aortic stenosis (Fig. 19) and mitral regurgitation (Fig. 20). These valvular lesions are the two most associated with altered pressure and volume loading, respectively. Correction of the valvular lesion produced the expected effects on systolic pressure and ejection fraction (Figs. 19 and 20). Despite these dramatic changes in ventricular loading and total shortening, however, stroke work-end diastolic volume curves remained relatively constant, confirming the load-insensitivity of this relationship in patients with heart disease. Current efforts are being directed toward defining basic differences in myocardial performance between the various valvular disorders.

In summary, our understanding of basic cardiac function has improved significantly in recent decades, as investigators have continued the tradition of direct experimental observation begun by William Harvey. It is reasonable to predict that knowledge will expand even more rapidly in coming years. Because of better measurement techniques together with more practical and relevant physiological models, medical science now is on the verge of significant breakthroughs in the area of clinical ventricular function, especially as related to pathophysiologic mechanisms in ischemic and valvular disorders. Associated diagnostic and therapeutic innovations should translate ultimately into improved care of patients with cardiac disease.

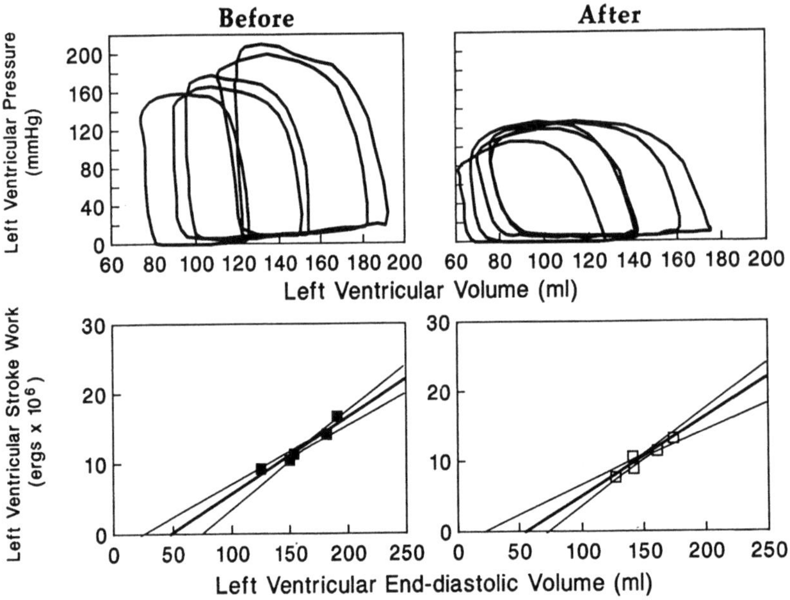

Fig. 19. Left ventricular pressure volume and stroke work-end diastolic volume characteristics before and after aortic valve replacement in a patient with aortic stenosis

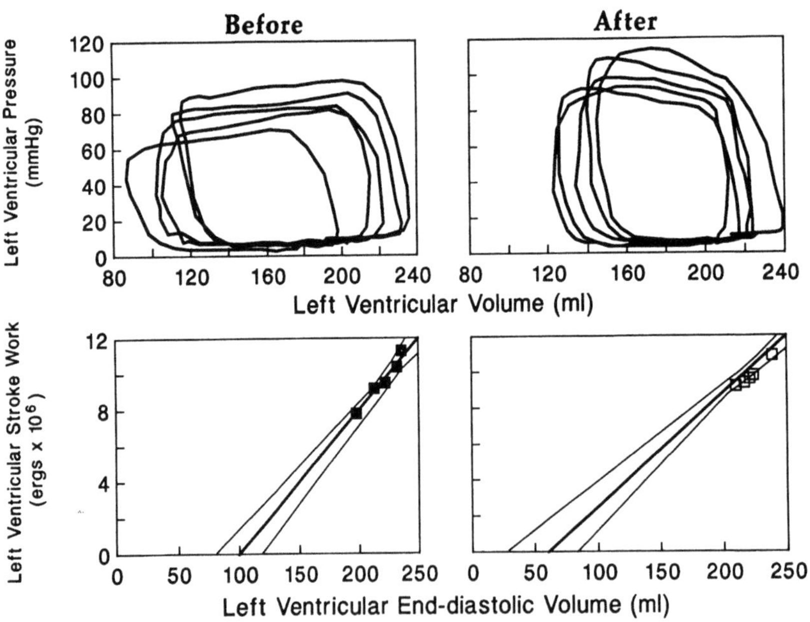

Fig. 20. Left ventricular pressure volume and stroke work-end diastolic volume characteristics before and after mitral valve replacement in a patient with mitral regurgitation

References

1. Cournand A (1964) Air and blood. In: Fishman AP, Richards DW (eds) Circulation of the blood. Oxford University Press, New York, pp 3-7
2. McDonald DA (1974) Blood flow in arteries. Williams & Wilkins, Baltimore, pp 1-16
3. Hamilton WF, Richards DW (1964) The output of the heart. In: Fishman AP, Richards DW, (ed) Circulation of the blood. Oxford University Press, New York, pp 71-125
4. Harvey W (1928) Exercitatio anatomica de motu cordis et sanguinis in animalibus. (translated from the Latin by CD Leake) Thomas, Springfield
5. Williams FA, Dry TJ (1948) A history of the heart and circulation. W.B. Saunders, Philadelphia
6. Bost J (1981) Chaveau and Marey's accomplishment; the first intracardiac pressure records. In: Snellen HA, Dunning AJ, Arntzenius AC (eds) History and perspective of cardiology. Leiden University Press, The Hague, pp 7-9
7. Chapman CB, Mitchell JH (1965) Starling on the heart. Dawsons of Pall Mall, London
8. Sarnoff SJ, Berglund E (1954) Starling's law of the heart studied by means of simultaneous right and left ventricular function curves in the dog. Circulation 9: 706
9. Sarnoff SJ, Mitchell JH (1961) The regulation of the performance of the heart. Am J Med 30: 747
10. Ross J Jr, Covell JW, Sonnenblick EH, Braunwald E (1966) Contractile state of the heart characterized by force-velocity relations in variably afterloaded and isovolumic beats.Circ Res 18: 149
11. Sagawa K, Maughan L, Suga H et al. (1988) Cardiac contraction and the pressure-volume relationship. Oxford University Press, New York
12. Rankin JS, McHale PA, Arentzen CE (1976) The three-dimensional dynamic geometry of the left ventricle in the conscious dog. Circ Res 39: 304
13. Rankin JS, Olsen CO (1980) The diastolic filling of the left ventricle. Eur Heart J 1: 95
14. Rankin JS (1980) The chamber dynamics of the intact left ventricle. In: Baan J, Arntzenius A, Yellin E (eds) Cardiac dynamics. Martinus Nijhoff, The Hague
15. Olsen CO, Rankin JS, Arentzen CE et al. (1981) The deformational characteristics of the left ventricle in the conscious dog. Circ Res 49: 843
16. Olsen CO, VanTrigt P, Rankin JS (1981) Dynamic geometry of the intact left ventricle. Fed Proc 40: 2023
17. Olsen CO, Glower DD, Lee KL et al. (to be published) Diastolic anisotropic properties of the left ventricle in the conscious dog. Circ Res
18. Arts T, Veenstra PC, Reneman RS (1982) Epicardial deformation and left ventricular wall mechanics during ejection in the dog. Am J Physiol 243: H379
19. Olsen CO, Tyson GS, Maier GW et al (1983) Dynamic ventricular interaction in the conscious dog. Circ Res 52: 85
20. Feneley MP, Olsen CO, Glower DD, Rankin JS (1989) Effect of acutely increased right ventricular afterload on work output from the left ventricle in conscious dogs. Circ Res 65: 135
21. Elbeery JR, Owen CH, Savitt MA et al. (submitted) Effects of the left ventricular assist device on right ventricular function. J Thorac Cardiovasc Surg
22. Rankin JS, Arentzen CE, McHale PA et al (1977) Viscoelastic properties of the diastolic left ventricle in the conscious dog. Circ Res 41: 64
23. Glower DD, Spratt JA, Snow ND (1985) Linearity of the Frank-Starling relationship in the intact heart: the concept of preload recruitable stroke work. Circulation 71: 994
24. Gaynor JW, Feneley MP, Maier GW (1987) Left ventricular adaptation to chronic aortic regurgitation in conscious dogs. Circulation 76: IV-888
25. Feneley MP, Gaynor JW, Maier GW (1988) In vivo estimation of left ventricular wall volume in volume-overloaded canine hearts. Am J Physiol 255: H1399
26. Feneley MP, Maier GW, Gaynor JW et al. (1987) An ellipsodial shell subtraction model for determining right ventricular volume. J Am Coll Cardiol 9: A214
27. Rankin JS, Gaynor JW, Feneley MP et al. (1987) Diastolic myocardial mechanics and the regulation of cardiac performance. In Grossman W, Lorell BH (eds) Diastolic relaxation of the heart. Martinus Nijhoff, Boston

28. VanTrigt P III, Olsen CO, Pellom GL (1980) Do acute hemodynamic interventions alter diastolic mechanics of the left ventricle? Circulation 62 (Suppl III): 778
29. Rankin JS, Arentzen CE, Ring WS et al. (1980) The diastolic mechanical properties of the intact left ventricle. Fed Proc 39: 141
30. McHale PA, Greenfield JC Jr (1973) Evaluation of several geometric models for estimation of left ventricular circumferential stress. Circ Res 33: 303
31. Squire J (1981) The structural basis of muscular contraction. Plenum Press, New York
32. Grossman W, Lorell BH (1988) Diastolic relaxation of the heart. Martinus Nijhoff, Boston
33. Yellin EL, Hori M, Yoran C et al. (1986) Left ventricular relaxation in the filling and nonfilling intact heart. Am J Physiol 250: H620
34. Nikolic S, Feneley MP, Pajaro O et al. (1989) On important volumes in the left ventricle. J Am Coll Cardiol 13: 86A
35. Nikolic S, Feneley MP, Pajaro O et al. (1988) Effect of transient filling perturbations on the stroke work-end diastolic volume relationship. Circulation 78 (Suppl II): 274
36. Feneley MP, Maier GW, Gaynor JW (1987) Comparison of elastic and non-elastic predictive models of myocardial oxygen consumption in conscious dogs. Circulation 76: IV-543
37. Maier GW, Owen CH, Feneley MP et al. (1988) The mechanical determinants of myocardial oxygen consumption in the conscious dog. Circulation 78 (Suppl II): 266
38. Elbeery JR, Owen CH, Savitt MA et al. (1989) Myocardial metabolic to mechanical energy transfer characteristics from single vena caval occlusions in conscious dogs. J Am Coll Cardiol 13: 87A
39. Glower DD, Schaper J, Kabas JS et al. (1987) The relationship between reversal of diastolic creep and recovery of systolic function after ischemic myocardial injury in the conscious dog. Circ Res 60: 850
40. Glower DD, Spratt JA, Kabas JS et al. (1988) Quantification of regional myocardial function after acute ischemic injury. Am J Physiol 255: H85
41. Kabas JS, Spratt JA, Davis JW et al. (submitted) Effects of dopamine on myocardial functional recovery after reversible ischemic injury. J Thorac Cardiovasc Surg
42. Elbeery JR, Williams RF, Rankin JS et al. (submitted) The effects of systolic loading on myocardial after ischemic injury. Circulation
43. Gall SA Jr, Maier GW, Gaynor JW et al. (1988) Repetitive ischemia delays recovery of regional myocardial function in awake dogs. J Am Coll Cardiol 11: 94A
44. Spratt JA, Olsen CO, Tyson GS Jr et al. (1983) Experimental mitral regurgitation: physiologic effects of correction on left ventricular dynamics. J Thorac Cardiovasc Surg 86: 479
45. Rankin JS, Ring WS, Arentzen CE et al. (1977) The functional reserves of the left ventricle in chronic volume overload. Circulation 56 (Suppl II): 906
46. Ross J Jr (1974) Adaptations of the left ventricle to chronic volume overload. Circ Res 34 and 35 (Suppl II): 64
47. Gaynor JW, Feneley MP, Maier GW et al. (1988) Time course of left ventricular hypertrophy and wall stress in aortic regurgitation. J Am Coll Cardiol 11: A92
48. Gaynor JW, Savitt MA, Feneley MP et al. (1988) Effects of volume overload on ventricular diastolic mechanics in conscious dogs. Circulation 78 (Suppl II): 276
49. Feneley MP, Skelton TN, Kisslo KB et al. (1987) Comparison of preload recruitable stroke work and end-systolic pressure-volume relationships in man. Circulation 76: IV-163
50. Gall SA Jr, Olsen CO, Reves JG et al. (1988) Beneficial effects of endotracheal extubation on ventricular performance. J Thorac Cardiovasc Surg 95: 819
51. Harpole DH, Rankin JS, Wolfe WG, Smith LR, Young WG, Clements FM, Jones RH (in Press) Assessment of left ventricular functional preservation during isolated cardiac valvular operations. Circulation 80 (Suppl III)

Ventricular Pressure-Volume Relations Demonstrate Positive Inotropic Effect of Increased Arterial Impedance

JAN BAAN and ENNO T. VAN DER VELDE[1]

Summary. The influence of different types of loading intervention on the end-systolic pressure-volume relation (ESPVR) of the left ventricle (LV) in situ was investigated in anesthetized open-chest dogs. LV volume was measured by conductance catheter. Two loading interventions were applied: a pressure intervention (INp) by gradually occluding the descending aorta, and a volume intervention (INv) by rapidly infusing 180 ml blood into the vena cava. The P-V loops during an intervention always showed a linear ESPVR, the slope of which, E_{es}, was calculated. Results from 16 dogs show that E_{es}(INp) was always larger (1.37 ± 0.45 kPa/ml) than E_{es}(INv) (0.73 ± 0.32 kPa/ml) ($P < .001$). This difference was enhanced by beta-blockade through metoprolol. The same phenomenon was found for the slope (SdPV) of the relation between dP/dt_{max} and EDV: SdPV(INp) = 45.17 ± 22.63 kPa/ml per s and SdPV(INv) = 20.55 ± 11.13 kPa/ml per s ($P < .001$).

Since INv increases stroke volume while INp decreases it, the results can be explained in part by shortening deactivation. However, the similar behavior of SdPV as compared to E_{es} suggests that INp leads to an increased inotropic state with respect to INv. We speculate that an intrinsic myocardial mechanism related to changes in calcium turnover forms the basis of this phenomenon.

Introduction

The traditional concept of an index of the inotropic state or contractility of cardiac muscle requires that it be independent of preload and afterload. Studies in isolated muscle over the past years, recently reviewed by Lakatta [1], have shown the muscle-length dependence of calcium turnover, which puts into question any consideration of contractility as a preload-independent property. Afterload dependence of the inotropic state, however, has been studied only sparsely in muscle fibers [2, 3]. When considered at the level of the total ventricle, this situation is not much different, which is largely due to the long-claimed insensitivity of the end-systolic pressure-volume relationship (ESPVR) to loading conditions [4]. Pressure-volume loops of the left ventricle may be employed to characterize and diagnose a number of cardiac function abnormalities [5], but the slope (E_{max} or E_{es}) and intercept (V_d) of the ESPVR determined from these loops have received most attention in their use for the assessment of contractility. Such usage has been based mostly on the behavior of

[1] Laboratory of Clinical Physiology, Department of Cardiology, Leiden University Hospital, Leiden, The Netherlands

the ESPVR in the isolated canine heart, a preparation in which ventricular volume could be measured accurately [4]. The development of the conductance catheter method in our laboratory [6], however, has brought measurement and usage of pressure-volume relations to the clinic [5, 7] as well as the intact animal lab [8]. The study described here was set up to investigate the possible dependence of the ESPVR on afterload, defined by arterial impedance in the intact dog. Increasing this impedance by gradual occlusion of the descending aorta resulted in a much steeper relation than increasing pressure and volume by venous loading.

Materials and Methods

Studies were performed in anesthetized mongrel dogs, with body masses ranging from 15–35 kg. After premedication with 5 ml Hypnorm (10 mg fluanisone and 0.2 mg fentanyl per ml) intramuscularly and 0.25 mg atropine subcutaneously, anesthesia was initiated with sodium-pentobarbital (Nembutal, 15 mg/kg), and maintained with an infusion of methadone (5 mg/h) and dehydrobenzperidol (25 mg/h). Ventilation was maintained with a 3:1 mixture of N_2O and O_2 using a Dräger-Pulmonat respirator. ECG (lead II) was recorded. The chest was opened via a midline sternotomy and a dual sensor microamanometer catheter (Millar PC-780B or Honeywell MTC) was introduced through the left carotid artery to measure aortic and left ventricular pressures. An 8-electrode conductance catheter (custom-made by Cordis Europa, Roden, The Netherlands) with appropriate electrode distance to match the LV long axis was introduced into the left ventricle via the right common carotid artery to measure LV volume as previously described [6]. Catheter tip position in the apex was verified by palpation before measurements, and post-mortem by opening the LV. Briefly, the method is based on measuring the time-varying electrical conductance of 5 segments of blood in the left ventricle, from which total left ventricular volume is calculated. The generation of a low alternating current, applied between the proximal and distal electrodes, measurement of conductances, and analog computations to obtain the desired volume variable, V(t), are performed by a model Sigma-5 signal-conditioner-processor (Leycom, Oegstgeest, The Netherlands). Calibration for the conductivity of blood was done at least every half hour and after infusion of fluids which might alter blood conductivity. The slope constant, α, in the relation between conductance and volume was taken as 1.0 in all experiments, although the average value found both in vivo [6] and in the isolated heart [9] is approximately 0.8. Because the present study is primarily concerned with comparison of ESPVR under different loading conditions within individual dogs, the value of α has no influence on the results. The value of the parallel conductance formed by tissue surrounding the LV cavity (myocardium, right ventricle contents, etc.) was determined in each experiment by injection of 1–2 ml of hypertonic saline (6 M), administered through a balloon-flotation catheter in the pulmonary artery [6]. To prevent reflex changes in heart rate bilateral vagotomy was performed, accompanied as needed by administering atropine (1 ml/kg).

Volume loading was effectuated by increasing end-diastolic volume through rapid infusion of 180 ml of homologous blood (37°C) or a mixture of glucose and dextran (also at 37°C) in a ratio to match the conductivity of blood. Pressure loading was effectuated by increasing the aortic impedance through gradual (5–10 s) inflation of a catheter-mounted balloon in the descending aorta, just distal to the aortic arch. The volume intervention will be referred to as INv, the pressure intervention as INp.

To investigate whether or not the behavior of the ESPVR under different loading conditions was affected by the beta-adrenergic system, we administered a beta-blocker, metoprolol (0.1 mg/kg) and a beta-sympathicomimeticum, dobutamine (1-6 mg/kg per hour) in three experiments. To rule out the influence of respiration on LV volume and/or conductance all data were collected during periods of suspended respiration of maximally 40 s. Ventilation was stopped at end-inspiration. Data collection occurred both during gradually changing loading conditions and at steady states, to test the possible influence of slowly working control mechanisms.

We used an iterative method to determine the ESPVR from the pressure-volume loops. With a starting value for the volume intercept (V_d) of 0 ml, in each cardiac cycle of the loading run the point at which $E(t) = P(t)/[V(t)-V_d]$ is maximal was determined, followed by linear regression analysis on these P-V points. From the resulting slope and intercept, a new V_d was calculated as: $V_d = -$ pressure intercept/slope. With this new value substituted in the equation for E(t), the process was repeated until the difference between the most recent V_d used and the last V_d calculated was smaller than a preset error value of 0.1 ml. The value of the slope of the resulting ESPVR is termed E_{es}, and its volume intercept V_d. The value of E_{es} generally differs from the maximum of E(t).

To investigate whether another index of contractile performance would reveal the same behavior under different loading conditions as the ESPVR, we calculated the slope of the relation between maximal LVdP/dt (dP/dt_{max}) and end-diastolic volume (EDV), in the manner proposed by Little [10]. Since dP/dt_{max} generally occurs before ejection, this offered the possibility to compare a pre-ejection index to the E_{es} index which occurs close to end-ejection. dP/dt was calculated by computer from LV pressure using a smoothed 3 point Lagrange derivative.

Results

Fig. 1 shows typical LV pressure-volume data resulting from a pressure intervention (INp) and a volume intervention (INv). There is a clear difference between the effect of INp and INv, resulting in an E_{es}(INp) of 1.29 kPa/ml, with V_d(INp) = 13.8 ml, and an E_{es}(INv) of 0.66 kPa/ml with V_d(INv) = 4.1 ml. The end-systolic P-V points during any intervention always correlated highly (median r = 0.995) and linearly. The P-V loops shown indicate that a volume intervention by venous loading not only increased EDV but also raised systolic pressure, which is to be expected in the intact circulation, since peripheral resistance remains approximately constant. Conversely, pressure loading by occluding the descending aorta increased aortic impedance, and as a consequence, stroke volume decreased. Probably because of excess filling volume with respect to ejected volume, pressure loading also raised EDV, albeit to a lesser degree than by the venous loading intervention. Establishing a steady state after the gradual intervention did not change the position or the slope of the ESPVR appreciably.

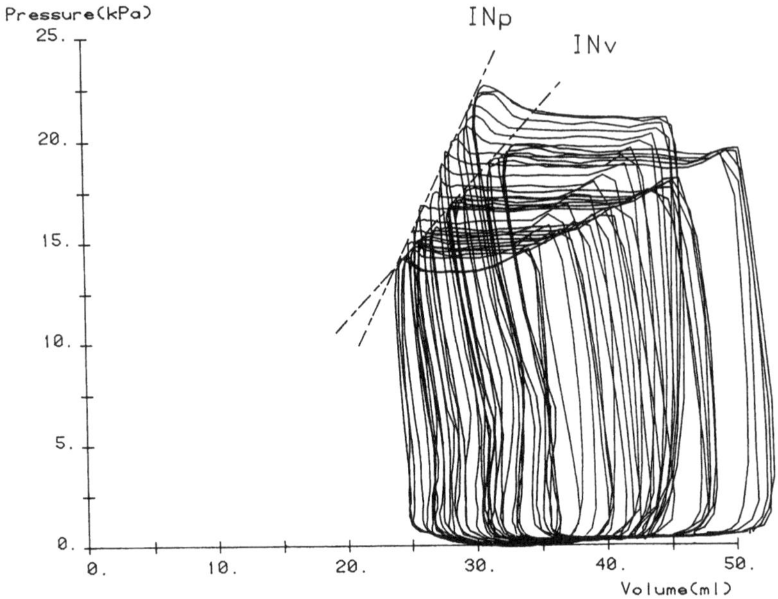

Fig. 1. Pressure-volume loops in a dog during pressure intervention (*steep* end-systolic *line*, marked *INp*, $E_{es} = 1.29$ kPa/ml) and during volume intervention (*INv*, $E_{es} = 0.66$ kPa/ml). E_{es}, end-systolic elastance. Redrawn from [23] with permission of authors and publisher

In Fig. 2 an example is shown of the effect of INp versus INv on E_{es} (Fig. 2a) and the slope (SdPV) of the dP/dt_{max}-EDV relation (Fig. 2b). Notice the similarity of the ratios between the two slopes: $E_{es}(INp)/E_{es}(INv) = 2.67$ (a) and SdPV(INp)/SdPV(INv) = 2.66 (b). Such a large similarity was not present in all experiments, however. The E_{es} and SdPV results of 22 pairs of INp versus INv interventions are displayed in Fig. 3a and b respectively, from a total of 16 dogs. The mean value (± SD) of $E_{es}(INp)$ is 1.37 (± 0.45) kPa/ml; the mean value (± SD) of $E_{es}(INv)$ is 0.73 (± 0.32) kPa/ml. The difference between these values is statistically highly significant with $P < 0.001$. Likewise, the average values for V_d for the two interventions are significantly different with $P < 0.002$: $V_d(INp)$ is 9.01 (±7.98) ml and $V_d(INv)$ is 0.33 (±8.65) ml.

The average values for the slope, SdPV, of the relation between dp/dt_{max} and EDV in the 22 measurement sets are: SdPV(INp) = 45.17 (± 22.63) kPa/ml per s and SdPV(INv) = 20.55 (± 11.13) kPa/ml per s, having a statistically significant difference with $P < 0.001$ (Fig. 3b). So, the phenomenon of a considerable dependence on the type of loading intervention, observed for the parameter E_{es}, was also found for the parameter SdPV. There was a good linear correlation between $E_{es}(INp)$ and $E_{es}(INv)$ ($r = 0.71$, $P < 0.001$) as well as between SdPV(INp) and SdPV(INv) ($r = 0.76$, $P < 0.001$). Moderately correlated linear relations were found between $E_{es}(INp)$ and SdPV(INp) ($r = 0.45$, $P < 0.05$), and between $E_{es}(INv)$ and SdPV(INv) ($r = 0.67$, $P < 0.005$). Heart rate was constant to within 2% per experiment.

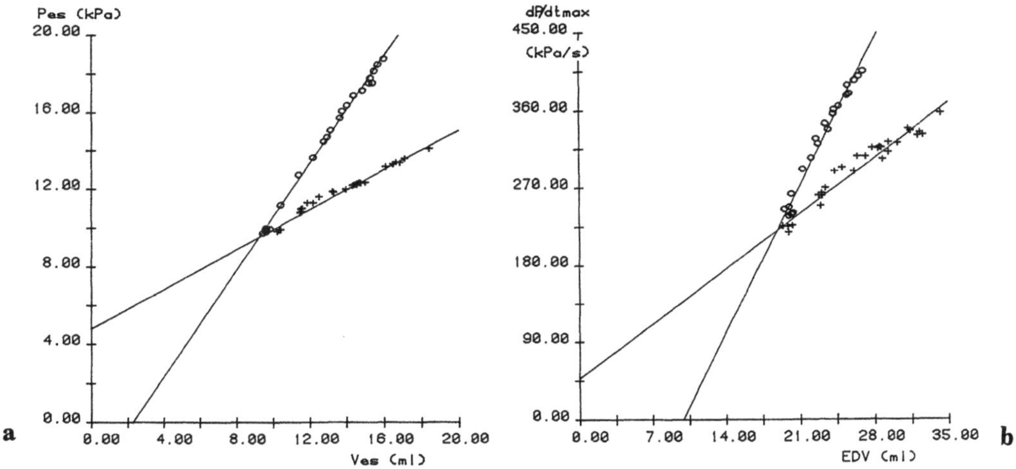

Fig. 2a. End-systolic pressure (P_{es}) versus end-systolic volume (V_{es}) (E_{es}(INp) = 1.38 kPa/ml, E_{es}(INv) = 0.52 kPa/ml), and **b** dP/dt_{max} versus end-diastolic volume (*EDV*) (slope SdPV(INp) = 24.43 kPa/ml per s, SdPV(INv) = 9.19 kPa/ml per s in a 23 kg dog during pressure intervention (*circles*) and during volume intervention (*crosses*). Reprinted from [8] with permission of authors and publisher

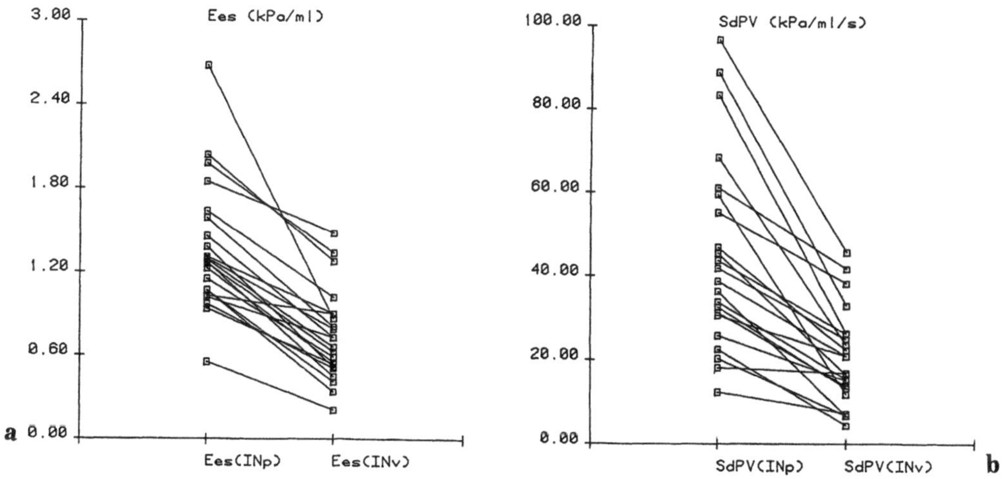

Fig. 3a. Values of end-systolic elastance (E_{es}) from pressure intervention (*INp*) and from volume intervention (*INv*) in 16 dogs (22 measurements). **b** Values of the slope (*SdPV*) of maximal dP/dt versus end-diastolic volume (EDV) from pressure intervention and volume intervention in the same set of data. Reprinted from [8] with permission of authors and publisher

In Table 1 the results are summarized from experiments where metoprolol and dobutamine were administered. The results show that E_{es} strongly decreases after metoprolol, probably due to its negative inotropic action, and increases back towards control values with subsequent administration of dobutamine, as expected.

Discussion

The results of this study indicate that the end-systolic pressure-volume relation is clearly dependent on the way in which loading conditions are altered. E_{es} is consistently and substantially larger for a pressure intervention than for a volume intervention. This difference is further enhanced by blocking the beta-receptors.

As demonstrated previously, the results obtained are not dependent on measurement properties of the conductance method, since LV pressure-diameter relations secured with ultrasonic crystals displayed a behavior after loading interventions similar to that of the P-V loops [8]. Moreover, comparable effects of changes in afterload on the ESPVR have been found recently in isolated canine hearts using the intracardiac balloon method as discussed later.

The mean value of E_{es} by volume intervention, 0.73 ± 0.32 kPa/ml, converts to 5.5 ± 2.4 mmHg/ml. This value compares well with those found in the isolated heart: 3.2 mmHg/ml [4] and 6.6 mmHg/ml [11], and also the intact dog (obtained with vena cava occlusion): 7.7 ± 2.9 mmHg/ml [10] and 5.8 ± 3.6 mmHg/ml [12]. In contrast, E_{es} found by aortic occlusion, an intervention not studied by others, was 1.37 ± 0.45 kPa/ml or 10.3 ± 3.4 mmHg/ml, exceeding all of the values reported in the literature.

The average value of the volume intercept, V_d, upon INv is much lower than its value upon INp, which is in concordance with the lower slope value, E_{es}. We do not attach much significance to this finding nor to the occasionally found negative values for V_d. Namely, V_d thus calculated merely represents a mathematical intercept with the volume axis obtained by extrapolation of a portion of the "total" ESPVR which happens to be rather straight. Obviously, a low E_{es} value will lead to a negative V_d value as also found by others [10, 13], a physical impossibility if V_d is considered to

Table 1. E_{es} values (in kPa/ml) for metoprolol/dobutamine experiments

	Control	Metoprolol	Dobutamine
Experiment 1			
E_{es} (INp)	1.02	0.61	1.39
E_{es} (INv)	0.73	0.27	0.84
Experiment 2			
E_{es} (INp)	1.30	0.76	0.87
E_{es} (INv)	0.79	0.39	0.55
Experiment 3			
E_{es} (INp)	2.43	1.30	2.08
E_{es} (INv)	0.95	0.30	1.26

E_{es}, slope of end-systolic pressure-volume relation; INp, pressure intervention; INv, volume intervention represent LV volume at zero pressure. In addition, such a physical value of V_d should not depend on the intervention used to determine the ESPVR: in all likelihood both relationships will curve toward the volume axis at low pressures [14] and must meet in the same point. This point, however, cannot be obtained in the intact circulation without controlled coronary perfusion. Rather than using V_d we recently advocated the use of a volume intercept ('V18') at a pressure occurring within the physiological range of any intervention (e.g., 18 kPa or 135 mmHg) [15].

The manifest dependency of the ESPVR on the type of loading intervention for the heart in situ as found by us cannot be explained on the basis of external control mechanisms. Firstly, such mechanisms would probably be exerted through the beta-adrenergic system, but beta blockade appears to enhance the afterload dependence of the ESPVR. Also, occlusion of the descending aorta increases the pressure in the baroreceptors in the aortic arch and carotid arteries, which, rather than lead to an enhanced inotropic state as observed, should result in a depressed inotropism. Secondly, similar results as reported here have been found in the isolated—obviously denervated—heart preparation [16, 17]: a large stroke volume (with ejection fraction $> 50\%$) led to an end-systolic P-V point below the isovolumic ESPVR, called pressure deficit [16], while a very small stroke volume (with ejection fraction $< 20\%$) resulted in a P-V point above the isovolumic ESPVR [17]. A similar dependence on the magnitude of stroke volume and ejection fraction for the slope of the ESPVR was found by us in a right heart bypass preparation in vivo [8]. It can be argued that the phenomenon of the pressure deficit and the lower value of E_{es} with increasing stroke volume reflects the basic property of isolated muscle known as shortening deactivation [18], while the phenomenon of the left-and-upward shift of the end-systolic P-V point and the larger value of E_{es} with decreasing stroke volume reflects some sort of hyperactivation due to moderate shortening [19]. Modeling such phenomena by incorporating a resistance term (to mimic flow-dependent pressure changes) in the time-varying elastance concept, however, can explain the observed effects only partially and qualitatively [20]. Other mechanisms than shortening- or flow-related deactivation are probably involved.

A clue to these mechanisms might be found in the circumstance that the afterload dependence of the ESPVR is reflected in the behavior of dP/dt_{max}, a variable known to occur before ejection sets in. Moreover, the behavior of E_{es} is mimicked also by the hypothetical line connecting the P-V points at the onset of ejection (Fig. 1). This indicates that the change in afterload affects the LV behavior during the entire systolic period rather than affecting shortening alone. The observed afterload dependence is reminiscent of the well-known Anrep effect, but we emphasize that in our experiments aortic pressure is increased during both the pressure and the volume intervention. Thus, rather than an increase in aortic pressure, it appears to be the increase in aortic impedance, and in particular the value of the ratio of stroke volume/systolic pressure known as E_a [21], which is the main factor responsible for the apparent increase in inotropic state.

It is likely that these inotropic changes occur at the cellular level of the muscle. The basis of this hypothesis is that an increase in dP/dt_{max} at a given constant EDV reflects a similar increase of maximal dF/dt (F = force) at the sarcomere level. As shown by Yue [22], dF/dt is linearly related to intracellular calcium concentration measured by aequorin luminescence, under a wide variety of stimulation patterns. The observed changes in dP/dt_{max} and ESPVR therefore probably reflect corresponding changes in intracellular calcium turnover. It remains unknown which specific factors provide the link between changes in calcium turnover and changes in the ratio force/shortening velocity—which is essentially the sarcomere equivalent of the ratio pressure/flow. Meanwhile, we consider the findings reported here mostly as evidence that the ESPVR indeed reflects the inotropic state of the myocardium. In the event that aortic impedance is increased, the ventricle adapts its performance by an enhanced inotropic state, a feature which seems independent of cardiac adrenergic innervation. Whether or not this effect is accompanied by changes in myocardial perfusion and metabolism remains to be investigated. The described phenomenon appears to enable the heart to adapt and maintain its stroke volume (and cardiac output if rate remains fixed) against an increased resistive load.

References

1. Lakatta EG (1987) Starling's law of the heart is explained by an intimate interaction of muscle length and myofilament calcium activation. J Am Coll Cardial 10: 1157-1164
2. Parmley WW, Brutsaert DL, Sonnenblick EH (1969) Effects of altered loading on contractile events in isolated cat papillary muscle. Circ Res 24: 521-532
3. Paulus WJ, Claes VA, Brutsaert DL (1980) End-systolic pressure-volume relation estimated from physiologically loaded cat papillary muscle contractions. Circ Res 47: 20-26
4. Sagawa K, Suga H, Shoukas AA, Bakalar KM (1977) End-systolic pressure/volume ratio: a new index of contractility. Am J Cardiol 40: 748-753
5. Kass DA, Maughan WL (1988) From 'Emax' to pressure-volume relations: a broader view. Perspective. Circulation 77: 1203-1212
6. Baan J, Van Der Velde ET, De Bruin HG, Smeenk GJ, Koops J, Van Dijk AD, Temmerman D, Senden PJ, Buis B (1984) Continuous measurement of left ventricular volume in animals and humans by conductance catheter. Circulation 70: 812-823
7. Thormann J, Kramer W, Kindler M, Kremer P, Schlepper M (1987) Bestimmung der Wirkkomponenten von Amrinon durch kontinuierliche Analyse der Druck-Volumen-Beziehungen; Anwendung der Conductance (Volumen-) Kathetertechnik und der schnellen Laständerung durch Ballonokklusion der Vena cava inferior. Z Kardiol 76: 530-540
8. Baan J, Van Der Velde ET (1988) Sensitivity of left ventricular end-systolic pressure-volume relation to the type of loading intervention in dogs. Circ Res 62: 1247-1258
9. Burkhoff D, Van Der Velde ET, Kass D, Baan J, Maughan WL, Sagawa K (1985) Accuracy of volume measurement by conductance catheter in isolated, ejecting canine hearts. Circulation 72: 440-447
10. Little WC (1985) The left ventricular dP/dt_{max}-end-diastolic volume relation in closed-chest dogs. Circ Res 56: 808-815
11. Suga H, Sagawa K, Shoukas A (1973) Load independence of the instantaneous pressure-volume ratio of the canine left ventricle and effects of epinephrine and heart rate on the ratio. Circ Res 32: 314-322
12. Kass DA, Yamazaki T, Burkhoff D. Maughan WL, Sagawa K (1986) Determination of left ventricular end-systolic pressure-volume relationships by the conductance (volume) catheter technique. Circulation 73: 586-595
13. Spratt JA, Tyson GS, Glower DD, Davis JW, Muhlbaier LH, Olsen CO, Rankin JS (1987) The end-systolic pressure-volume relationship in conscious dogs. Circulation 75: 1295-1309

14. Burkhoff D, Sugiura S, Yue DT, Sagawa K (1987) Contractility-dependent curvilinearity of end-systolic pressure-volume relations. Am J Physiol 252: H1218–1227
15. Van der Velde ET, Burkhoff D, Steendijk P, Sagawa K, Baan J (1988) Afterload sensitivity and nonlinearity of the end-systolic pressure-volume relation in dogs. Circulation 78: II-68
16. Suga H, Yamakoshi K (1977) Effects of stroke volume and velocity of ejection on end-systolic pressure of canine left ventricle. End-systolic volume clamping. Circ Res 40: 445–450
17. Igarashi Y, Goto Y, Yamada O, Ishii T, Suga H (1987) Transient vs steady state end-systolic pressure-volume relation in dog left ventricle. Am J Physiol 252: H998–H1004
18. Meiss RA, Sonnenblick EH (1972) Controlled shortening in heart muscle: velocity-force and active state properties. Am J Physiol 222: 630–639
19. Hunter WC, Burkhoff D, Oikawa R, Sagawa K (1985) Evidence for a mechanism opposing deactivation during ejection in canine left ventricles. Fred Proc 44: 1736
20. Van Der Linden LP, Van Der Velde ET, Bruschke AVG, Baan J (1988) Identifiability of left ventricular end-systolic pressure-volume relationships. Am J Physiol 254: H1113–H1124
21. Sunagawa K, Maughan WL, Burkhoff D, Sagawa K (1983) Left ventricular interaction with arterial load studied in isolated canine ventricle. Am J Physiol 14: H773–H780
22. Yue DT (1987) Intracellular [Ca^{2+}] related to rate of force development in twitch contraction of heart. Am J Physiol 252: H760–H770
23. Baan J, Van Der Velde ET, Van Der Linden LP (1987) Aortic occlusion leads to larger left ventricular elastance (Ees) and DP/DT_{max} than venous loading in dogs. Automedica 9: 238

Physiological and Pharmacological Interventions on Coupling of Heart and Arterial Load

NICO WESTERHOF, G. PAUL TOOROP, GIJS ELZINGA, and VINCENT J.A. SCHOUTEN[1]

Summary. Cardiac output (F) increases, but the mean pressure (P) decreases with decreasing peripheral resistance. This inverse relationship is called the pump function graph, which characterizes the heart as a fluid pump. Power output (P × F) depends on the properties of the heart and of the arterial load, and when plotted as a function of cardiac output shows a relative maximum at a certain flow (F_p). This power maximum is relative since it depends not only on flow but also on the "setting of the heart," i.e., contractility, heart rate, and end-diastolic volume. The actual values of pressure and flow measured in the intact animal define working point (P_w and F_w). A consistent finding in anesthetized, open-thorax cats is that the heart works at the relative maximum of power output ($F_w = F_p$).

This paper discusses the influence of experimental and pathological conditions on power output. For variations in heart rate and volume loading and under influence of halothane the heart remained working at maximum power output ($F_w = F_p$). Noradrenalin and the vasodilator Hydralazine caused the working point to shift away from the maximum ($F_w > F_p$ and $F_w < F_p$, respectively). These results are compatible with the hypothesis that the heart is regulated to work at maximum power output, and that noradrenalin and hydralazine somehow interfere with the regulatory mechanisms.

Introduction

Analysis of mechanical behaviour of the ventricles has been approached mainly in two ways. The first approach has been developed by Suga, Sagawa, and Shoukas [1] and is known as the varying elastance concept. It is based on the instantaneous relationship between left ventricular pressure and volume. Using this concept simple equations were derived for stroke volume as a function of afterload (aortic pressure), preload (end-diastolic volume) and contractility (slope of the end-systolic pressure-volume relationship) [2]. The second approach, developed in our laboratory, is to characterize the ventricle by the pump function graph: the inverse relationship between mean left ventricular pressure (P) and cardiac output(F) [3].

The arterial system can be characterized by the relationship between mean aortic pressure and mean flow, from which the peripheral resistance can be calculated. Since the ratio of mean left ventricular pressure to mean aortic pressure is rather constant over a wide range and under varying conditions [4], the mean of either pressure can be used to evaluate the heart and the periphery. Thus their charac-

[1] Laboratory for Physiology, Free University, 1081 BT Amsterdam, The Netherlands

teristics can be plotted in the same coordinate system. The working point is where the pressure-flow relationships of heart and periphery intersect (Fig. 1). A fundamental assumption in our analysis is that the coupling between heart and vascular system can be described adequately in terms of mean pressure and mean flow, thus circumventing complex phase relationships of the pulsatile components in pressure and flow.

Pump Function

Experimental data in this report were recorded from isolated cat hearts and the in situ heart of anesthetized, open-thorax cats (for detailed descriptions of the techniques, see [5, 6]. The isolated cat heart was connected to an artificial hydraulic load of which the resistive and capacitive properties could be varied. In the in situ experiments, a pneumatic clamp for rapid occlusions was fixed around the ascending aorta, and close to the heart a large-diameter cannula was inserted into the aorta. The cannula was equipped with a pneumatically driven valve and an adjustable resistor. Thus, when the clamp is open and the valve is shut the left ventricle faces the normal load of the vascular bed. Occlusion of the aorta results in an isovolumic beat (zero flow) that produces maximum ventricular pressure (P_{max} in Fig. 1). Occlusion of the clamp and opening of the valve with different values of the resistor allows for variation of the artificial load. Even very low loads may be obtained, so that the maximum flow (F_{max} in Fig. 1) is approximated, whereas ventricular pressure is virtually zero. Single artificially loaded beats are studied in an otherwise steady-state situation to maintain a constant setting of the heart (i.e., heart rate, contractility, and end-diastolic volume). Variation of the artificial load between the two extremes results in a curve as shown in Fig. 2a, which is representative for the behavior of intact [6, 7] and isolated heart of cat [3, 8] and dog [9].

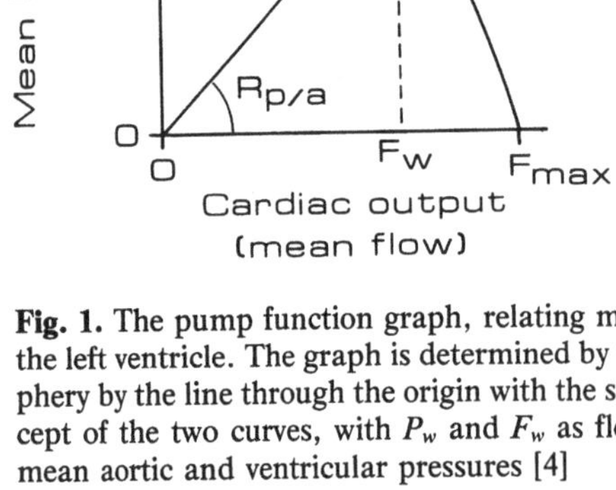

Fig. 1. The pump function graph, relating mean pressure and cardiac output, characterizes the left ventricle. The graph is determined by the two parameters P_{max} and F_{max}, and the periphery by the line through the origin with the slope $R_{p/a}$. The working point (W) is at the intercept of the two curves, with P_w and F_w as flow and pressure values. The constant a relates mean aortic and ventricular pressures [4]

The mean pressure–mean flow relationship presented here is related to the inverse relationship between force and shortening velocity in isolated trabeculae [10, 11]. Isovolumic (zero flow) pressure, P_{max}, is proportional to the isometric force generated by the muscle fibres, i.e., to contractility. According to the force-length relationship of the fibres, P_{max} is also dependent on the end-diastolic volume of the ventricle. Maximum flow, F_{max}, is equal to heart rate times ejected volume per beat (stroke volume). Maximal stroke volume is mainly determined by end-diastolic volume and hardly by contractility [1, 9].

It was found that the pump function graph can be fitted well with a parabola:

$$P = P_{max}\,(1 - F^2/F_{max}{}^2). \tag{1}$$

This experimental result implies that the heart can be characterized by two parameters: P_{max} and F_{max}.

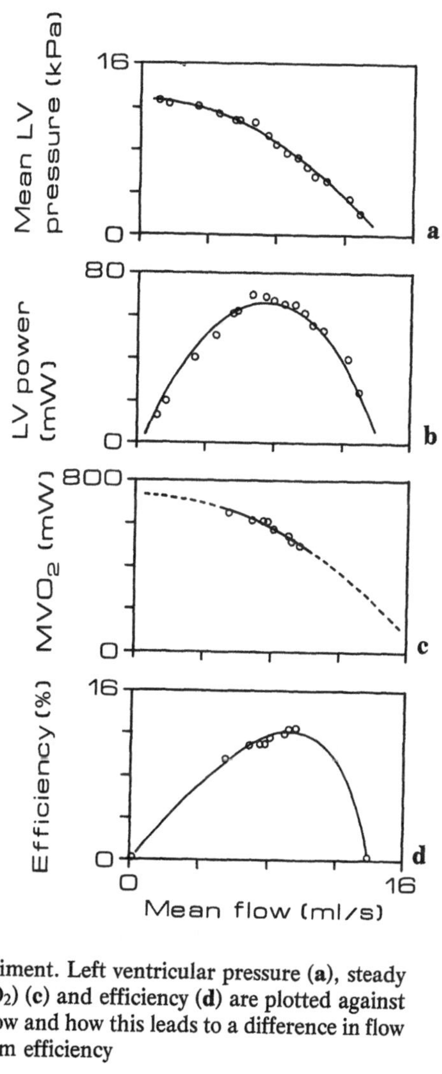

Fig. 2a–d. Example of an isolated cat heart experiment. Left ventricular pressure (**a**), steady power (**b**), myocardial oxygen consumption (MVO_2) (**c**) and efficiency (**d**) are plotted against flow. Note how MVO_2 decreases with increasing flow and how this leads to a difference in flow values at maximum steady power and at maximum efficiency

Power Output and Efficiency

The external work (mechanical energy) produced by the ventricle is pressure times volume change ($P \times \Delta V$) and the mean external power output is the work per unit of time, and is calculated from the instantaneous pressure and flow:

$$\text{Power} = (1/T)\int_0^T P(t)F(t)dt \quad (2)$$

where T is the heart period. We have shown previously that power calculated as the product of mean pressure and mean flow (steady power) is proportional to the mean external power calculated with Eq. 2 [4]. It should be clear that steady power is zero when either pressure or flow is zero while in between these extremes power output finds a maximum as shown in Fig. 2b. From Eq. 1, and using the fact that steady power is mean pressure times mean flow, we can derive the result that the maximum power output is found at $P_p = (2/3)P_{max}$ and $F_p = F_{max}/\sqrt{3}$. Thus, the two parameters of the pump function graph predict the pressure and flow values of maximum power output.

In order to supply the tissues with oxygen and metabolites, the heart has to generate pressure as well as flow, and thereby produces power. There is as yet, however, no thorough theoretical ground for deciding which combination of pressure and flow (i.e., which point on the pump function graph and the power-flow curve) would be most valuable to the organism [2, 7]. Nevertheless, in the hearts in situ, we always found the working point at or near the maximum power output [6]. This empirical finding has led to the idea of the "matching principle" which basically states that control mechanisms regulate heart function such that the working point coincides with the maximum of the power curve [4, 6, 10, 13].

Oxygen consumption (MVO_2) of the (isolated) heart depends predominantly on developed pressure whereas the influence of ejection is much less [1, 9]. Accordingly, pressure and MVO_2 decrease with increasing flow (Fig. 2c). Oxygen consumption may be taken as a measure of the energy used by the heart [5]. Efficiency (energy output/energy input) is represented by the ratio of power output and oxygen consumption. Efficiency as a function of flow (Fig. 2d) is zero when power output is zero, and maximum efficiency occurs at a larger flow value than maximum power output [6, 14]. Our experimental findings indicate that maximum efficiency is found at a flow value of $F_e = 0.7F_{max}$ and at a pressure of $P_e = 0.5P_{max}$ [6]. Thus, once we know the values of P_{max} and F_{max}, the point of maximum efficiency can be calculated.

It is now simple to determine if the working point (F_w, P_w) coincides with maximum power output (F_p, P_p) or maximum efficiency (F_e, P_e), from the fitting of the pump function graph with the parameters F_{max} and P_{max} only. When working point and maximum power output coincide we consider the heart and arterial system matched.

Variations in Pump Function and Working Point

It is not readily clear how a control mechanism could regulate the heart to maximize power output since there is no such thing as a power receptor. On the other hand, the circulation is submitted to various regulatory mechanisms for pressure and (local) flow, and their interaction and integration at the level of the central nervous system may have power matching as a resultant effect.

An interesting question is whether conditions can be found where the ventricle does not operate at maximum power output. We have tested the influence of a change in circulating blood volume and heart rate, as well as the effects of adrenalin halothane, and hydralazine (vasodilation) (for a summary of the data see Table 1). As shown in Fig. 3, in the open-thorax, anesthetized cat, the heart works at the relative maximum of power output in all five groups under control conditions. Infusion of 15–20 ml of dextran solution (Haemacel) led to an increase in P_{max} (27% ± 18%) and F_{max} (8% ± 5%). Nevertheless, the working point remained at maximum power output (Fig. 3a). The increase in P_{max} and F_{max} was probably due to increased end-diastolic volume. It has been shown also by others, that changes in end-diastolic volume markedly affect P_{max} and F_{max} while the working point remained at maximum power output [13].

An increase of heart rate from 160 to 180 min^{-1} led to a reduction of both P_{max} (− 10% ± 6%) and F_{max} (− 12% ± 4%). These effects were not expected since in the isolated cat heart [5] an increase of heart rate caused an increase in F_{max} and P_{max}. The shorter time for filling of the atrium at high rates may, however, have caused a smaller end-diastolic volume, as indicated by the lower end-diastolic pressure in the ventricle [7], which in turn would explain the decrease in P_{max} and F_{max}. Again, the working point remained at maximum power output (Fig. 3b). Noradrenalin predominantly enhanced P_{max} (37% ± 16%) and to a lesser extent F_{max} (8% ± 8%). These changes are consistent with the well-known positive inotropic effect of catecholamines. Peripheral resistance increased by 90% ± 49%, indicating a strong overall vasoconstrictive action by noradrenalin. The working point was shifted to a smaller flow value and was significantly different from the point of maximum power output. The working point moved also away from the point of maximum efficiency (Fig. 3c). We propose as tentative explanation that an important part of matching is regulated via sympathetic nerve terminals in the myocardium and vascular wall, i.e., infusion of extra noradrenalin may have disabled these control mechanisms.

Table 1. Maximum mean pressure (P_{max}) and flow (F_{max}), and mean pressure (P_w) and flow (F_w) in the working point (mean ± standard deviation of the mean) in control hearts and after various interventions

	P_{max}	F_{max}	P_w	F_w
Control (n = 8)	10.2 ± 0.7	6.7 ± 0.5	6.8 ± 0.5	3.7 ± 0.2
Volume loading	12.6 ± 0.4	7.3 ± 0.5	8.3 ± 0.5	4.1 ± 0.2
Control (160 min^{-1}) (n = 8)	10.6 ± 0.6	7.5 ± 0.4	7.6 ± 0.6	4.2 ± 0.2
Heart rate (180 min^{-1})	9.6 ± 0.6	6.6 ± 0.3	6.4 ± 0.4	3.6 ± 0.2
Control (n = 8)	8.9 ± 0.3	8.8 ± 0.8	6.1 ± 0.3	4.6 ± 0.3
Noradrenalin	12.0 ± 0.5	9.4 ± 0.7	9.5 ± 0.6	4.0 ± 0.3
Control (n = 6)	10.7 ± 0.5	7.0 ± 0.5	7.1 ± 0.5	3.9 ± 0.3
Halothane	6.9 ± 0.2	7.0 ± 0.5	4.9 ± 0.1	3.7 ± 0.3
Control (n = 6)	15.0 ± 0.7	14.8 ± 0.3	14.3 ± 0.8	9.1 ± 0.3
Hydralazine	16.0 ± 1.5	15.2 ± 1.1	9.9 ± 1.0	10.8 ± 1.3

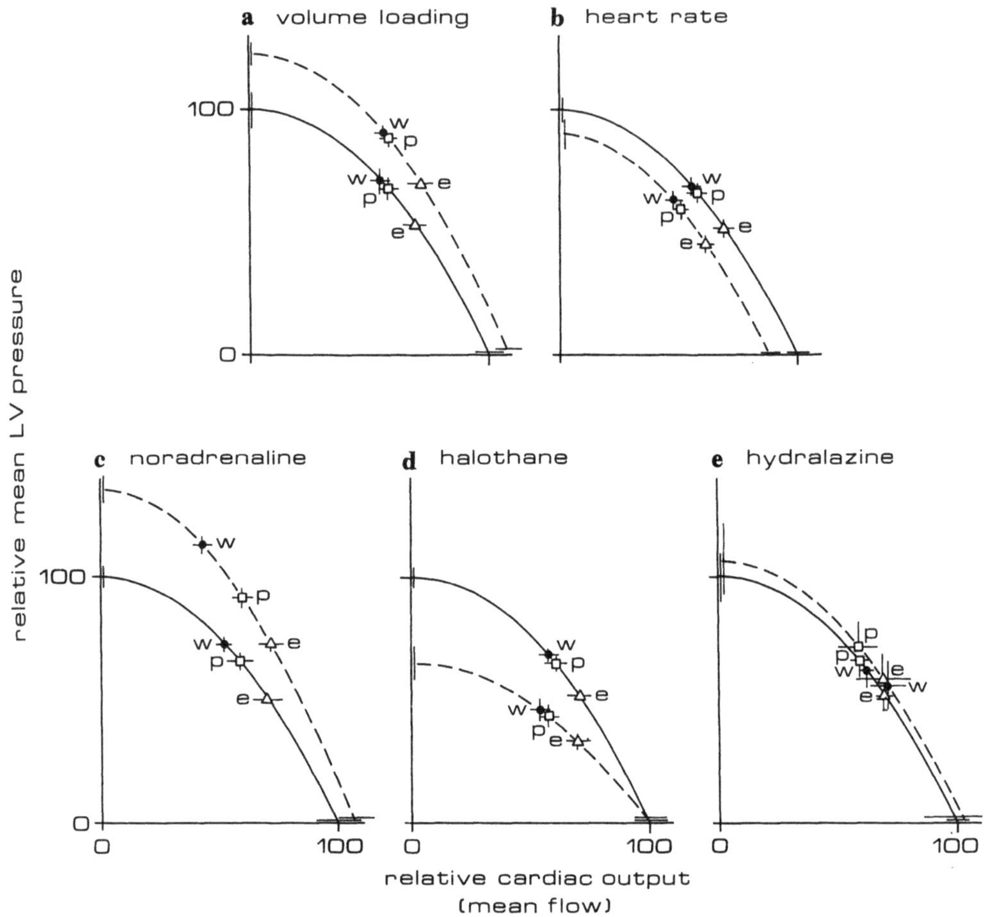

Fig.3. Pump function graphs in control (*solid lines*) and during interventions (*dashed lines*). For each graph, averaged F_{max} and P_{max} obtained from the experiments (number of animals is given in Table 1) allow for the calculation of maximum power and maximum efficiency (F_p, P_p) and (F_e, P_e) respectively. The working points are indicated by *w*, and the points of maximum power and efficiency by *p* and *e*, respectively. Each curve was normalized to the control curves ($P_{max} = 100$, $F_{max} = 100$)

Addition of halothane to the inhalation mixture selectively reduced P_{max} ($-35\% \pm 3\%$) and had no effect on F_{max}; i.e., halothane had a negative inotropic effect. As can be seen in Fig. 3d the working point remained at maximum power output during this intervention.

The vasodilator hydralazine had no significant effect on P_{max} or F_{max} indicating no effect on properties of the ventricle. Flow in the working point increased by $7\% \pm 15\%$, and the reduction of the peripheral resistance was 22%. Thus, matching was lost also in the case of vasodilation (Fig. 3e) and the working point was close to the point of maximum efficiency. It is unclear whether this result should be explained again in terms of disabled control mechanisms, or should be taken as evidence against the hypothesis of regulation of power output. At present we cannot give an answer because it is not known whether the effect of hydralazine was attenuated by regulatory mechanisms; i.e., the effect might have been greater if sympathetic, parasympathetic, and other control mechanisms had been blocked.

Burkhoff and Sagawa [14] advocated the hypothesis that the heart normally works at maximum efficiency, but there is neither theoretical nor experimental evidence in support of this hypothesis. So far, however, experimental studies concerned resting and anesthetized animals, and efficiency may turn out be more a more relevant parameter than power output when oxygen supply becomes a limiting factor, as during severe exercise.

The working point is at the intersection of the pump function graph and the pressure-flow relationship of the periphery. According to our results, the working point under control conditions coincides with maximum power output (Fig. 3) (matching). Any intervention or drug that affects contractility or peripheral resistance tends to separate the working point from maximum power output, and may thereby interfere with matching. Hence, knowledge of matching and of the regulatory mechanisms involved may advance optimal application of drugs.

References

1. Suga H, Sagawa K. Shoukas AA (1973) Load independence of the instantaneous pressure-volume ratio of the canine left ventricle and effects of epinephrine and heart rate on the ratio. Circ Res 32: 314-322
2. Sunagawa K, Maughan WL, Sagawa K (1985) Optimal arterial resistance for the maximal stroke work studied in isolated canine left ventricle. Circ Res 56: 586-595
3. Elzinga G, Westerhof N (1973) Pressure and flow generated by the left ventricle against different impedances. Circ Res 32: 178-186
4. Van den Horn GJ, Westerhof N, Elzinga G (1984) Interaction of heart and arterial system. Ann Biomed Eng 12: 151-162
5. Elzinga G, Westerhof N (1980) Pump function of the feline left heart: changes with heart rate and its bearing on the energy balance. Cardiovasc Res 14: 81-92
6. Toorop GP, Van der Horn GJ, Elzinga G, Westerhof N (1988) Matching between feline left ventricle and arterial load: optimal external power or efficiency. Am J Physiol 254: H279-285
7. Van den Horn GJ, Westerhof N, Elzinga G (1986) Feline left ventricle does not always operate at optimum power output. Am J Physiol 250: H961-H967
8. Piene H (1987) Matching between right ventricle and pulmonary bed. In: Yin FCP (ed): Ventricular/vascular coupling. Springer-Verlag, New York, pp 180-202
9. Weber KT, Janicki JS, Reeves RC, Hefner LL, Reeves TJ (1974) Determinants of stroke volume in the isolated canine heart. J Appl Physiol 37: 742-747
10. Chiu YC, Ballou EW, Ford LE (1987) Force, velocity, and power changes during normal and potentiated contractions of cat papillary muscle. Circ Res 60: 446-458
11. Daniels M, Noble MIM, ter Keurs HEDJ, Wohlfart B (1984) Velocity of sarcomere shortening in rat cardiac muscle: relationship to force, sarcomere length, calcium and time. J Physiol 355:367-381
12. Wilcken DEL, Charlier AA, Hoffman JIE, Guz A (1964) Effects of alterations in aortic impedance on the performance of the ventricles. Circ Res 14: 283-293
13. Myhre ESP, Johansen A, Bjornstad J, Piene H (1986) The effect of contractility and preload on matching between the canine left ventricle and afterload. Circulation 73: 161-171
14. Burkhoff D, Sagawa K (1986) Ventricular efficiency predicted by an analytical model. Am J Physiol 250: R1021-R1027

Ventricular Matching with the Arterial System in Chronically Instrumented Dogs

KENJI SUNAGAWA, MASARU SUGIMACHI, KOJI TODAKA, and MOTOOMI NAKAMURA[1]

Summary. We investigated how exercise (physiologic stress) and cardiac dysfunction (pathologic stress) influence the mechanical energy transmission from the left ventricle to the arterial system. With exercise, end-systolic elastance (E_{es}) increased by 46% and effective arterial elastance (E_a) increased by 35%, while the ratio of E_a to E_{es} remained fairly constant. The estimated ratio (Q_{mch}) of external work to its theoretically derived maximal value was 0.95 ± 0.06 at rest and 0.93 ± 0.10 during maximal exercise. Similarly the ratio (Q_{mtb}) of estimated minimum oxygen consumption of the heart to that actually measured to achieve a given level of cardiac output and arterial pressure was 0.97 ± 0.02 at rest and 0.98 ± 0.02 during exercise. With the creation of left ventricular dysfunction, Q_{mch} was still close to unity (0.94 ± 0.07) in spite of a significant increase in E_a/E_{es} (1.42 ± 0.58). Q_{mtb}, on the other hand, was significantly lowered (0.81 ± 0.09). We conclude that external work of the left ventricle of normal dogs is at a near-maximal level during exercise as well as at rest without compromising metabolic efficiency. In the left ventricle with dysfunction, external work is still close to optimal, whereas metabolic efficiency has significantly deteriorated.

Goal of Cardiovascular Regulation

The purpose of the cardiovascular system is to provide adequate blood perfusion to peripheral tissues, and thus support the normal function of these tissues. Since adequate cardiac output for peripheral perfusion can be achieved by multiple combinations of ventricular contractility and loading conditions, it is likely that the cardiovascular regulatory system chooses a unique combination of them according to a preprogrammed optimization criterion. Although the optimization criterion is not yet known, the fact that myocardium utilizes a tremendous amount of metabolic energy throughout its lifetime suggests that maximizing the energy transmission from the left ventricle to the arterial system, and/or minimizing the energy cost to achieve required cardiac work, would be of major concern. Indeed, the mean stroke power output of the normal excised feline left or right ventricle was found to be maximum when the ventricle was loaded with normal arterial impedance [1]. Also, in the cat right ventricle, pump efficiency, defined as the ratio of external work to total usable energy, was found to be maximum at normal pulmonary arterial impedance [2]. In the canine heart, external work was expected to be nearly

[1] Research Institute of Angiocardiology and Cardiovascular Clinic, Kyushu University School of Medicine, 3-1-1 Maidashi, Higashi-ku, Fukuoka, 812 Japan

maximal under the normal loading condition [3, 4]. Although all these studies indicated that external work or efficiency was optimized in anesthetized animals, it is not known whether these optimization principles hold under physiologic stress such as exercise or pathologic stress such as left heart dysfunction.

The purpose of this investigation, therefore, is to clarify the influence of exercise and left heart dysfunction on the energy transmission of ventricular contraction to the arterial load.

Maximum External Work and Minimum Oxygen Cost

We evaluated mechanical optimality of ventricular ejection in terms of the ratio of external work to its maximal, theoretically derived, value. Assuming a constant left ventricular pressure during ejection (i.e., isobaric contraction), the external work (EW) of the left ventricle for a given end-diastolic volume (V_{ed}) can be approximated as

$$EW = \frac{E_a E_{es}^2}{(E_{es} + E_a)^2} (V_{ed} - V_0)^2 \quad (1)$$

where E_{es} is end-systolic elastance, E_a effective arterial elastance, and V_0 end-systolic unstressed volume [3, 6]. E_a can be estimated as the ratio of end-systolic pressure to stroke volume [3–5]. From Eq. 1, maximal external work (EW_{max}) will be obtained when E_{es} equals E_a, i.e.,

$$EW_{max} = \frac{E_{es}}{4} (V_{ed} - V_0)^2 \quad (2)$$

We indexed the mechanical optimality (Q_{mch}) by taking the ratio of EW to EW_{max} as

$$Q_{mch} = \frac{EW}{EW_{max}} \quad (3)$$

Substituting Eqs. 1 and 2 into Eq. 3 yields

$$Q_{mch} = \frac{4\ E_a/E_{es}}{(1 + E_a/E_{es})^2} \quad (4)$$

Therefore, once we know E_a and E_{es}, we can evaluate the mechanical optimality of the ventricular ejection by Eq. 4.

To evaluate metabolic optimality of left ventricular ejection, we estimated the optimal combination of ventricular contractility and heart rate that minimizes oxygen consumption (Vo_{2min}) to support a given cardiac output and arterial pressure. We define the metabolic optimality index (Q_{mtb}) by the ratio of Vo_{2min} to the actual oxygen consumption (Vo_2) as

$$Q_{mtb} = \frac{Vo_{2\,min}}{Vo_2} \quad (5)$$

We used the end-systolic pressure volume area framework to estimate ventricular oxygen consumption under various conditions of contractility and loading [6, 7].

Effect of Exercise

We used six chronically instrumented dogs preconditioned to run on a treadmill. After complete recovery from the surgery, we imposed on the dogs exercise stress with the treadmill. The stress varied stepwise in random order from complete standstill to maximal load (7 km/h with a slope of 0.2). With increases in exercise load, both heart rate and cardiac output were nearly doubled, while arterial pressure increased a little. End-stystolic elastance (E_{es}) increased by 46% and effective arterial elastance (E_a) also increased by 35%. The ratio of E_a to E_{es} remained unaltered [8].

The estimated mechanical optimality index (Q_{mch}) was 0.95 ± 0.06 at rest and 0.93 ± 0.10 during exercise. The metabolic optimality index (Q_{mtb}) was 0.97 ± 0.03 at rest and 0.98 ± 0.02 during exercise. This is to say that external work of the left ventricle at rest was nearly maximal and remained maximal in spite of tremendous changes in cardiac contractility and hemodynamics. The left ventricle achieves maximal external work with minimal oxygen cost [8].

Effect of Left Ventricular Dysfunction

Using the same dogs, we created left ventricular dysfunction by intracoronary injections of 15 micron microspheres. We repetitively injected the microspheres until left atrial pressure increased up to 25 mmHg. We evaluated hemodynamics of these animals one or two days after the creation of left ventricular dysfunction. Cardiac output decreased by 20%, heart rate increased by 60%, and arterial pressure remained relatively constant. E_{es} was relatively unaltered, whereas E_a was markedly increased. The increase in E_a was attributable to an increase in arterial resistance and heart rate. Thus the ratio of E_a to E_{es} increased from 0.74 ± 0.14 to 1.42 ± 0.58.

The estimated mechanical optimality index (Q_{mch}) was 0.97 ± 0.03 in the control condition and 0.94 ± 0.07 after the creation of left ventricular dysfunction. The metabolic optimality index (Q_{mtb}) decreased significantly from 0.92 ± 0.05 to 0.81 ± 0.09. It seems that even in the left ventricle with dysfunction, external work was hardly impaired. Metabolic efficiency, however, was obviously compromised.

Physiologic Significance

We have demonstrated that external work of the left ventricle was nearly maximal during exercise as well as at rest in normal dogs. External work was maintained at a near-maximal level even in the presence of mild cardiac dysfunction. As we indicated, external work is a function of the ratio of E_a to E_{es}. The mechanical optimality index is the highest (i.e., unity) when E_a/E_{es} is unity. It remains higher than 0.9 as long as E_a/E_{es} is between 0.52 and 1.92. The actual E_a/E_{es} values under the stresses we imposed were within this range. Therefore, the afterloading conditions are always chosen to derive maximal external work for a given level of ventricular contractility during exercise. Even a mild degree of ventricular dysfunction will not alter this relation.

The mechanical optimality index (Q_{mch}) quantifies how efficiently mechanical energy of the left ventricle is transferred to the afterloading system. In other words, we indexed by Q_{mch} the efficiency of the afterload in terms of the mechanical energy transmission for a given level of ventricular contractility. The metabolic optimality index (Q_{mtb}), on the other hand, dealt with the overall efficiency of the ventricle to achieve a given level of cardiac output and arterial pressure. Since the amount of cardiac output and the level of arterial pressure are determined in response to peripheral demand, we are quantifying with Q_{mtb} the overall energy efficiency of the left ventricle to carry out its required tasks.

Obviously the ventricle can choose multiple combinations of contractility, heart rate, and end-diastolic volume for a given level of peripheral demand. A theoretical analysis indicated that for a given heart rate and arterial resistance, Q_{mtb} becomes maximal at a particular combination of ventricular contractility and end-diastolic volume. Q_{mtb}, however, increases monotonically with a decrease in heart rate and increase in end-diastolic volume. In the naturally operating left ventricle, the ventricular volume is relatively constant even during exercise. If we accept this approximation, one can choose a particular combination of heart rate and ventricular contractility which maximizes Q_{mtb} (and thus minimizes cardiac oxygen consumption) for a given level of peripheral demand.

In the normal dogs, Q_{mtb} was nearly maximal during exercise as well as at rest. Thus, the normal left ventricle is perfusing peripheral tissue with minimal oxygen cost. At the same time, the afterload is chosen to derive maximal external work from the left ventricle. In the dogs with left ventricular dysfunction, although external work is well-maintained for a given ventricular contractility, the oxygen cost is no longer minimal.

To summarize: we investigated how exercise and heart dysfunction influence mechanical and metabolic optimality of the left ventricular ejection. We demonstrated that in the normal heart ventricular ejection is close to optimal, mechanically and metabolically. In the heart with mild dysfunction, mechanical optimality is maintained, but metabolic optimality is compromised. Needless to say, however, whether the maximization of those indices is the preprogrammed criterion of the regulatory system remains to be seen.

References

1. Elzinga G, Piene H, Jong JP (1980) Left and right ventricular pump function and consequences of having two pumps in one heart. A study on the isolated cat heart. Circ Res 46: 564–574
2. Piene H, Sund T (1982) Does normal pulmonary impedance constitute the optimum load for the right ventricle? Am J Physiol 242: H154–H160
3. Sunagawa K, Maughan WL, Burkhoff D, Sagawa K (1983) Left ventricular interaction with arterial load studied in isolated canine ventricle. Am J Physiol 245: H773–H780
4. Sunagawa K, Maughan WL, Sagawa K (1985) Optimal arterial resistance for the maximal stroke work studied in isolated canine left ventricle. Circ Res 56: 586–595
5. Burkhoff D, Sagawa K (1986) Ventricular efficiency predicted by an analytical model. Am J Physiol 250 (Regulatory Integrative Comp Physiol 19): R1021–R1027
6. Suga H (1979) Total mechanical energy of a ventricle model and cardiac oxygen consumption. Am J Physiol 236 (Heart Circ Physiol 5): H498–H505
7. Burkhoff D, Yue DT, Oikawa RY, Franz MR, Schaefer J, Sagawa K (1985) Influence of contractility on ventricular oxygen consumption. Circulation 72: III-298
8. Hayashida K, Sunagawa K, Noma M, Sugimachi M, Ando H, Suyama A, Tomoike H, Taijimi T, Nose Y, Nakamura M (1986) Mechanical matching of the left ventricle with the arterial system in exercising dogs. Jpn Circ J 51: 182

Right Ventricular Function in Relation to Pulmonary Arterial Impedance

HROAR PIENE[1]

Summary. The mechanical function of the right ventricle (RV) and the input impedance of the pulmonary arterial (PA) tree define in combination the RV-PA hydraulic power production and dissipation. Mutual RV-PA adaptation may be discussed in the light of the efficiency of the hydraulic energy transfer. In pump-perfused rabbit lungs, characteristic PA impedance and pulsatile (= wasted oscillatory) power were minimal when PA pressure was normal, and this seemed to be associated with a slight degree of vasoconstriction. Stimulation of sympathetic nerves to intact cat lungs resulted in increased PA impedance; however, this effect diminished if RV flow was increased. The flow and pressure of an isolated supported and working cat heart RV preparation showed characteristic dependence of the PA load impedance. The produced hydraulic power exhibited a distinct maximum, indicating the existence of an optimum load impedance. Analysis of RV flow and pressure over broad ranges of load impedance indicated that the relation: pressure = $E(t, \text{volume}) \cdot (\text{volume} - V_d) + Pc$ (where V_d and Pc are constants, t is time, and $E(t, \text{volume})$ is time- and volume dependent cavity volume elastance), described RV pressure according to its volume (and flow). The above function may also be expressed by a three-dimensional surface with coordinates of pressure, volume and time, i.e., Pressure = F(Volume, t). Such RV descriptions were coupled to PA impedance in computer computations, which gave flows and pressures close to those observed. Also, analysis depending on such computations, together with the assumption of maximum efficiency in terms of external and produced RV hydraulic power, indicated that optimum efficiency is obtained when the PA impedance is close to normal.

Analysis of the RV-PA system indicated, roughly, that the end-systolic pressure-volume relation E_{max}, heart rate (HR), and PA peripheral resistance (Rp) should be related as $E_{max}/HR = Rp$ for optimum power output. In the cat, RV E_{max} was 1.2×10^4 dyne. cm^{-5}. Rp may be around 4×10^3 dyne. s cm^{-5} and HR 3 s^{-1}, which would balance the above relation and assure that power was around optimum. Although the evidence is only indirect, this and the other observations point to mutual RV-PA adaptation to achieve efficient power transfer from heart to vessel bed.

Introduction

Pulmonary arterial (PA) impedance in relation to right ventricular (RV) function defines how energy developed during RV wall contraction is utilized for propelling blood through the pulmonary vasculature, and PA-RV interaction is discussed here

[1] Department of Biomedical Engineering, Faculty of Medicine, University of Trondheim, Trondheim, Norway

primarily in this perspective. By RV contraction the hydraulic energy in the RV is raised sufficiently to overcome the pressure head in the PA system, leading to opening of the pulmonary valves, and to net forward blood motion in the pulmonary vessels. The hydraulic power in the pulmonary blood-stream is consumed through the perfusion of the lungs and the filling of the left side of the heart. If the RV were a non-oscillatory flow pump drawing blood from a zero-pressure reservoir, the demand on the RV for hydraulic power production would simply be (nonoscillatory) PA pressure multiplied by RV flow. However, due to its cyclic function the RV has to produce surplus power which is consumed by oscillatory blood motion in the pulmonary vessel system. In addition to that, a fraction of the potential hydraulic energy developed in the RV cavity during the systolic contraction is, according to the pressure-volume-area (PVA) concept of Suga and co-workers [1], not utilized for blood movement, but released as heat during ventricular relaxation.

This article will discuss indications that the RV and the PA system may be mutually adapted to keep such power "waste" at a low level. As implicit in the term "mutually" such adaptation seems to involve both the pulmonary bed and the RV, and their functions will therefore be discussed both separately and in connection.

Pulmonary Arterial Impedance

Opposition to oscillatory blood motion in the pulmonary bed depends on resistance, blood inertia, and wall compliance, and also on their distribution, which influences pressure and flow wave transmission and reflection. All these factors are combined in the impedance description of the PA bed, i.e., impedance is expressed by a relation involving oscillatory pressure and flow, and whatever influences their interrelation. Splitting the pressure and flow waves into their respective sets of harmonic (sinusoidal) oscillations, the impedance modulus is calculated as the ratio of pressure to flow oscillation amplitudes at each harmonic, and the impedance phase as pressure minus flow harmonic phase. An example of recordings of PA input impedance (in cat lungs) is given in Fig. 1 [2]. Impedance of the feline PA bed dropped from the zero frequency value (i.e., mean pressure divided by mean flow) to a fairly constant value above 3–4 Hz. The phase was negative below about 6 Hz and positive above. By impedance analysis one may examine the possible influence of adrenergic nerves to the lung: as seen in Fig. 1, maximal electric stimulation of the adrenergic nerves to the lung (*curve* NS) caused a significant increase of the impedance modulus from the first harmonic and upwards, i.e., the *characteristic* impedance was increased. Alpha and beta adrenoceptor blockade (*curves* AB and BB of Fig. 1) did not change impedance from control (*curve* C). By combining such data with analog models of the lung, one may deduce the probable influence of adrenergic or other vascular smooth muscle in terms of vascular resistance and compliance changes [3, 4].

The hydraulic power linked to the pressure and flow oscillations (oscillatory power, Wp) is, when a small kinetic term is omitted, given by the equation:

$$Wp = \frac{1}{2}\,\Sigma(F_f^2 \times Z_f \times \cos\phi_f) \qquad (1)$$

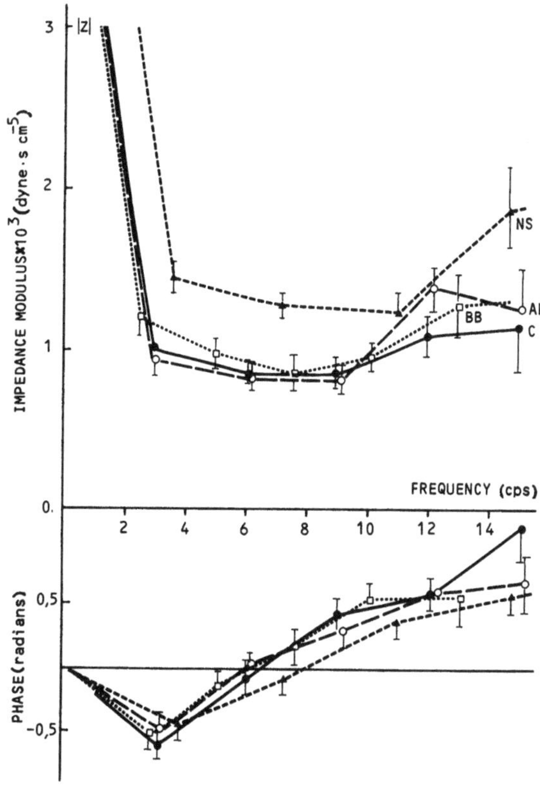

Fig. 1. Pulmonary arterial input impedance vs oscillation frequency of cat lungs. *C*, (●) control conditions; *NS*, (▲) sympathetic nerve stimulation; *AB*, (○) alpha receptor blockade; *BB*, (□) beta receptor blockade; *Z* , modulus of impedance. From [2]

where F_f is the flow amplitude of the (harmonic) oscillation of frequency f, and Z_f and ϕ_f are the impedance modulus and phase at f. To examine the influence of impedance on Wp it may be convenient to calculate the real part of the complex impedance, i.e., *real* impedance = $Z_{r,f} = Z_f \times \cos \phi_f$. According to Eq. 1 $Z_{r,f}$ relates directly to Wp.

Experiments in pump-perfused rabbit lungs [5], employing a pulsatile flow pump which was not influenced by alterations in lung vessel pressure and hence had constant F_f, demonstrated directly the effect of Wp by showing change in Z_f. By progressive vasoconstriction from a completely relaxed state Wp exhibited a minimum when the PA pressure had risen to 15–20 cm H_2O and rose both at lower and higher pressures (Fig. 2). The associated alterations in impedance are shown in Fig. 3. By slight vasoconstriction, the impedance modulus fell in the frequency range 2–8 Hz and the phase ϕ increased both in negative and positive values. Hence, Z_r decreased significantly in the low frequency range which carries most of the oscillatory power.

This was the first observation that indicated minimization of the oscillatory power losses (and impedance) at normal vascular pressure, which is about 20 cm H_2O in the pulmonary artery. Similar observations have subsequently been obtained in the

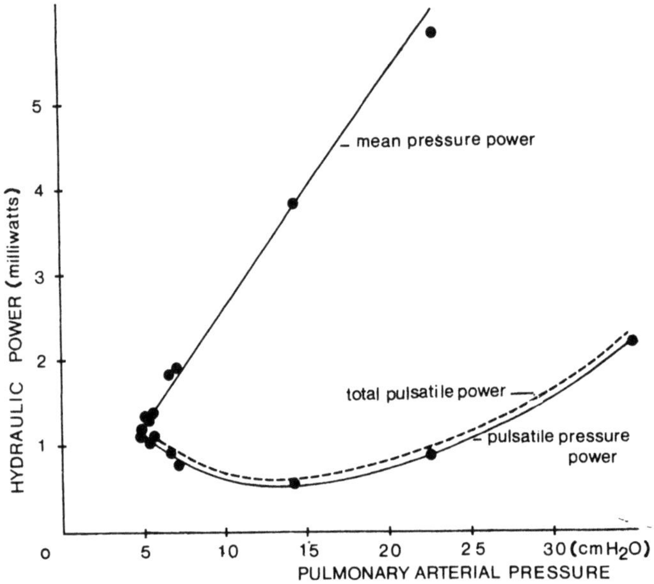

Fig. 2. Hydraulic power of isolated rabbit lung perfused by pulsatile blood pump with constant setting. Hydraulic (pressure) power is split into pulsatile (oscillatory) and mean power (mean pressure × mean flow). Total pulsatile power contains, in addition to the pressure power, a small kinetic power term linked to the velocity of oscillatory blood motion. Pulmonary arterial pressure was raised by adding a vasoconstrictor agent to the lung perfusate. Note minimum oscillatory power at normal arterial pressure. From [5]

systemic bed [6]. One possible explanation for the above observation was that, with an increase of the PA pressure from subnormal to normal levels, the main pulmonary arteries, which influence impedance most [5], were expanded and central arterial blood inertia and resistance were thereby reduced. This might reduce Wp even though the PA compliance was reduced as well. High compliance is traditionally considered to be the most important factor for keeping Wp at a low level. At higher PA pressures the continued reduction of compliance could increase Wp and Z_T. These assumptions were tested by calculating the 4 components of a lumped model of the pulmonary bed, i.e., central resistance and inertia (R', L), total compliance (C), and peripheral resistance (R") (Fig. 4). The calculations supported the notion that low Wp was associated with low values of the central R' and L while total C seemed less important. This was further supported by a more complete model simulation of the distributed pulmonary vascular system. The model study demonstrated that vascular expansion reduces Z_T and Wp even at reduced C [4].

PA pressure, i.e., the systolic load on the RV, is also influenced by the alveolar and left atrial pressures (Fig. 5). Expressing the load as the hydraulic power associated with pulmonary flow, experiments in the isolated rabbit lung showed that oscillatory (pulsatile) power was largely unaffected by the alveolar and left atrial pressures, although the power associated with mean pressure rose progressively.

In these isolated lung experiments, the flow pulse was essentially maintained constant in size and shape irrespective of impedance. This is not the case in vivo, where

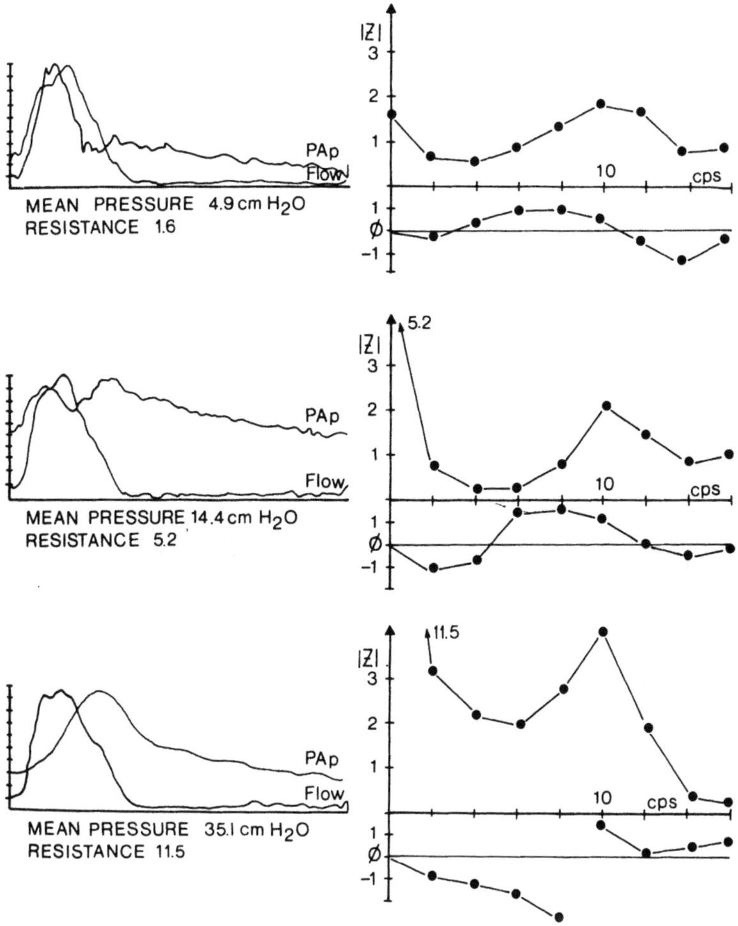

Fig. 3. Recordings of pressure and flow, and calculated impedance (modulus, Z, in 10^3 dyne.s. cm^{-5}; phase, ϕ, in radians), of same experiment as in Fig. 2, at three levels of pulmonary artery pressure (4.9, 14.4, and 35.1 cm H_2O). Note the depression of impedance modulus and the increase of impedance phase magnitude from low (*top panels*) to medium (*center panels*) pressure level. Both these changes contribute to low Z_r and oscillatory power

impedance influences the RV output, and vice versa, flow through the pulmonary bed influences the impedance by the effect of pressure on vessel distension. One could hypothesize that at high flow, it would be favourable if impedance were lowered in order to minimize Wp and thereby the necessary power output of the RV. In terms of mean power (mean pressure × mean flow) this is certainly the case since pulmonary resistance decreases with flow [2]. In an open-chest cat preparation [2] it was shown that Z_r was not influenced by flow (and the associated change in PA pressure) at control conditions; hence, Wp was related to the square of total flow (or cardiac output, CO) in line with Eq. 1. Consequently, the ratio Wp/CO which expresses oscillatory power waste per ml/min) blood flow increased almost linearly with CO (Fig. 6a). This situation changed, however, during adrenergic stimulation.

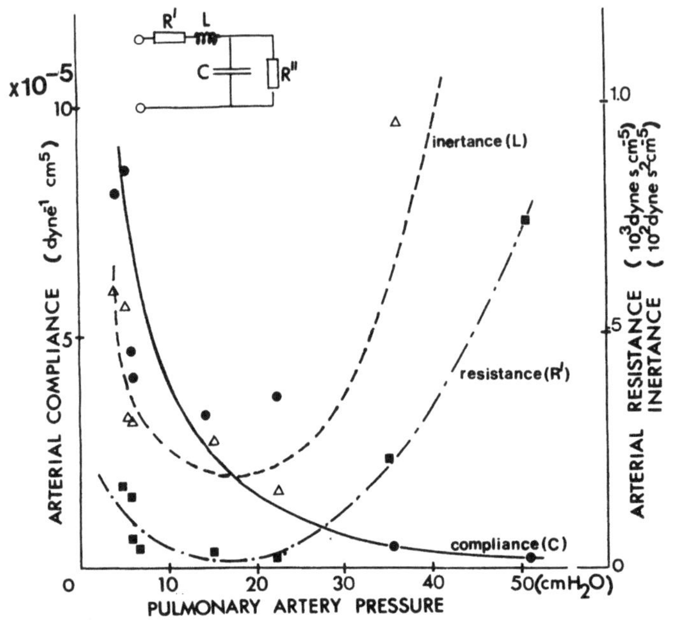

Fig. 4. Central resistance (*R*'), inertia (*L*), and compliance (*C*) of the pulmonary arterial bed during vasoconstriction at constant pulsatile flow, according to a lumped vascular model. (Experiment as in Figs. 2 and 3). Note coincidence of low oscillatory power (shown in Fig. 2) and low central resistance and inertance

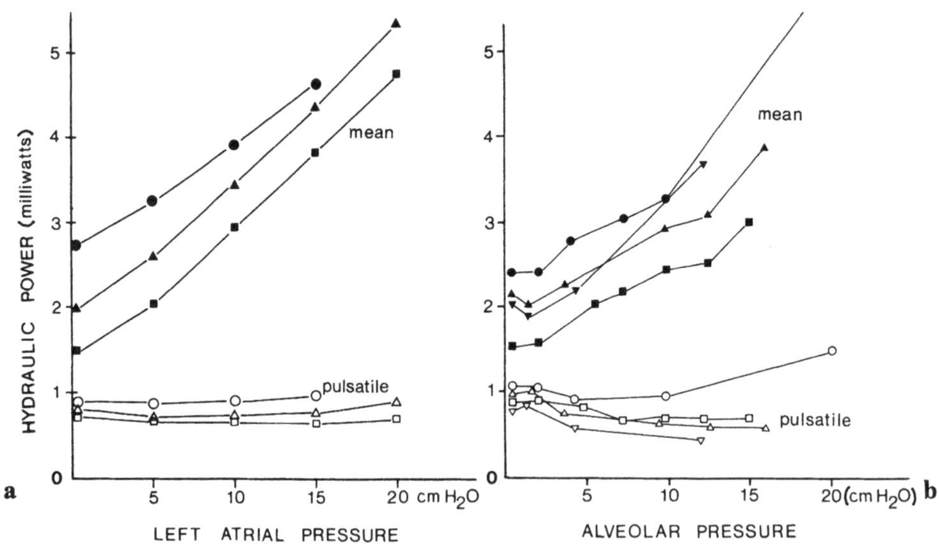

Fig. 5a, b Pulsatile (oscillatory) and mean hydraulic power vs left atrial pressure (**a**) and alveolar pressure (**b**) of isolated pump-perfused rabbit lugs. Note the approximately constant levels of oscillatory power

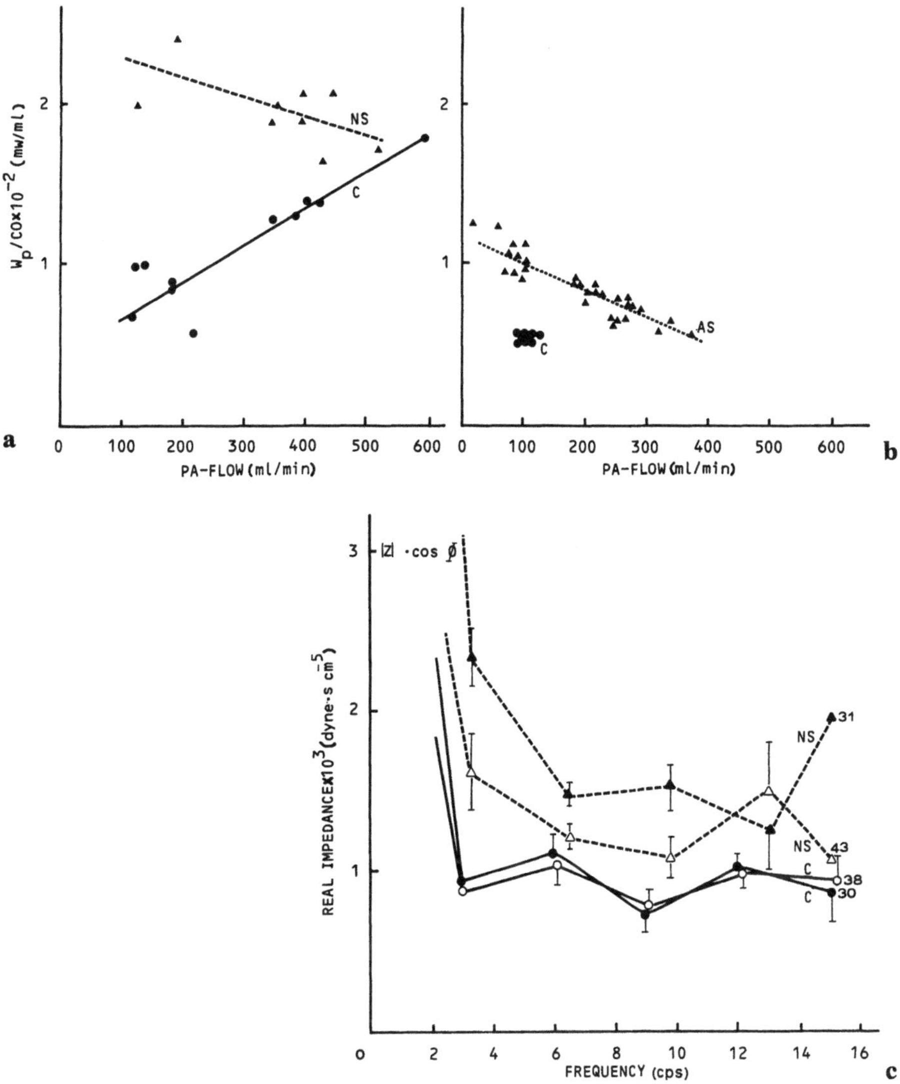

Fig. 6a, b Ratio of oscillatory power to cardiac output (*Wp/CO*) vs pulmonary arterial (*PA*) flow at control conditions (*C*) and during stimulation of sympathetic nerves to the lung (*NS*) (**a**) or administration of epinephrine (*AS*)(**b**). Note reduction of Wp/CO by PA flow in the stimulated conditions. **c.** Impedance(real part, *Z . cos* ϕ, see text) at low (*filled symbols*) and high (*open symbols*) flow levels at control (*C, circles*) and sympathetic nerve stimulation (*NS, triangles*). Note the depression of impedance by increasing flow during NS, resulting in reduction of Wp/CO, as demonstrated in Fig. 6a. *PAP*, pulmonary artery pressure [2]

(Fig. 6b) In this condition Wp/CO fell with increasing CO from an initially high level at low CO, to approach at high CO the Wp/CO level observed at control. The Wp/CO reduction with flow during adrenergic stimulation was associated with a progressive reduction of Z_T as flow (and PA pressure) increased (Fig. 6c). Both at low and high flow, sympathetic stimulation reduced PA compliance. Thus, in the "fight and flight reaction", characterized by high general sympathetic stimulation level and flow, the animal may make effective use of the lung as a blood reservoir by reducing PA compliance without extra load on the RV.

More recent experiments in intact dogs have shown that similar depression of the impedance level occurs during chronically increased pulmonary flow [7].

Influence of Impedance on RV Output

In the above studies the RV was primarily considered as a flow source, and the influences on impedance by variations of pressures (arterial, left atrial, alveolar), vessel tone, and flow were examined. To understand the mutual adaptations of the system more completely a series of experiments was performed where isolated perfused RV of cat hearts pumped into hydraulic models which simulated the pulmonary impedance [8, 9]. It was observed that RV load impedance had a major influence on the configuration of the pressure and flow pulses, and on the power produced by the RV. Fig. 7 shows how alterations in load compliance (C) and peripheral resistance (Rp) altered RV pressure, RV flow, and PA pressure [9]. The changes were qualitatively similar to observations of the left side of the heart [10]. Calculations of the hydraulic power output vs mean flow, which was altered by beat-to-beat variations in load pressure at constant filling, heart rate (2 Hz), and inotropy, demonstrated distinct optima (maxima) in external power output (Fig. 8). The external power vs flow curves were of parabolic shape and approximately symmetrical about a line through the power optimum and vertical to the flow axis. If maximum ventricular flow is assumed to occur at zero load pressure, then flow at the power optimum seemed to be about 50% of maximum flow.

Since RV pressure and flow changed in characteristic patterns with the load parameters it was hypothesized that PA impedance coupled to a description of the ventricle could predict these changes. A description of the RV which incoporated time, t, as an explicit variable, was however needed. It was examined whether RV pressure could be expressed as

$$\text{RV pressure} = \text{F}(t, \text{RV volume}, \text{RV flow}) \tag{2}$$

where F was an arbitrary functional relationship, and t the time after the onset of systole. Calculating RV volume of the isolated cat heart as an arbitrary (fixed) end-diastolie value minus the instantaneous volume reduction (obtained by integration of measured RV flow), the above relation was examined [11]. By manipulation of the load at constant inotropy and diastolic filling pressure, one could obtain a wide dispersion of pressure vs volume and flow during systole. In the first experiments we could not discover any direct relationship to flow, hence:

$$\text{RV pressure} = \text{F}(t, \text{RV volume}) \tag{3}$$

Fig. 7. Changes in RV pressure, PA flow, and PA pressure by changes in load compliance (*C*) and resistance (*Rp*). Isolated cat hearts pumped into a hydraulic model (with variable C and Rp) of the pulmonary bed. From [9]

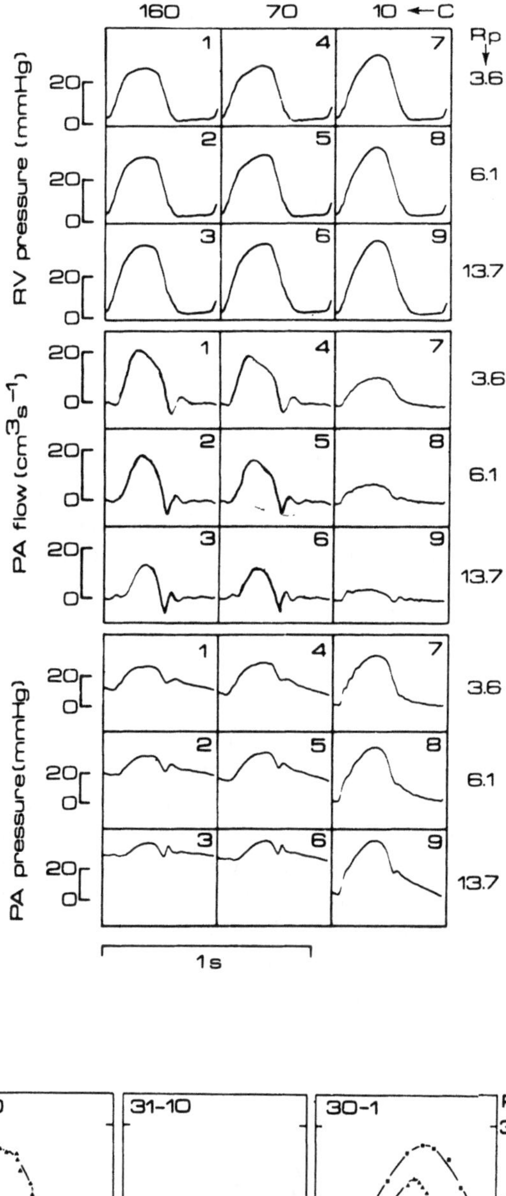

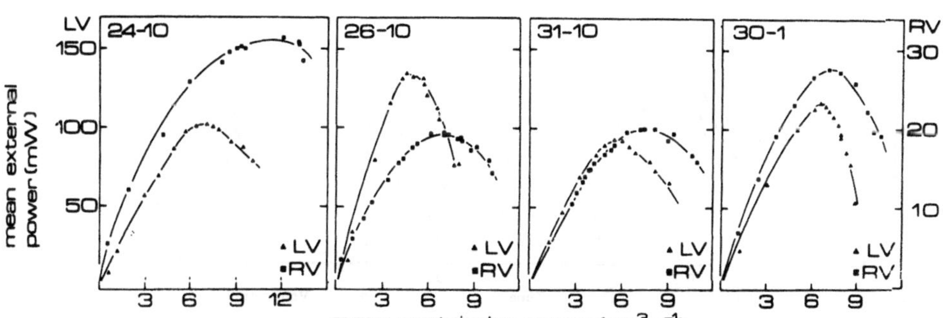

Fig. 8. External power output of the right and left ventricles (*RV, LV*) of isolated cat hearts vs ventricular output. From [9]

seemed sufficient as a first-order description of the assumed functional relationship. The function F was described as a set of curves describing RV volume vs t at constant RV pressure (isobar chart) [11], but could equally well be described as pressure vs t at constant volume, or pressure vs volume at constant t (isochronic chart) [12] (Fig. 9). From later experiments it was suggested that Eq. 2 could be written as

$$\text{RV pressure} = E(t, \text{volume}) \times (\text{volume} - V_d) + P_C \tag{4}$$

(where V_d and P_C are constants and $E(t, \text{volume})$ is cavity volume elastance), which is not very different from the expression suggested by Suga et al. [13] for the left ventricle (LV):

$$\text{LV pressure} = E(t) \times (\text{LV volume} - V_0) \tag{5}$$

except for the nonlinear relation between pressure and volume which is included in Eq. 4.; e.g., Eq. 5 does not include the possibility that maximum $E(t)$, E_{max}, may occur at different values of t depending on load, and hence, does not predict that the time of maximum pressure of isovolumic pressure curves may depend on the volume, as is often observed.

By assuming an alternative relation:

$$\text{RV pressure} = E(t) \times (\text{volume} - V_d) + R_v \times \text{pressure} \times \text{flow}, \tag{6}$$

where R_v represents pressure-dependent internal viscous losses [14], Sund and Piene [11] calculated R_v to be 1.2×10^{-4} s/ml in the isolated cat heart. That is, at maximum RV pressure and flow, which were about 20 mmHg and 20 ml/s, respectively, the contribution of the second term was less than 1 mm Hg and therefore almost negligible. Eq. 4 gave a significantly better fit to the experimental data than any other proposed method of describing ventricular pressure vs t, volume, and flow [12]. By employing a ventricular description according to Eq. 3, and a 4-component description of the arterial load (Rc, L, C and Rp) similar to the lumped model of Fig. 2, pressures and flow were calculated by a computer algorithm; the calculations included an estimation of the arterial load component values to obtain a similar

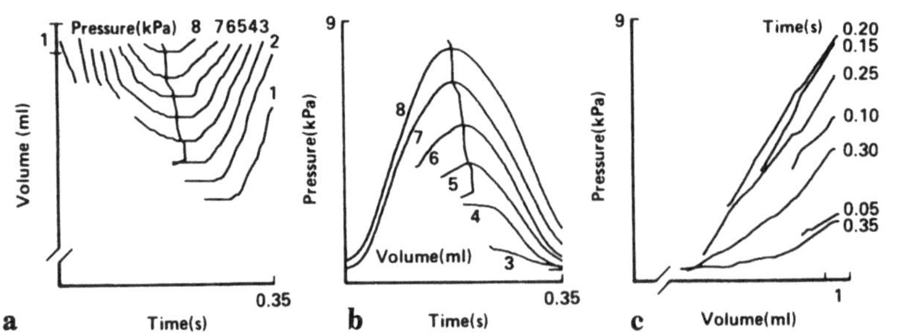

Fig. 9a-c. Relations among RV pressure, time after onset of systole, and RV volume (pressure-volume-time relation in three projections: **a** into the volume-time plane, **b** into the pressure-time plane, and **c** into the pressure-volume plane

impedance in the low-frequency range as that observed in experiments. An example of the resulting calculated pressure and flow curves is shown in Fig. 10, and compared to those observed at identical impedance [11]. Due to a rather crude analytical representation of Eq. 3 in these computations, the calculated pressures deviated somewhat from the observed, but the calculated flow pulses were strikingly similar to those obtained in the experiments, at least for the first half of the ejection period. As this was the first attempt to demonstrate the feasibility of obtaining pressures and flow from separate descriptions of ventricle and load, little emphasis was put on reproducing pressure and flow in high detail. Another computation technique, combining the pressure-volume-*t* (Eq. 3) representation of the ventricle and the impulse response (Fourier transform) representation of the input impedance, supported the notion that pulsatile pressure and flow can be accurately rederived by combining the independently described RV and PA systems [15].

Adaptation Among PA Impedance and RV

Experiments in isolated cat hearts suggested that the RV, at normal end-diastolic filling pressure (about 2 mmHg), heart rate (about 2 Hz), and inotropy (no artificial inotropic stimulation or depression), worked at peak external power output [8, 9]. Since oscillatory power normally is only around 20% of total external power, one may as a first approximation assume that total power, Wt, is given by:

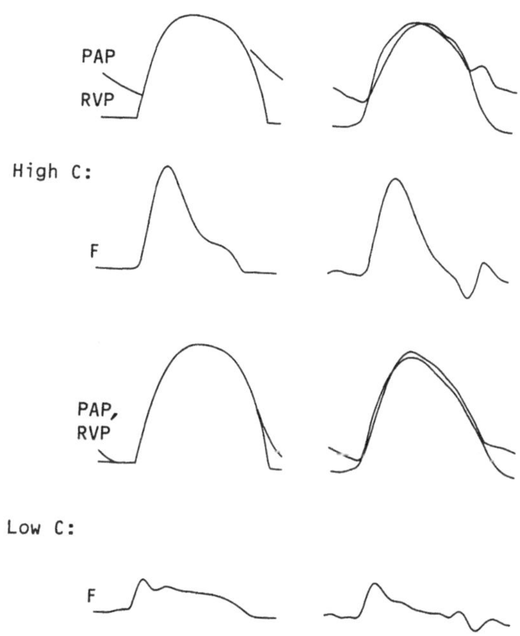

Fig. 10. Calculated (*left panels*) and measured (*right panels*) RV pressure (*RVP*), PA pressure (*PAP*), and flow (*F*) at high and low load compliance (*C*). Note the close similarity of the calculated and measured flow pulses, particularly during the first parts of ejection

$$Wt = SV \times HR \times PAP \tag{7}$$

where SV is stroke volume, PAP is PA pressure, and HR is heart rate. According to Piene [16] and others [17], maximum power is achieved when

$$E_{max} = P_{es}/SV = PAP/SV = Rp \times HR, \tag{8}$$

(P_{es}: end-systolic pressure), or written:

$$E_{max}/HR = Rp. \tag{9}$$

In Eq. 9 the left and right sides belong to the heart and load, respectively. It predicts how the ventricle and the vessel bed should be adapted to achieve maximum power output.

Looking at data pertinent to the cat heart-lung system, E_{max} in the isolated heart was around 12×10^3 dyne. cm^{-5} [12]. Assuming HR of 3 s^{-1}, Rp should be about 4×10^3 dyne. s.cm^{-5}, which accords well with the normal lung vascular resistance in the cat [2]. Looking at dogs, E_{max} for the RV of isolated hearts is, according to Maughan et al. [18], 2.9 mmHg/ml. At a physiologic HR of 1.5 s^{-1}, Rp should be at 2 mmHg. s/ml to obtain maximum power. This is approximately twice the value expected assuming a (physiologic) cardiac output of about 35 ml/s (2 1/min) at a mean PAP of 15–20 mmHg. Thus, the present data do not equivocally support the proposition that Eq. 9 is fulfilled, i.e., that RV normally pumps at maximum power in the dog. There is, however, a lack of simultaneously obtained data for the RV and the PA bed at physiologic conditions. One reason for this is the difficulty of measuring RV volume and E_{max} in more intact preparations.

Myocardial oxygen consumption has been connected with the pressure-volume-area (PVA) of a ventricle [1]. The PVA is the area in the P-V diagram confined within the end-diastolic and end-systolic pressure-volume relationships, and the pressure-volume curve described by the ventricular contraction from end-diastole to end-systole. PVA represents the static hydraulic energy build-up in the ventricle. Ventricular external (stroke) work is part of the PVA; the PVA is, however, larger by an amount not used for blood ejection, but assumed to be released as heat during relaxation. In this perspective, it would be of advantage to prolong the ejection phase to make use of this energy for blood movement. The ejection phase is influenced by the peripheral resistance, compliance, and the level of the characteristic impedance of the arterial load, among other factors. Measurements of pulmonary pressure and flow regularly show that closure of the pulmonary valves occurs well below peak RV pressure and is delayed in time relative to the closure of the aortic valves. Hence, it is possible that ventricular energy is more efficiently used for blood forward movement in the RV than in the LV.

To examine the effect of load parameters on the utilization of the PVA for external work, experimentally obtained pressure-volume-*t* descriptions of RV performance of isolated cat hearts according to Eq. 4 [12] were connected by a computer algorithm to a 4-component lumped model of the pulmonary arterial bed, i.e., a central resistance (Rc) simulating the characteristic impedance, an inductance (L) representing arterial blood inertia, compliance (C), and peripheral resistance (Rp). This model was similar to that of Fig. 2. The ratio Q = external work/PVA was used as indicator of pump efficiency [19]. According to these calculations Q was improved by increasing C, but exhibited little sensitivity to C above the physiologic level; Q also showed an optimum for Rc in the physiologic range (Fig. 11). Hence, the normal impedance graphs for the cat lung [2] were in good agreement with optimum load impedance according to these calculations (Fig. 11). In particular, the calculations indicated that the normal characteristic impedance was important for effective utilization of the ventricular energy represented by the PVA.

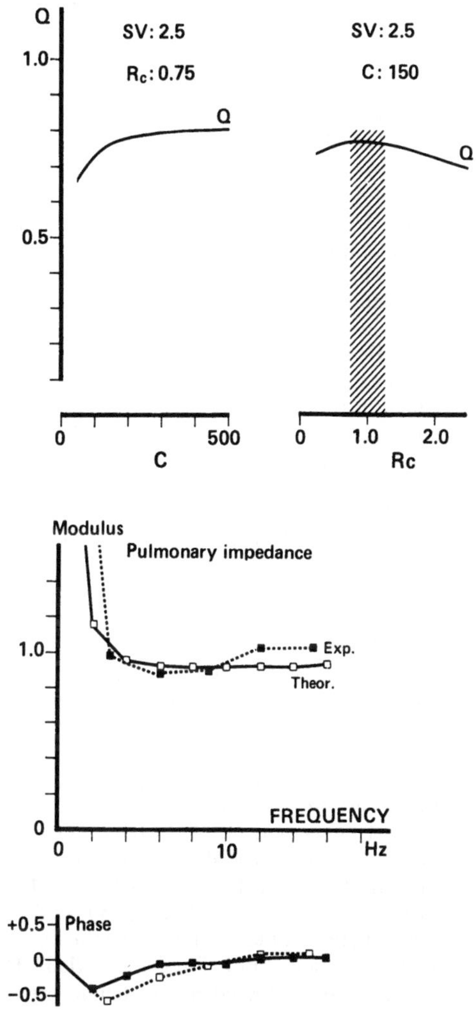

Fig. 11. Dependence of the ratio (*Q*) of external power and pressure-volume area with load compliance (*C*) and charcteristic resistance (*Rc*) (*top panels*). Note the relatively stable Q above C 150–200 (in 10^{-6} dyne^{-1} cm^5) and the Q optimum at Rc about 1.0 (in 10^3 dyne.s.cm^{-5}). The *lower panels* demonstrate impedance (modulus in 10^3 dyne.s.cm^{-5} and phase in radians) at optimum Q (*Theor* . □), and the experimentally obtained impedance of cat lungs (*Exp.* ■).[19]

Concluding Remarks

Attacking the RV-PA impedance functional relationships from different points of view led to results which indicated that there are mutual adaptation mechanisms which serve to optimize RV hydraulic power production. Their true physiologic meaning is, however, still obscure since the ventricular power output is only a fraction of the chemical energy expended by myocardial contraction. Also, how such adaptation is brought about is completely unknown at present.

References

1. Suga H, Hayashi T, Shirahata M, Suehiro S, Hisano R,(1981) Regression of cardiac consumption on ventricular pressure-volume area in dog. Am J Physiol 240: H320–H325
2. Piene H (1976) The influence of pulmonary blood flow rate on vascular input impedance and hydraulic power in the sympathetically and noradrenaline stimulated cat lung. Acta Physiol Scand 98: 44–53
3. Piene H (1976) Some physical properties of the pulmonary arterial bed deduced from pulsatile arterial flow and pressure. Acta Physiol Scand 98: 295–306
4. Piene H (1976) Influence of vessel distension and myogenic tone on pulmonary input impedance. A study using a computer model of rabbit lung. Acta Physiol Scand 98: 55–66
5. Piene H, Hauge A (1976) Reduction of pulsatile hydraulic power in the pulmonary circulation caused by moderate vasoconstriction. Cardiovasc Res 10: 503–513.
6. Dujardin JPL, Stone DN, Paul LT, Piper HP (1980) Response of systemic arterial input impedance to volume expansion and hemorrhage. Am J Physiol 238: H902–H908
7. Hopkins RA, Hammon JW, McHale PA, Smith PK, Anderson RW (1979) Pulmonary vascular impedance analysis of adaptation to chronically elevated blood flow in the awake dog. Circ Res 45: 267–274
8. Piene H, Sund T (1979) Flow and power output of right ventricle facing load with variable impedance. Am J Physiol 237: H125–H130
9. Elzinga G, Piene H, DeJong JP (1980) Left and right ventricular pump function and consequences of having two pumps in one heart. A study on the isolated cat heart. Circ Res 46: 564–579
10. Elzinga G, Westerhof N, (1973) Pressure and flow generated by the left ventricle against different impedances. Circ Res 32: 178–186
11. Piene H, Sund T (1980) Performance of the right ventricle: a pressure plane analysis. Cardiovasc Res 14: 217–222
12. Sund T, Piene H (1983) Right ventricular mechanics: a comparison of models. Cardiovasc Res 17: 320–330
13. Suga H, Sagawa K (1974) Instantaneous pressure-volume relationships and their ratio in the excised supported canine left ventricle. Circ Res 35: 117–126
14. Schroff SG, Janicki JS, Weber KT (1985) Evidence and quantification of left ventricular systolic resistance. Am J Physiol 249: H358–H370
15. Piene H (1980) Interaction between the right heart ventricle and its arterial load: a quantitative solution. Am J Physiol 238: H932–H937
16. Piene H (1984) Impedance matching between ventricle and load. Ann Biomed Eng 12: 191–207
17. Sunagawa K, Maughan WL, Sagawa K (1985) Optimal arterial resistance for the maximal stroke work studied in isolated canine left ventricle. Circ Res 56: 586–595
18. Maughan WL, Sunagawa K, Sagawa K (1987) Ventricular systolic interdependence: volume elastance in isolated canine hearts. Am J Physiol 253: H1381–H1390
19. Piene H, Sund T (1981) Does the pulmonary impedance constitute the optimum load for the right ventricle? Am J Physiol 242: H154–H160

Dynamics of Left Ventricular Filling

EDWARD L. YELLIN[1]

Summary. The laws of physics indicate that transmitral filling patterns are uniquely determined by the impedance of the mitral apparatus and the atrioventricular pressure difference. This paper reviews the results of the simultaneous measurement of left atrial and ventricular pressures and mitral flow in the experimental laboratory. The data support the theory: the pressure difference is due to accelerative and dissipative forces. It is now important to determine how disease and stress acutely and chronically affect the properties of both the ventricle and the atrium. With this knowledge, we will be better able to relate the non-invasive measurement of transmitral filling patterns to the active and passive properties of both chambers of the left heart.

Introduction

The last decade has witnessed a growing interest among cardiac physiologists in left ventricular filling dynamics and diastolic function (for a thorough review of relaxation and diastolic function see [1] and p 103, this volume). This current interest is due in part to the growing recognition that diastolic dysfunction frequently precedes systolic dysfunction and to the observation that cardiac failure may be present despite apparently normal indices of systolic performance. It has thus become clear that inadequate venous return and/or inability of the ventricle to fill adequately can lead to pump failure. Partial credit for the current focus on diastole, and for these important observations, must be given to the rapid development of non-invasive technology, in particular to echocardiography and the ability to combine ultrasonic imaging with pulsed-Doppler measurement of transmitral flow [2]. The interpretation and understanding of data derived by these non-invasive methods has been helped considerably by data derived from studies in the animal laboratory, and this is the focus of this paper.

We will review the highly invasive but highly accurate canine studies from our laboratory and show how these investigations have led to the development of a conceptual framework in which to analyze transmitral flow patterns and their relation to the active and passive properties of both the ventricle and atrium. We will also present some new and interesting unpublished data comparing Doppler and elec-

[1] Department of Cardiothoracic Surgery and Department of Physiology and Biophysics, Albert Einstein College of Medicine, Bronx, NY 10461, USA

tromagnetic measurement of transmitral flow. (The focus on reports from our laboratory is not meant to imply that others have not also made important contributions; it is meant only to improve the didactic value of this brief review.)

Rather than describing experimental methods (the details can be found in the literature) we will instead concentrate on discussing the results and applying them to the interpretation of currently used Doppler-derived indices of diastolic function. We will also focus on the raw oscillographic records, i.e., on original "cath-lab" type data, rather than on compiled results. Our intent is to clarify the physiology and the dynamic character of ventricular filling by describing and analyzing the phasic patterns of pressure and flow.

Discussion of Experimental Results

Determination of the Transmitral Pressure-Flow Relation

Fig. 1 shows a schematic of the instrumented dog heart in the anesthetized, acute, open-chest preparation that we used in our early studies [3]. With this approach we were able to measure the phasic patterns of transmitral flow; the atrioventricular pressure difference that provided the motive force for the flow; and the motion of the mitral valve [4–7]. More recently, we have succeeded in chronically implanting the mitral flow probe along with atrial and ventricular micromanometers, and in studying diastolic function in the conscious dog [8]. It is particularly important to note that the electromagnetic flow probe measures volume flow rate (ml/s) and is

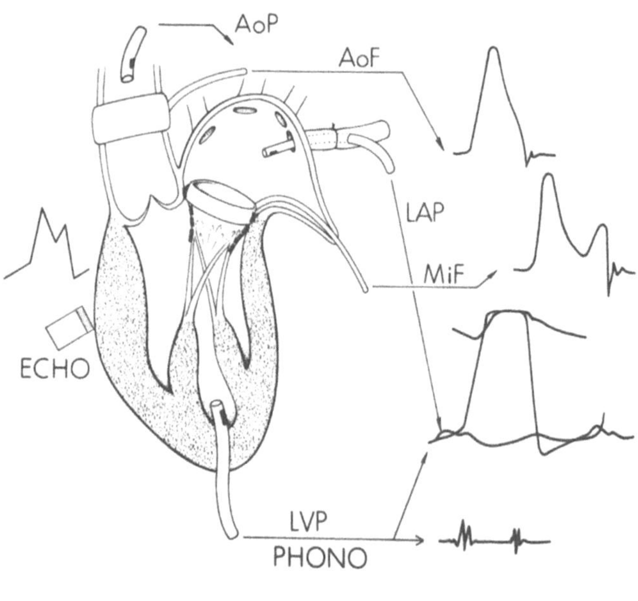

Fig. 1. Schematic of the instrumented canine heart and recorded waveforms. *AoP*, aortic pressure; *AoF*, aortic flow; *LAP*, left atrial pressure; *MiF*, mitral flow; *LVP*, left ventricular pressure. Redrawn from [3] with kind permission of authors and publisher

independent of changes in orifice size. It is therefore a true indicator of LV volume change. The electromagnetic flow patterns are comparable to those obtained by pulsed-Doppler [9, 10].

Fig. 2 is an oscillographic record from a chronically instrumented conscious dog that illustrates all the pertinent relationships; and Fig. 3 defines the measured parameters that are related to diastolic function. The quietly resting dog tends to have large respiratory variations in sinus rhythm, hence the short RR interval in cycle 2, Fig. 2, and in cycle 1, Fig. 3. These figures thus provide a useful illustration of the effect of variations in diastolic filling time. We now direct our attention to the temporal events that describe the dynamic determinants of ventricular filling and diastolic function.

Left ventricular pressure (LVP) rapidly falls during deactivation and at the moment of crossover with the left atrial pressure (PCO), the mitral valve opens and flow starts. The initial slow upward deflection in mitral flow is due to motion of the closed leaflets toward the ventricle; no forward flow or volume change has occurred—there is only a shape change. LVP continues to fall, but at a decreased rate because deactivation has slowed and because filling relengthens the myocardial structures, thereby increasing wall stress. LVP_{min} is reached at a value determined by the interaction of active decay of force and passive stretch and is, therefore, a function of rate and duration of relaxation, and of end systolic volume (ESV determines where on its passive pressure-volume (P-V) curve filling starts). At the same

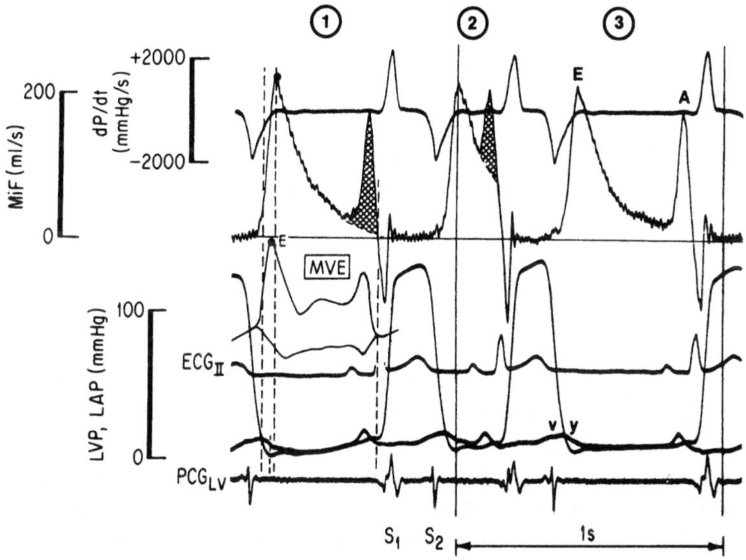

Fig. 2. Oscillographic record from a conscious chronically instrumented dog. A mitral valve echo (*MVE*) has been superimposed. Because of sinus respiratory variations in rhythm, the heart rate in cycle 2 is faster than in cycles 1 and 3 (127 compared to 86 and 82/min). *S1*, *S2*, first and second heart sounds; PCG_{LV}, intracardiac phonogram; *MiF*, mitral flow; *LVP*, left ventricular pressure; *LAP*, left atrial pressure; *V*, increase in *LAP* due to atrial filling; Y, decline in *LAP* due to atrial emptying. The *dashed vertical lines* in cycle 1 denote onset of filling peak AV pressure difference, peak flow rate, and end of filling. Redrawn from [22] with kind permission of authors and publisher

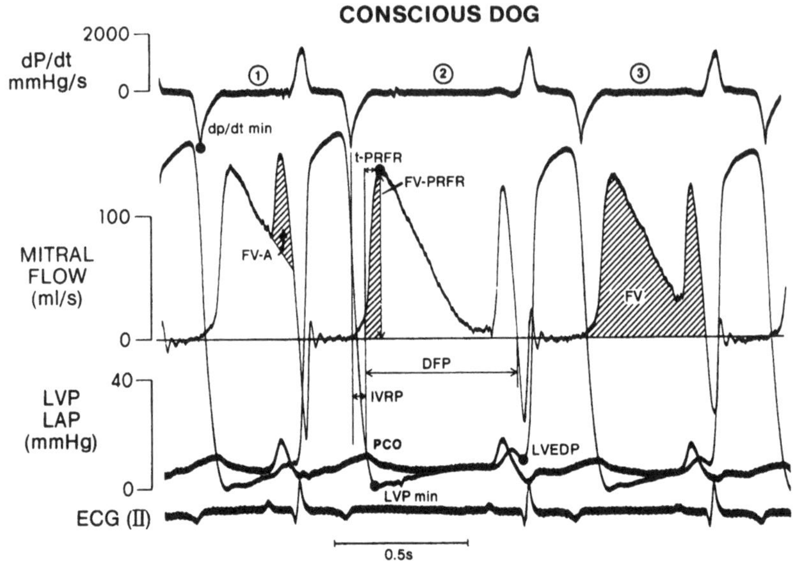

Fig. 3. Oscillographic record from a conscious dog, defining the indices of diastolic function. *FV*, filling volume; *FV-A*, filling volume due to the atrial contraction; *PRFR*, peak rapid filling rate; *t-PRFR*, time from onset of flow to PRFR; *IVRP*, isovolumic relaxation period; *DFP*, diastolic filling period; *PCO*, atrioventricular pressure crossover; LVP_{min}, minimum *LVP; LVEDP*, left ventricular end diastolic pressure. Unpublished observation, courtesy of Drs. T. Tamura and K. Tamura

time, left atrial pressure falls (Y descent) because it empties faster than it fills via the pulmonary veins. In fact, pulmonary venous flow is quite large in early diastole so that the atrium acts as both a reservoir and conduit during early ventricular filling [11].

The small (c. 3-5 mmHg) early atrioventricular (AV) pressure difference provides the motive force for the rapid acceleration of mitral flow to its peak (E-point, or peak rapid filling rate, PRFR). Flow then rapidly decelerates, and despite the fact that the pressure gradient soon becomes zero, sometimes even reversing slightly (Fig. 4) [9, 12], forward mitral flow continues because of inertia. The kinetic energy in the blood must, however, be dissipated because the ventricular chamber is unable to convert kinetic energy into pressure energy. The AV pressure difference may therefore be presumed to be related to acceleration and resistance, as discussed later. The rapid deceleration of mitral flow due to AV pressure reversal is illustrated in Fig. 4. The significance of small differences in AV pressures should not be underestimated; they may lead to large changes in transmitral flow patterns! The importance of fluid inertia must also be stressed; it explains the temporal delay (c. 30 ms) between the atrioventricular pressure crossover at end diastole and the cessation of inflow. the closure of the valve, and the onset of the first heart sound S1 (Figs. 2,3,5,6) [4, 6, 13].

The atrial contraction then reaccelerates mitral flow to a second peak (A-point) from which flow decelerates and pressure falls as the atrium empties and relaxes.

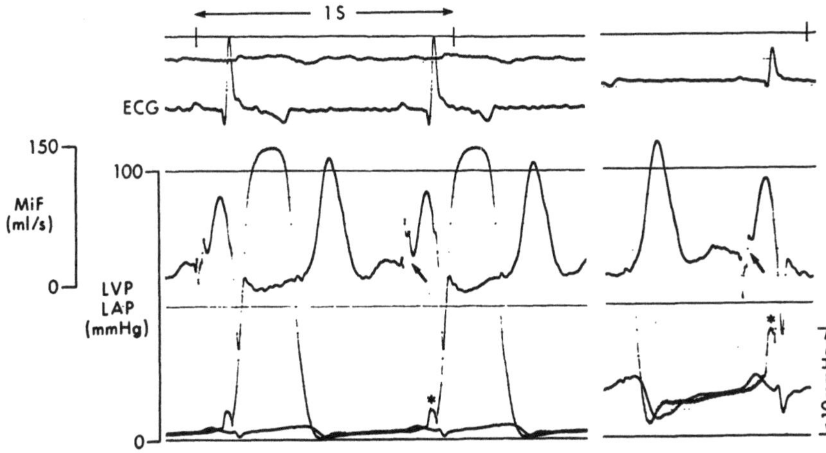

Fig. 4. Oscillographic record from a chronically instrumented dog, illustrating the response of mitral flow (*MiF*) to small changes in atrioventricular (*AV*) pressure difference. The high-gain pressure traces in the *right panel* accentuate the relationship. The "rebound" in flow following the rapid deceleration and preceding the atrial contraction is frequently seen in the electromagnetic measurement of flow. We suspect it is also present in subjects who have rapid decelerations, but is masked by signal filtering. The * denotes an artifact in the pressure trace; the *arrow* indicates electrical interference between the P-wave of the ECG and the flow signal. *LVP*, left ventricular pressure: *LAP*, left atrial pressure. Unpublished observation courtesy of Mr. O. Pajaro

Flow reaches zero and the mitral valve closes competently either as a result of atrial pressure fall or ventricular contraction [6]. Note that the large, short duration, negative deflection in the mitral flow trace is due to the relative motion between the rapidly closing valve and the flow probe; it does not reflect retrograde flow [14]. Note also that this closing "spike" and its subsequent oscillation about zero is synchronous with S1 (Figs. 2, 5, 6). These data, along with simultaneous recording of mitral valve motion (Figs. 2, 5), have led us to conclude that the first heart sound originates in the vibration of the closed valve and its structures following the impulsive force of ventricular contraction [4]. The *shaded areas* in Fig. 2 and in Fig. 3, cycle 1, define the contribution to filling of the atrial contraction as: the filling volume during the atrial contraction that would not have entered the ventricle if the atrium had failed to contract (assuming that the ventricular contraction occurred at the normal RR interval). With this definition, we have determined that the atrium contributes approximately 15% of the total filling volume in the resting consious dog, and that it does not vary with respiratory variations in heart rate [15]. We hesitate to extend this number (i.e., 15%) to humans because the PR interval and duration and strength of atrial contraction are not similar in dogs and humans. But we do think that the lack of variation with heart rate is applicable. We also think that it is inappropriate to discuss the E/A relation at this time.

Conceptual formulation. The simultaneous measurement of phasic mitral flow along with left atrial and ventricular pressures has revealed the sensitivity of flow to small pressure differences; the importance of inertia; and the role of dissipative energy losses. The transmitral pressure-flow relations can thus be described mathematically [16-18]:

$$\Delta P = (L)dQ/dt + (R)Q^2$$

where ΔP is the atrioventricular pressure difference, Q is the volume flow rate, L is an inertial coefficient inversely related to the annulus area, and R is a resistive (i.e., dissipative) coefficient inversely related to the square of the area. The nonlinear combination of L and R determines the impedance to flow; it is clear that small decreases in flow area lead to large increases in impedance. The formulation of an equation of fluid motion not only helps us understand the dynamic relations between pressure and flow; it also provides a basis for the development of cardiac models [8, 17, 18].

Relation Between Mitral Flow and Valve Motion

An interesting oscillographic record that reveals the relation between transmitral flow and valve motion, and that also provides additional insight into many of the points discussed above, is shown in Fig. 5. The record shows the pressures and flow with a simultaneous mitral valve echo during a transient episode of complete heart block at a sinus rate of 100/min ("a" waves) and an idioventricular rate of 64/min. The *vertical broken lines* in cycles 1 and 3 enclose the onset and end of mitral flow; they also reveal that the mitral valve opens and closes at these times. (The *solid vertical lines* are one-second time lines.) Note that the valve closes c. 40 ms after the

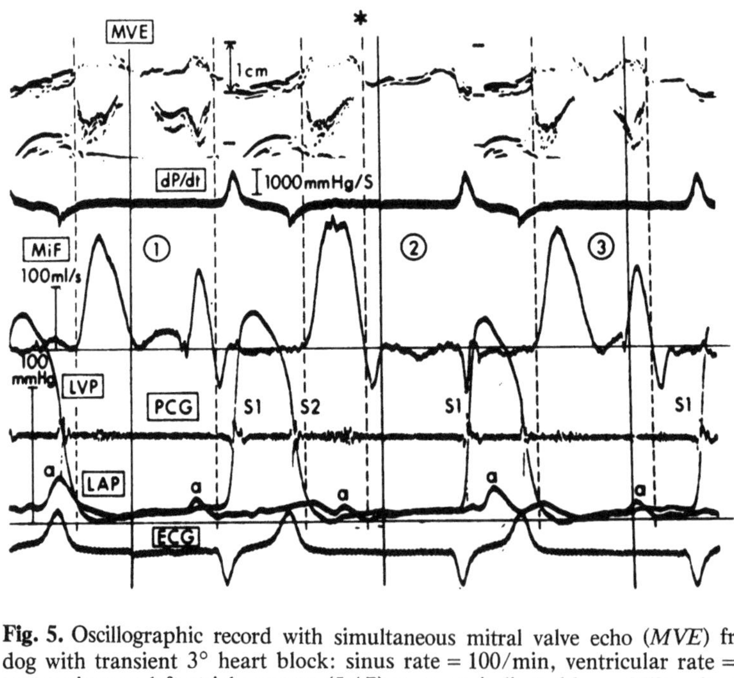

Fig. 5. Oscillographic record with simultaneous mitral valve echo (*MVE*) from a conscious dog with transient 3° heart block: sinus rate = 100/min, ventricular rate = 64/min. Atrial contractions on left atrial pressure (*LAP*) trace are indicated by *a*. *MiF*, mitral flow; *LVP*, left ventricular pressure; *PCG*, phonocardiogram; *dP/dt*, rate of change in pressure with time; *S1*, *S2*, first and second heart sounds. Unpublished observation courtesy of Dr. J.S. Meisner

atrial pressure falls below the ventricular, and that peak amplitude of S1 occurs still later, during rapid ventricular pressure development. These results support our conclusions that: the negative deflection in the flow trace is indeed an "artifact" and does not indicate regurgitation; the mitral valve can close competently in the absence of an "appropriately" timed atrial contraction; atrial relaxation can decelerate flow and close the valve; and the first heart sound is due to vibration of the closed and tensed valve.

Cycle 2 in Fig. 5 clearly illustrates the ability of an atrial relaxation to create a negative AV pressure difference, thereby decelerating flow and closing the valve without a measurable S1 (*broken vertical line* denoted by *). Cycles 2 and 3 are also illustrations of mitral valve locking [19]: the valve closed and remained closed for the duration of diastole. The large reversed (i.e., negative) AV gradient in cycle 2 provided the force to decelerate flow more rapidly than the small, barely discernable gradient in cycles 1 and 3. In all three cycles the mitral valve closed, but in cycles 1 and 3, in contrast to cycle 2, pulmonary venous inflow to the atrium raised the left atrial pressure (LAP), and the rebound in flow reopened the valve prior to the atrial contraction.

The temporal relations between the mitral valve echo and mitral flow in Fig. 2 are based on the results of studies designed to determine the interaction between diastolic inflow and the mitral apparatus [6]. The data consistently show that the mitral valve reaches its maximum excursion (E-point on the mitral valve echo, MVE) and starts its mid-diastolic closing motion approximately 25 ms before peak flow (E-point on the mitral flow, MiF); that is, while flow is still accelerating, and before large circulating vortices have had a chance to form. These and other data have led us to conclude that the mitral valve is always under chordal constraint. During mitral flow acceleration the valve overshoots its equilibrium position and is brought toward closure by the chordae. The Venturi effect due to the configuration of the mitral apparatus serves to increase the local velocity of the blood and to lower the pressure within the leaflets; large circulating vortices can then form and participate in the final closure of the valve [6].

It is important to understand that the observations and conclusions described above are based on data from normal dog hearts. Changes in the morphology of the mitral apparatus, for example, may lead to mitral regurgitation rather than to complete valve closure, when there is a sudden reversal of the AV pressure difference, due, for example, to a premature atrial or ventricular contraction. Of importance to this presentation, however, is the recognition that the equation describing the pressure-flow relations remains valid; only the relative values of the impedance parameters will change in pathology. For example, in mitral stenosis the value of R becomes so large that L is physiologically insignificant, and the equation reduces to the Gorlin equation.

Determinants of Early Diastolic Filling

From the physical point of view, it is clear that diastolic inflow is determined solely by the atrioventricular pressure difference and the impedance of the mitral apparatus. A discussion of the determinants of filling must, therefore, concentrate on those *factors that determine the time variation of atrial and ventricular pressures.* We emphasize this point because we think it will lead to a more meaningful understanding of physiological mechanisms and interpretation of non-invasive data. Increased heart rate, for example, will certainly change filling patterns, particularly if the increase is due to exercise. But causality, the answer to *why* filling changes and why it changes differently among patients, can only be found by examining the effect of heart rate on: fall in LVP (a function of deactivation rate; related in turn to inotropic state); LVP during early filling (a function of elastic recoil and passive compliance; therefore also of end systolic volume); and on left atrial pressure (determined by LA compliance and venous return). Because chamber pressures depend on chamber properties, we must also determine if increases in heart rate make inappropriate metabolic demands that change myocardial properties. Clearly, the process is complex and the number of questions limitless. We now turn our attention to one study that illustrates the importance of examining left atrial pressure as well as left ventricular relaxation rate.

Effects of Increased Venous Return on LV Filling

In a recent study we changed loading conditions in the conscious chronically instrumented dog to examine the influence of both relaxation rate and filling pressure on early transmitral flow [8]. The importance of left atrial filling pressure as a determinant of flow is illustrated in Fig. 6. Volume infusion increased venous return and raised mean left atrial pressure; but more importantly for early filling, it increased the atrioventricular pressure difference at the time of peak flow. Although this is the pressure difference that uniquely determines the PRFR, for convenience of measurement we used the LAP at AV crossover (PCO) as an index of pressure difference. The time constant (T) of an assumed exponential decay of pressure also increased with volume loading. Central venous pressure apparently led to a reflex increase in sympathetic tone (note the increased heart rate), that maintained the end systolic volume and a low LVP_{min}. The increased atrioventricular pressure gradient produced an increased peak rapid filling rate, *despite the measured increase in T.* Although left atrial pressure and relaxation rate each separately had a statistically significant correlation with PRFR under a variety of loading conditions, multiple correlation analysis of T and PCO with PRFR reduced the variance and confirmed the combined importance of both parameters. This important conclusion from the experimental laboratory [8, 10] has now been confirmed in patient studies [20]. *Thus, a normal early filling pattern need not imply normal diastolic function, and conversely, depressed early filling may not imply diastolic dysfunction.*

Comparison of Flow Waveforms

The flow patterns of Figs. 2-6 all show a rapid acceleration to peak flow but differ in the deceleration phase. Figs. 4 and 5 illustrate the rapid deceleration toward zero that is frequently observed clinically by pulsed-Doppler measurements. Figs. 2, 3, and 6 show slower deceleration patterns more indicative of disease or aging, i.e., patterns usually attributed to impaired diastolic function. Figs. 2 and 6, on the other

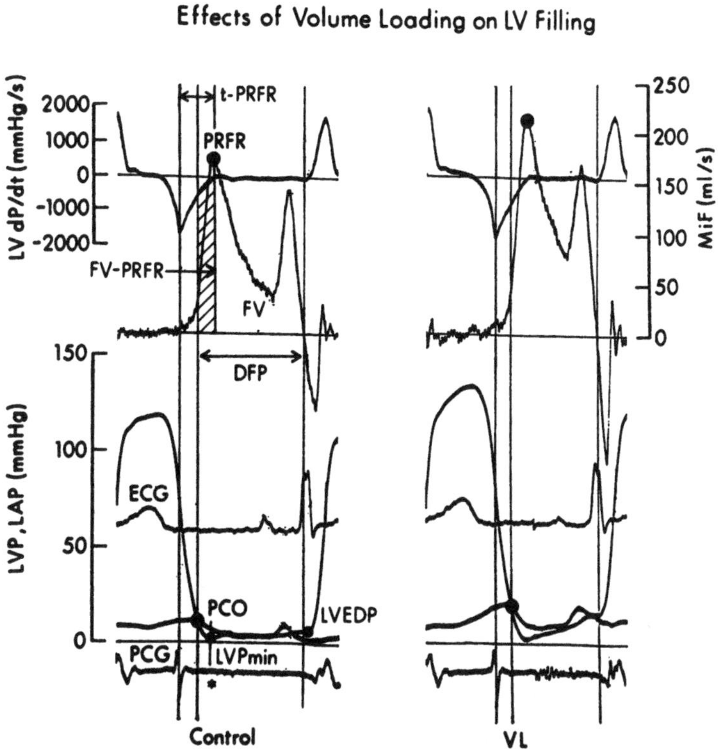

Fig. 6. Oscillographic record from a chronically instrumented dog showing the effects of increasing venous return by volume loading. Note the increased atrioventricular (*AV*) pressure difference and increased E-wave after volume (*VL*) (*right panel*). (In contrast to Fig. 2, t-PRFR in this record includes the IVRP.) *LAP*, left atrial pressure; *LVP*, left ventricular pressure; *MiF*, mitral flow; *PCG*, phonocardiogram; *PCO*, atrioventricular pressure crossover; *LVEDP*, left ventricular end-diastolic pressure; *PRFR*, peak rapid filling rate; *t-PRFR*, time from onset of flow to PRFR; *FV*, filling volume. Reprinted from [8] with kind permission of the authors and the American Heart Association

hand, have rapid early decelerations followed by a slower phase; only Fig. 3 has a sustained slow deceleration, perhaps due to a "functional" mild stenosis (note the relatively large AV pressure difference). A direct comparison of these waveforms with those obtained from patients is made difficult by the limitations of pulsed-Doppler measurements in the clinical setting: the use of filters that attenuate the low frequency signals; the "fuzziness" of the recordings; the movement of the sample volume and beam angle relative to the heart during a single diastole; and the phasic change in flow area. Some recent studies from the experimental laboratory that report simultaneous measurement of pulsed-Doppler flow and intracardiac pressures indicate that progress is being made in the ongoing effort to use invasive experimental methods for the understanding of clinical data obtained non-invasively [9, 10, 21]. We now present data from our laboratory comparing mitral flow (electromagnetic) with mitral velocity (pulsed Doppler) and with mitral valve motion (M-mode).

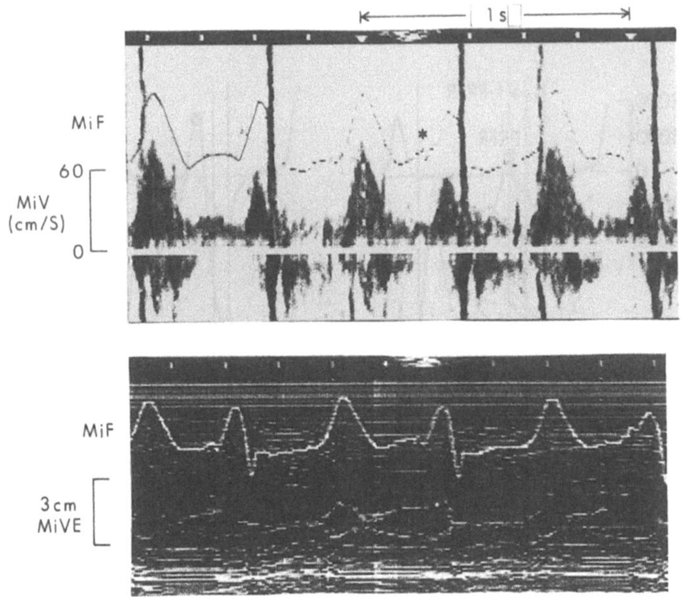

Fig. 7. Simultaneous recordings of mitral flow (*MiF*, uncalibrated) and velocity (*MiV*), *upper panel*, and mitral flow and echo (*MiVE*), *lower panel*, in a dog lightly anesthetized with isoflurane. Note the transient closure of the mitral valve at the end of the rapid deceleration phase of flow (or velocity). Note also that the re-opening of the valve is accompanied by a mid-diastolic rebound in flow. (The MiF trace in the first cycle, *upper panel*, and the MiVE trace in the second cycle, *lower panel*, have been "touched up" for clarity. The * in the MiF trace, second cycle *upper panel*, indicates an artifact due to P-wave interference. Unpublished observation courtesy of Mr. O. Pajaro and Drs. L. Biasucci and M. Nana)

Fig. 7 is from a lightly anesthetized dog (isoflurane) with an implanted electromagnetic flow probe. The *upper panel* demonstrates the similarity between the volume flow and velocity waveforms. Note particularly the rapid deceleration from peak early filling followed by a mid-diastolic rebound in flow. The mitral valve echo in the *lower panel* shows a transient closure of the valve at the end of deceleration followed by a re-opening corresponding to the flow rebound. These patterns are similar to those shown in Figs. 4 and 5, (see also [9, 21]). We suspect that the mid-diastolic rebound in flow is usually not recorded because of filtering. The physiological and clinical significance of changes in mid-diastolic flow patterns require further studies. Further work is also necessary to determine the conditions under which the valve may not close competently at the end of rapid deceleration, thus leading to regurgitation.

Conclusion

The experimental laboratory has provided important insights into the dynamics of left ventricular filling. Since it is clear that transmitral filling patterns in health and disease are determined by the time variations in the atrioventricular pressure difference, it is also clear that we must direct our experimental efforts toward determining how changes in chamber properties of both the atrium and ventricle change the respective pressures. This implies, of course, that we must determine how disease and stress acutely and chronically affect chamber properties. We will then be in a stronger position in our attempts to use non-invasively obtained mitral flow patterns as indices of chamber properties and diastolic function.

Acknowledgments. The work reviewed in this paper was done in collaboration with the many students, fellows, and colleagues that appear in the Reference section, and with the skilled technical assistance of Messrs. Leon, Bon, Rivera, and Wasserman.

References

1. Grossman W, Lorell BH (eds) (1988) Diastolic relaxation of the heart. Martinus Nijhoff, Boston
2. Spencer MP (ed) (1983) Cardiac Doppler diagnosis. Martinus Nijhoff, Boston
3. Kennish A, Yellin EL, Frater RWM (1975) Dynamic stiffness profiles in the left ventricle. J Appl Physiol 39: 665-671
4. Laniado S, Yellin EL, Miller H, Frater RWM (1973) Temporal relation of the first heart sound to closure of the mitral valve. Circulation 47: 1006-1014
5. Laniado S, Yellin EL, Kotler M, Levy L, Stadler J, Terdiman R (1975) A study of the dynamic relations between the mitral valve echogram and phasic mitral flow. Circulation 51: 104-113
6. Yellin EL, Peskin CS, Yoran C, Koenigsberg M, Matsumoto M, Laniado S, McQueen D, Shore D, Frater RWM, (1981) Mechanism of mitral valve motion during diastole. Am J Physiol 241 (Heart Circ Physiol 10): H389-H400
7. Meisner JS, McQueen DM, Ishida Y, Vetter HO, Bortolotti U, Strom JA, Frater RWM, Peskin CS, Yellin EL (1985) Effects of timing of atrial systole on LV filling and mitral valve closure: Computer and dog studies. Am J Physiol 249 (Heart Circ Physiol 18): H604-H619
8. Ishida Y, Meisner JS, Tsujioka K, Gallo JI, Yoran C, Frater RWM, Yellin EL (1986) Left ventricular filling dynamics: influence of left ventricular relaxation and left atrial pressure. Circulation 74: 187-196
9. Courtois M, Kovacs SJ Jr, Ludbrook PA (1988) The transmitral pressure-flow velocity relationship: The importance of regional pressure gradients in the left ventricle. Circulation 78: 661-671
10. Choong CY, Abascal VM, Thomas JD, Guerrero JL, McGlew S, Weyman AE (1988) Combined influence of ventricular loading and relaxation on the transmitral flow velocity profile in dogs measured by Doppler echocardiography. Circulation 78: 672-683
11. Keren G, Sherez J, Megidish R, Levitt B, Laniado S (1985) Pulmonary venous flow pattern—its relationship to cardiac dynamics. Circulation 71: 1105-1112
12. Van de Werf F, Minten J, Carmeliet P, De Geest H, Kesteloot H (1984) The genesis of the third and fourth heart sounds: a pressure-flow study in dogs. J Clin Invest 73: 1400-1407
13. Nolan SP, Dixon SH, Fisher RD, Morrow AG (1969) The influence of atrial contraction and mitral valve mechanics on ventricular filling. Am Heart J 77: 784-791
14. McQueen DM, Peskin CS, Yellin EL (1982) Fluid dynamics of the mitral valve: physiological aspects of a mathematical model. Am J Physiol 242 (Heart and Circ Physiol 11): H1095-H1110
15. Meisner JS, Ishida Y, Tsujioka K, Gallo JI, Yoran C, Frater RWM, Yellin EL (1984) Effect of spontaneous variation in heart rate on atrial contribution to left ventricular

filling in the conscious dog: analysis by electromagnetic phasic mitral flowmetry. (abstract) Circulation 70: II-237
16. Yellin EL, Peskin CS (1975) Large amplitude pulsatile water flow across an orifice. J Dynamic Systems, Measurement and Control, Trans ASME, vol 97, series G, no 1, pp 92-95
17. Meisner JS, Pajaro OE, Yellin EL (1986) Investigation of left ventricular filling dynamics: development of a model. Einstein Quart J of Biol and Med 4: 47-57
18. Thomas JD, Weyman AE (1989) Fluid dynamics model of mitral valve flow: Description with in vitro validation. J Am Coll Cardiol 13: 221-233
19. David D, Michelson EL, Naito M, Chen CC, Schaffenburg M, Dreifus LS (1983) Diastolic locking of the mitral valve: the importance of atrial systole and intraventricular volume. Circulation 67: 640-645
20. Choong CYP, Herrmann HC, Weyman AE, Fifer MA (1987) Preload dependence of Doppler-derived indexes of left ventricular diastolic function in humans. J Am Coll Cardiol 10: 800-808
21. Courtois M, Vered Z, Barzilai B, Ricciotti NA, Perez JE, Ludbrook PA (1988) The transmitral pressure-flow velocity relation: effect of abrupt preload reduction. Circulation 78: 1459-1468
22. Yellin EL, Yoran C, Frater RWM (1984) Physiology of mitral valve flow. In: Duran C, Angel WW, Oury JH, Johnson AD (eds) Recent progress in mitral valve disease. Butterworths Scientific, London, pp 47-59

Ventricular Interaction and the Role of the Pericardium: Effects of Right Ventricular Loading*

JOHN V. TYBERG and ELDON R. SMITH[1]

Summary. For several years, the concepts that left ventricular diastolic compliance and systolic performance might be influenced by alterations in right ventricular hemodynamics (ventricular interaction) or by the constraining effect of the pericardium have been recognized. However, to a considerable degree, these concepts have been examined and discussed in isolation. That is, it has often been implicitly assumed that ventricular interaction and pericardial constraint are independent phenomena. In this review of recent experimental studies which involve right ventricular dilation (pulmonary embolization, volume loading, and right ventricular infarction), we attempt to present an integrated view—increased right ventricular diastolic pressure both moves the septum leftward and, by raising pericardial pressure over the ventricles, augments pericardial constraint, thus diminishing left ventricular preload. Therefore, the pressure gradients both across the ventricular septum and across the left ventricular free wall are diminished. This concerted dual action is consistent with the definition of left ventricular transmural pressure proposed by Mirsky and Rankin.

Introduction

In the past, ventricular interaction and the constraint exerted by the normal pericardium have sometimes been considered separately in a way that implies a considerable degree of independence between the two phenomena. Although this was not invariably the case [1] and although a model for the intergration of ventricular interaction and pericardial effects was proposed several years ago [2], recent experimental work involving interventions which result in dilation of the right ventricle (RV) has served to emphasize that ventricular interaction and pericardial constraint are inextricably related phenomena.

Theoretical and Experimental Background

In 1979, Mirsky and Rankin set forth a simple but useful approach to integrating the effects of the RV and the pericardium on left ventricular (LV) diastolic filling [2].

* Studies reviewed in this paper were supported by Grants from the Alberta Heart and Stroke Foundation.

[1] Alberta Heritage Foundation for Medical Research, Alberta, Canada

[2] Departments of Medicine and Medical Physiology, The University of Calgary, Alberta, Canada

They assumed that the structural properties of the LV were uniform around its circumference (i.e., the stress-strain characteristics of the septum were the same as those of the free wall). Their formulation was based on the simple, semiquantitative observation that the "external" surface of the LV was comprised of the RV and the pericardium in the approximate proportion of $\frac{1}{3}:\frac{2}{3}$. Intuitively, since transmural pressure (Plv*) is the difference between intracavitary pressure (Plv) and the effective pressure or stress applied to the external surface of the LV, it follows that the components of the external pressure (Prv and Pp) should be weighted in proportion to the fraction of the total surface area over which they apply. Hence,

$$\text{Plv}^* = \text{Plv} - \{\frac{1}{3}\text{Prv} + \frac{2}{3}\text{Pp}\}.$$

Recent data from our laboratory has shown that, except in the presence of pericardial effusion, the pressure of the fluid contained in the pericardial cavity does not accurately reflect the degree of pericardial constraint [3]. While this assertion was, at first, extremely controversial, it is supported by the analogy of the knee joint [4]. Although the pressure in the synovial fluid is negligible, the stress (average force/area) on the weight-bearing surface of the tibia is equivalent to a pressure of approximately 10 atmospheres. Thus, it is obvious that the pressure of the synovial fluid is not fundamentally related to the weight-bearing characteristics of the knee joint and, by analogy, the pressure in the pericardial fluid does not define the stress that the pericardium applies to the LV epicardium. When the pericardium contains only a normal amount of liquid, the constraining effect of the pericardium can be measured only with a transducer designed to integrate compressive contact stress over an area of epicardium. In this review, pericardial "pressure" (Pp) is defined as the output of such a stress transducer, expressed as an equivalent pressure. While it is recognized that this stress is not a pressure in the strict physical sense, the term pressure is retained in deference to conventional usage, even though some risk of confusion continues.

The magnitude and variability of Pp is of note. In both laboratory and clinical experiments [5-7], Pp has been shown to be approximately equal to Prv during diastole. Significant exceptions have only been demonstrated under rather extreme and transient circumstances [8]. Thus, in most instances, Pp is closely paralleled by diastolic Prv and, as such, Pp is both greater in magnitude and more variable than was previously appreciated. In this context, one must acknowledge the remarkably prescient comment of L.N. Katz who, in 1955, wrote that neither end-diastolic Plv nor even changes in LV end-diastolic pressure (Plved) can be assumed to reflect LV preload because of the unknown effect of the pericardium [9].

Experimental Pulmonary Embolization

Clinically, pulmonary embolism is a life-threatening condition in which thrombi, usually formed in the veins of the legs, become dislodged, migrate through the right heart, and finally stop in the branches of the pulmonary artery, thus obstructing pulmonary artery flow to a variable degree. In severe cases, this is followed by a fall in systemic blood flow and pressure and, sometimes, death. The fact that the rise in right atrial pressure is sometimes associated with an increase in LV filling pressure and fall in cardiac output has led some to believe that this represents "right heart failure causing left heart failure."

We studied these phenomena recently [10]. Our hypothesis was that, even though LV filling pressure may be elevated after pulmonary embolization, LV transmural pressure decreases and, *pari passu*, LV end-diastolic volume (Vlved) decreases. If this is the case, the cardiovascular collapse might be explained, at least in part, on the basis of the Frank-Starling mechanism.

Using anesthetized dogs, we measured intracavitary pressures, pericardial "pressures" with balloon transducers on the lateral aspects of the LV and RV, and LV minor-axis diameters and RV septum-to-free wall "diameter" by means of sonomicrometry [10]. LV systolic performance was measured as the area of the pressure-volume loop (i.e., LV stroke work, LVSW) using the product of the two minor-axis diameters as an estimate of volume. Systolic performance and diastolic "compliance" (i.e., as defined by the position as well as the slope of the diastolic pressure-dimension relationship [11] were evaluated over a range of dimensions effected by transfusion and phlebotomy. As pulmonary embolization proceeded incrementally (autologous clots were infused intravenously and, intermittently, cardiac volume was manipulated by transfusion and phlebotomy), RV end-diastolic pressure (Prved) increased. Pp over both ventricles and Prved increased almost equally [12]. During the course of the experiment, the differences between Plved and Prved (i.e., the transseptal gradient) and between Plved and Pp decreased and, in the end, became negative. Prved was equal to Pp except during the later stages of the experiment, when it was slightly greater. As we had shown earlier [13], the ventricular septum moved progressively leftward as the normal transseptal pressure gradient (Plv > Prv) diminished and reversed. Fig. 1a shows that LV function (i.e., LVSW as a function of Plved) apparently diminishes following pulmonary embolization. Despite a considerable increase in Plved, LVSW decreased considerably. Thus, by the conventional clinical definition, contractility clearly decreased. Fig. 1b shows that LV diastolic compliance decreased concomitantly [11]. That is, during progressive pulmonary embolization and intermittent volume loading, Plved increased or stayed the same while Vlved (as estimated by the product of the minor-axis diameters) decreased. However, when transmural LV end-diastolic pressure (Plved*) was used, it became apparent that there was no change in ventricular contractility, in that the relationship between LV systolic performance and preload was unchanged by embolization (Fig. 2a). (The relationship between LVSW and Vlved was also unchanged by embolization.) Similarly, there was no change in "true" LV compliance in that there was no change in the Plved*-Vlved relationship (Fig. 2b). Thus, the hemodynamic effects of pulmonary embolization proved to be another example of the mechanism outlined by Glantz and Parmley in 1978 [14]. They indicated that when apparent diastolic compliance decreases (i.e., Plved increases with no change or a decrease in Vlved), LV systolic performance appears to diminish (i.e., LVSW decreases or stays the same while Plved increases), even when contractility (as defined by the relationship between LVSW and Vlved or Plved*) remains unchanged. In this model, pulmonary embolization and intermittent volume loading eventually produced systemic cardiovascular collapse despite an increase in Plved (i.e., LV failure). However, this deterioration could be almost entirely explained by the Frank-Starling mechanism—there was no significant deterioration of intrinsic myocardial performance.

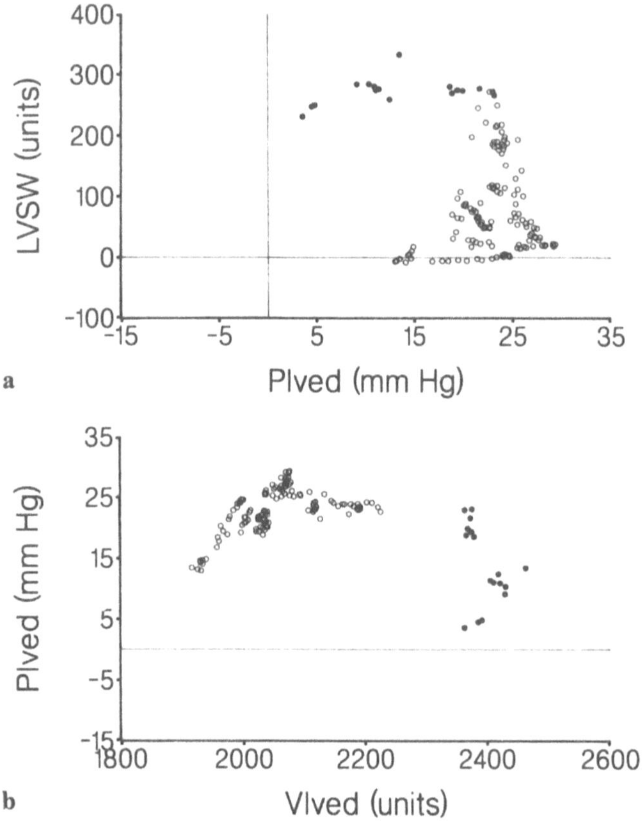

a

b

Fig. 1.a. A conventional ventricular function curve showing that pulmonary embolism and intermittent volume loading appear to decrease contractility, i.e., LV performance (stroke work) decreases at a given end-diastolic pressure. Control data are shown as *filled circles*. Data are taken from a representative experiment. *LVSW*, LV stroke work; *Plved*, LV end-diastolic pressure. **b.** Pulmonary embolism and intermittent volume loading apparently decrease LV compliance in that these pressure-volume points lie to the *left* and *above* the control data (*filled circles*). LV end-diastolic volume (*Vlved*) was estimated as the product of the two minor-axis diameters

This study strongly supports the Mirsky-Rankin approach to transmural pressure. It supports, but does not prove it, in that the relationships shown in Figs. 2a and 2b are very similar to those constructed using the alternative measure of transmural pressure (i.e., Plv – Pp), since Prv = Pp during diastole in this study. In this review, we indicate a theoretical preference for the Mirsky-Rankin formulation because it integrates the concerted effects of septum and pericardium and because defining LV transmural pressure in terms of only pericardial pressure does not resolve the problem of comparing pericardial constraint to septum-mediated ventricular interaction. Whether this will be supported by experimental data remains to be determined. For example, pulmonary hypertension leads to an increase in RV transmural pressure and, perhaps, to a disparity between Prv and Pp. However, concomitant

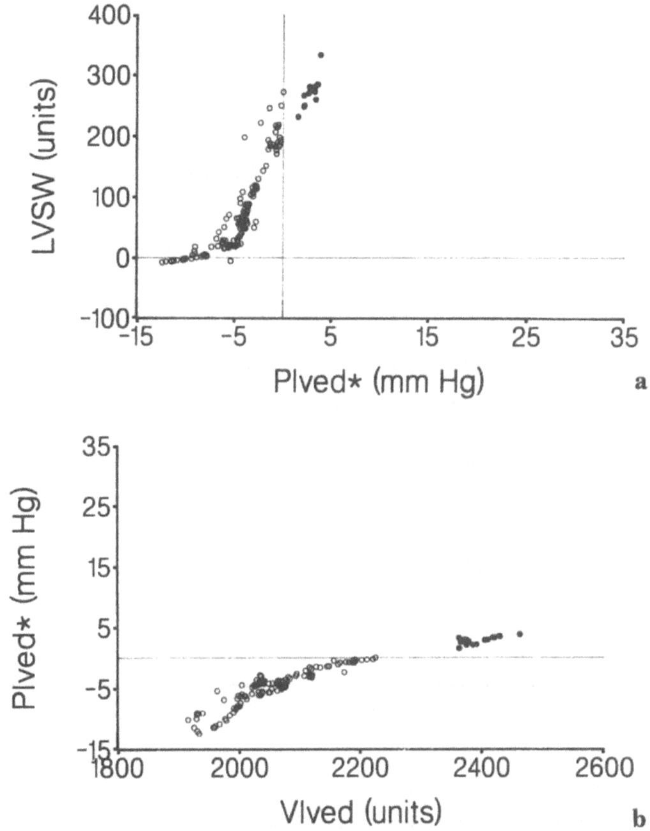

Fig. 2.a. A plot of LV performance (stroke work, *LVSW*) against transmural pressure [*Plved** = Plved — ($\frac{1}{3}$ Prved + $\frac{2}{3}$ Pp)] shows that there is no decrease in contractility when transmural pressure is used as a measure of preload. The data are the same as those shown in Fig. 1a. *Prved*, right ventricular end-diastolic pressure; *Plved*, left ventricular end-diastolic pressure; *Pp*, pericardial "pressure". **b.** A plot of LV transmural pressure (*Plved**) against end-diastolic volume (*Vlved*) indicates that there is no change in "true" ventricular compliance. The smaller end-diastolic volumes observed after pulmonary embolism and volume loading are commensurate with the reduction in transmural pressure. *Filled circles* show control data. The data are the same as those plotted in Fig. 1b

changes in the structure of the septum might complicate the interpretation of such an experiment.

Pulmonary embolization primarily increases resistance to RV ejection, which results in decreased systolic emptying and increased Prved. This results in a decrease in the normal transseptal end-diastolic gradient so that the septum shifts leftward, thereby diminishing LV volume. However, at the same time, the increase in Prved causes Pp over the RV to increase by a similar magnitude, since RV transmural pressure is very small over the range of RV dimensions observed when the pericardium is intact [11]. As observed in this study, the rise in Pp over the RV was almost equal to that over the LV, as predicted previously [15]. Thus, pulmonary embolism-induced RV dilation is seen to "squeeze" the LV between itself and the pericardium:

the increased Prved both shifts the septum leftward and compresses the LV by causing the increase in Pp. That the Mirsky-Rankin formulation properly integrates these effects is supported by the coherence of the data.

The Hemodynamic Effects of Volume Loading

In the course of these studies on the mechanical effects of pulmonary embolization, we noticed a variable response to the intravenous transfusion used to define ventricular function and diastolic compliance throughout the experiment [16] (see Fig. 3). At the beginning of the experiment, volume loading had the expected effect—as Plved rose, Plved* also increased and, with it, LVSW. However, after several pulmonary embolizations had been performed, volume loading no longer improved LV performance via the Frank-Starling mechanism. With progressive embolization, for the same increase in Plved, Plved* and LVSW first increased less,

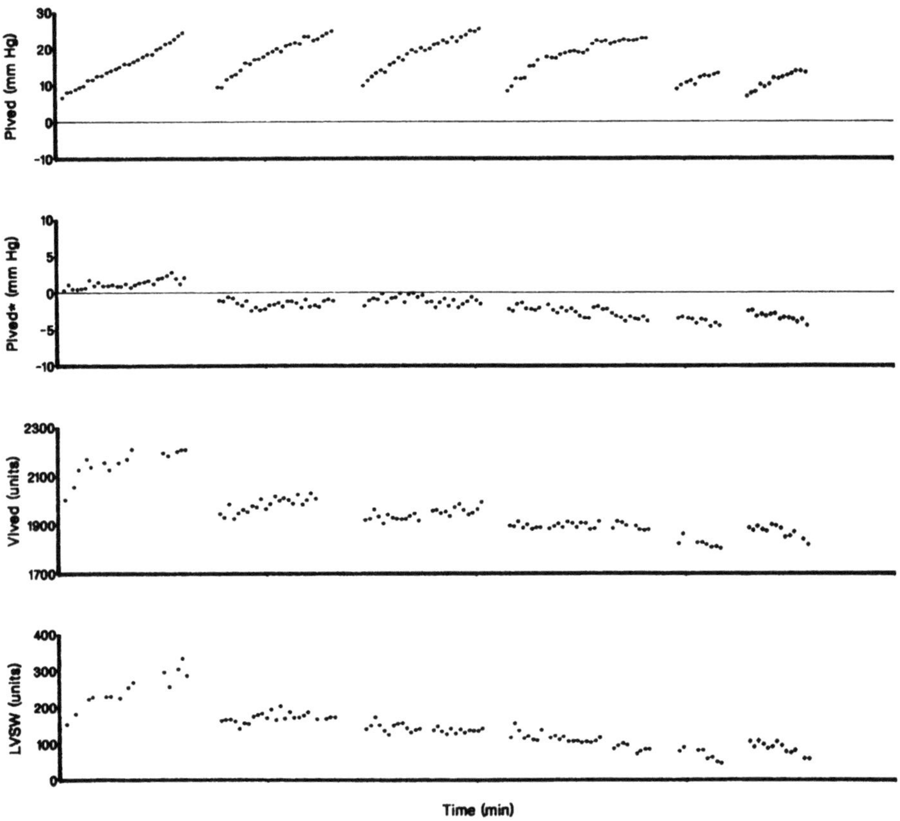

Fig. 3. Plots of LV end-diastolic pressure (*Plved*), transmural pressure (*Plved**), end-diastolic volume (*Vlved*), and LV stroke work (*LVSW*) against time. Panels indicate volume-loading episodes beginning with control. Autologous thrombi were infused before subsequent episodes. Note that volume loading which produced equal increases in Plved produced diminishing increments in Plved*, Vlved, and *LVSW* and, finally, actually decreased Plved*, Vlved, and *LVSW*

then remained unchanged, and finally they both decreased. Thus, volume loading, which is one of the standard clinical therapeutic options for pulmonary embolism, sometimes had markedly deleterious effects. Effectively, after several pulmonary embolizations, we found that volume loading decreased rather than increased LV preload. This result was mediated by exactly the same mechanisms as were observed with pulmonary embolization alone. As shown in Fig. 4a, volume loading apparently decreased compliance; however, the relationship between Plved* and Vlved was unchanged (Fig. 4b). Whether this experimental observation is applicable to man and the range of circumstances under which volume loading may be detrimental, remains to be determined.

These results also suggested that if volume loading increases Plved more than it increases Prved and Pp, LVSW will increase. The converse—when Plved increased less than Prved and Pp, LVSW decreased—was also true. While this may prove to have important therapeutic prognostic implications, it is also interesting theoreti-

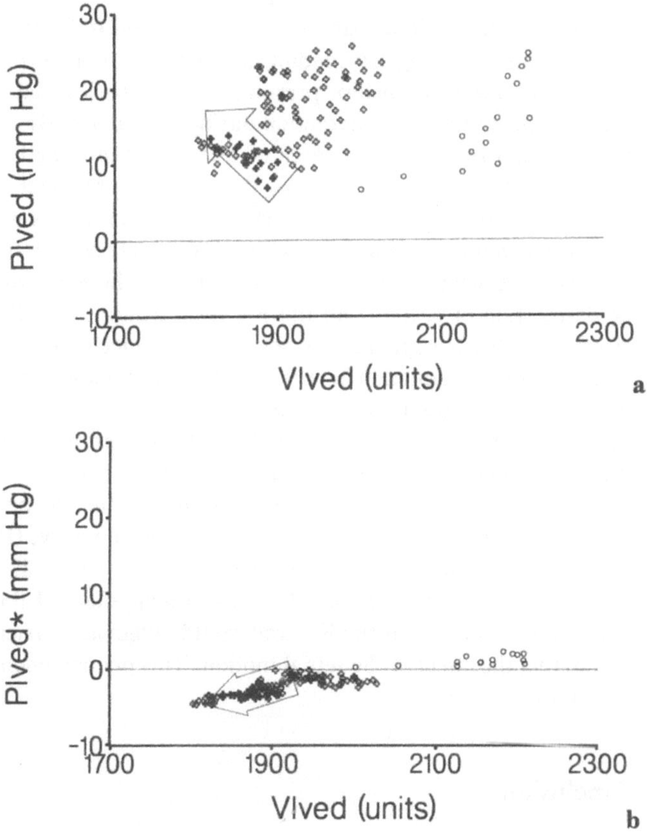

Fig. 4.a. Apparent LV compliance changes during the final volume loading shown in the final episode (see Fig. 3) are indicate by the *filled crosses* and the *arrow* indicating temporal progression. *Open circles*, control data; *open crosses*, other post embolization volume loading data. **b.** The volume loading data shown in **a** are *not* associated with a change in "true" ventricular compliance. *Plved*, LV end-diastolic pressure; *Vlved*, LV end-diastolic volume; *Plved**, transmural pressure

cally. It supports the utility of using Plved* as a measure of preload and is in keeping with the concept of a concerted effect of the RV and pericardium. Furthermore, since Prved and Pp were similar throughout this study, the role of ventricular interaction via the septum cannot be distinguished from that via the pericardium.

Diastolic and Systolic Aspects of Right Ventricular Infarction

Recently, in a series of experiments in anesthetized dogs in which RV infarction was produced by the injection of mercury in the right coronary artery, we have demonstrated other aspects of pericardium-mediated ventricular interaction. RV infarction produced RV dilation which resulted in increased Pp over both ventricles. The effect of increased Pp over the LV was an upward shift in the LV diastolic pressure-volume relationship [17]. This parallels the interaction observed earlier by Smiseth et al. during ischemic failure [18]. An opposite interaction (i.e., the LV affecting RV compliance) was clearly demonstrated recently by Dell'Italia and Walsh who showed, in patients, that changing LV loading by angiotensin or nitroprusside shifted the RV pressure-volume relationship, presumably via a pericardial mechanism [19].

Significant ventricular interaction during systole was also observed [20]. RV work was evaluated regionally (the area of a Prv-segment length loop) and globally (the area of a Prv-RV diameter loop). As expected, regional performance decreased toward zero after right coronary artery obstruction. However, global performance (i.e., stroke volume), while somewhat diminished, decreased much less than would have been predicted by the loss of contraction of the RV myocardium. Thus, despite the destruction of virtually all the RV myocardium, the RV may still pump blood effectively. This is explained by the following mechanism. With infarction, Prved increases with a reduction in the normal end-diastolic transseptal gradient. This causes a progressive leftward shift of the septum which, if sufficiently marked, results in rightward (paradoxic) motion of the septum during systole, leading to a positive counter-clockwise RV pressure-diameter loop. The work that is thus imparted to the blood appears to be primarily due to the contraction of the LV which drives the septum rightward by developing a positive (Plv > Prv) systolic transseptal pressure gradient. We postulate that this piston-like action of the septum would be less effective without the constraining effect of the pericardium; without this it would appear that the RV wall would bulge paradoxically during systole, thus attenuating the effect of the septal motion. This phenomenon obviously requires further study.

Conclusion

It is counter-productive to consider ventricular interaction and the constraining effect of the pericardium as independent phenomena. Rather, right ventricular dilation diminishes LV preload by a concerted dual effect: it forces the septum leftward during diastole and simultaneously increases pericardial pressure, over the LV as well as the RV. Thus, LV transmural pressure (preload) is decreased because of increases in both RV and pericardial pressures.

References

1. Maruyama Y, Ashikawa K, Isoyama S, Kanatsuka H, Ino-Oka E, Takishima T (1982) Mechanical interactions between four heart chambers with and without the pericardium in canine hearts. Circ Res 50: 86–100
2. Mirsky I, Rankin JS (1979) The effects of geometry, elasticity, and external pressures on the diastolic pressure-volume and stiffness-stress relations: how important is the pericardium? Circ Res 44: 601–611
3. Smiseth OA, Frais MA, Kingma I, Smith ER, Tyberg JV (1985) Assessment of pericardial constraint in dogs. Circulation 71: 158–164
4. Smiseth OA, Frais MA, Kingma I, White AVM, Knudtson ML, Cohen JM, Manyari DE, Smith ER, Tyberg JV (1986) Assessment of pericardial constraint: the relation between right ventricular filling pressure and pericardial pressure measured after pericardiocentesis. J Am Coll Cardiol 7: 307–314
5. Smiseth OA, Refsum H, Tyberg JV (1984) Pericardial pressure assessed by right atrial pressure: a basis for calculation of left ventricular transmural pressure. Am Heart J 108: 603–608
6. Tyberg JV, Taichman GC, Smith ER, Douglas NWS, Smiseth OA, Keon WJ (1986) The relation between pericardial pressure and right atrial pressure: an intraoperative study. Circulation 73: 428–432
7. Boltwood CM, Skulsky A, Drinkwater DC, Lang S, Mulder DG, Shah PM (1986) Intraoperative measurement of pericardial constraint: role in ventricular diastolic mechanics. J Am Coll Cardiol 8: 1289–1297
8. Smiseth OA, Scott-Douglas NW, Thompson CR, Smith ER, Tyberg JV (1987) Non-uniformity of pericardial surface pressure in dogs. Circulation 75: 1229–1236
9. Katz LN (1955) Analysis of the several factors regulating the performance of the heart. Physiol Rev 35: 91–106.
10. Belenkie I, Dani R, Smith ER, Tyberg JV (1988) Ventricular interaction during experimental acute pulmonary embolism. Circulation 78: 761-768
11. Braunwald E, Ross J Jr (1963) The ventricular end-diastolic pressure. Am J Med 34: 147–150 (editorial)
12. Traboulsi M, Scott-Douglas NW, Smith ER, Tyberg JV (1987) Measurement of right ventricular transmural pressure. (abstract) Clin Invest Med 10(5): C33
13. Kingma I, Tyberg JV, Smith ER (1983) Effects of diastolic transseptal pressure gradient on ventricular septal position and motion. Circulation 68: 1304–1314
14. Glantz SA, Parmley WW (1978) Factors which affect the diastolic pressure-volume curve. Circ Res 42: 171–180
15. Tyberg JV (1985) Ventricular interaction and the pericardium. In: Levine HJ, Gaasch WH (eds) The ventricle: basic and clinical aspects. Martinus Nijhoff, The Hague pp 171–184
16. Belenkie I, Dani R, Smith ER, Tyberg JV (1989) Effects of volume loading during experimental acute pulmonary embolism. Circulation 80: 178–188
17. Traboulsi M, Scott-Douglas NW, Belenkie I, Tyberg JV, Smith ER (1989) Right ventricular infarction decreases left ventricular compliance and performance by a pericardial mechanism. J Am Coll Cardiol 13: 235A.
18. Smiseth OA, Refsum H, Junemann M, Sievers RE, Lipton MJ, Carlsson E, Tyberg JV (1984) Ventricular diastolic pressure-volume shifts during acute ischemic ventricular failure in dogs. J Am Coll Cardiol 3: 966–977
19. Dell'Italia LJ, Walsh RA (1988) Right ventricular diastolic pressure-volume relations and regional dimensions during acute alterations in loading conditions. Circulation 77: 1276–1828
20. Traboulsi M, Scott-Douglas NW, Tyberg JV, Smith ER (1988) Right ventricular performance during extensive right ventricular infarction. Clin Invest Med 11(5): D85

Chapter D. Regional and Chamber Mechanics in Ischemic and Hypertrophic Hearts

HIROYUKI SUGA

The heart is an organ made of an assemblage of myocardial cells. As an organ, it serves as a pump and its performance can be assessed in terms of global variables such as internal pressure, dP/dt, volume, stroke volume, ejection fraction, E_{max}, etc. As an assemblage of myocardial cells, the whole chamber or regions of its wall can be treated as a lumped muscle mass and its average contractile properties can be assessed in terms of regional variables such as force (or tension or stress), length (or area), and shortening velocity calculated from the global variables. However, when the details of the contractile properties of the cardiac chamber and regions of its wall are of interest, the global variables and the average regional variables are no longer sufficient for the purpose, although the global variables are variably affected by regional contractile abnormalities and particularly so in the case of ischemic and hypertrophic hearts. In such cases, the chamber has to be treated as an assemblage of two or more regions of different contractile properties. Then, average force, tension, stress, length, and shortening velocity no longer represent the true values of the individual regions, and they must be assessed in each region despite technical difficulties in doing so.

Contributions from 6 representative groups working on several different aspects of the present theme are included in this chapter. Sugawara and Nakano summarized their new method to quantify regional work and elastic modulus (regional E_{max}) using wall stress and wall thickness. This theoretically sound method seems promising as a clinically applicable method. Tomoike et al. reviewed their studies of the effect of regional hypokinesis on the end-systolic pressure–regional length relation and found that the x-axis intercept of the relation shifted to the right with the degree of regional hypokinesis. Using wall thickness measured by echocardiography in place of regional length, the method could be applied clinically. Ishida et al. analyzed the factors that alter left ventricular early filling pattern in patients with hypertrophic cardiomyopathy. The results showed that the classification of the early filling pattern may be useful for assessing the degree of impaired ventricular relaxation as a result of hypertrophy. Kitabatake et al. studied the abnormalities of left ventricular diastolic filling in hypertensive patients and patients with hypertrophic cardiomyopathy by using pulse Doppler echocardiography. They found that the degree of diastolic filling abnormality could be assessed by peak velocity in the early filling phase and its ratio to peak velocity in the atrial contraction phase. Hirota et al. critically reviewed both global and regional indexes of ventricular systolic and diastolic properties and discussed their theoretical and methodological limitations. Finally, Gaasch and Izzi reviewed the recent advancement of the clinical diagnosis and management of left ventricular diastolic dysfunction, reexamining the concept of regional and global stiffness. All these contributions will surely provide a condense summary of the state-of-the-art in this field and are expected to greatly promote understanding of the theme of cardiac mechanics.

A New Method of Analyzing Regional Myocardial Function of the Ventricle

MOTOAKI SUGAWARA and KIYOHARU NAKANO [1]

Summary. We introduce a new method of defining regional work and regional elastic modulus of the myocardium of the ventricular wall using the relationship between mean wall stress (σ) and the natural logarithm of reciprocal of wall thickness ($\ln(1/H)$). The area surrounded by the loop described by the σ-ln $(1/H)$ relationship during a cardiac cycle is equal to regional work per unit volume of myocardium per stroke. The slope of the σ-ln $(1/H)$ relationship corresponds to Young's modulus of an incompressible Hookean material. The end-systolic σ-$\ln(1/H)$ data point moves on a rectilinear line during pre-and afterloadings. The slope of this line is not affected by alterations in pre-and afterloads, but changes singnificantly due to the changes in the contractile state. Therefore, the slope can be regarded as the normalized E_{max}, i.e., a normalized index of myocardial contractility index of myocardial contractility.

Introduction

The set of variables which expresses the work done by a small region of the ventricular wall in the greatest detail will be the stresses and strains in that region. However, at present, there are no reliable methods for measuring stresses in the ventricular wall directly. Therefore, we must rely on mathematical models to estimate the stresses. As there are both quantitative and qualitative discrepancies between the stress distributions calculated for the various ventricular models [1, 2], it is difficult to judge which is the most realistic representation of the actual stress distribution across the ventricular wall.

We have proposed the use of the relationship between the mean wall tension (T) and the area (A) of a region of interest of the ventricular wall to define regional work [3]. The instantaneous relation between the mean wall tension T and the regional area A describes a loop in the T-A plane during a cardiac cycle. The regional work is approximately equal to the area surrounded by this loop. The validity of our method based on the T-A diagram was also confirmed by Goto et al. [4–6].

Our next step involved comparison of the results derived from the T-A relationships for hearts with different area sizes and wall thicknesses. To normalize the results to unit volume of myocardium, we proposed the use of a new relationship between mean wall stress (σ) and the natural logarithm of A(areal strain), which is

[1] The Heart Institute of Japan, Tokyo Women's Medical College, 8-1 Kawada-cho, Shinjuku-ku, Tokyo, 162 Japan

derived directly from the T-A relationship [7, 8]. In clinical cases, however, the area A cannot be easily measured. On the assumption that the myocardium is incompressible throughout one cardiac cycle, the reciprocal of wall thickness ($1/H$) at a point on the epicardial surface is proportional to the regional area containing that point. Since wall thickness is readily measured by echocardiography, we used $1/H$ in place of A. To show the validity of our method for defining regional work and evaluating regional myocardial function, we investigated the σ-ln($1/H$) relationship in dogs under various conditions and in human subjects at the time of cardiac catheterization [7, 9].

In this article we explain the method of calculating regional work and clarify that the slope of the end-systolic σ-ln($1/H$) relationship line indicates the maximum elastic modulus of the myocardium of the region. Suga and Sagawa [10] and Sagawa [11] designated the slope of the end-systolic pressure-volume relationship as the maximum elastance of the ventricle, E_{max}; and several studies have shown that E_{max} is useful in evaluating the contractile state of the ventricle. However, E_{max} depends not only on the myocardial contractility, but also on the heart size. In this respect, E_{max} is not normalized, i.e., the values of E_{max} of hearts of differing sizes cannot be compared [12]. On the other hand, the slope of the end-systolic σ-ln(1/H) relationship does not include the effects of ventricular geometry as a parameter.

Normalization of Regional Work

The top left part of Fig. 1 shows a schematic illustration of the imaginary frustum of the myocardium cut from the ventricular wall. H is the wall thickness, and A is the area of the midwall layer of the frustum. Then, the volume of the frustum, V, is approximately given by

$$V = A \times H. \tag{1}$$

The volume of the frustum is constant throughout a cardiac cycle according to the incompressibility of the myocardium. The mechanical work RW done by the frustum is given by

$$RW = -\int T\,\mathrm{d}A, \tag{2}$$

where T is the average wall tension, and the integral is taken over a cardiac cycle. Division of regional work RW by regional myocardial volume, namely AH, yields

$$RWM = RW/AH = -(1/AH)\int T\mathrm{d}A \tag{3}$$

$$= -\int (T/H)(\mathrm{d}A/A) \tag{4}$$

$$= -\int (T/H)\mathrm{d}(\ln A) \tag{5}$$

$$= -\int \sigma \mathrm{d}(\ln A) \tag{6}$$

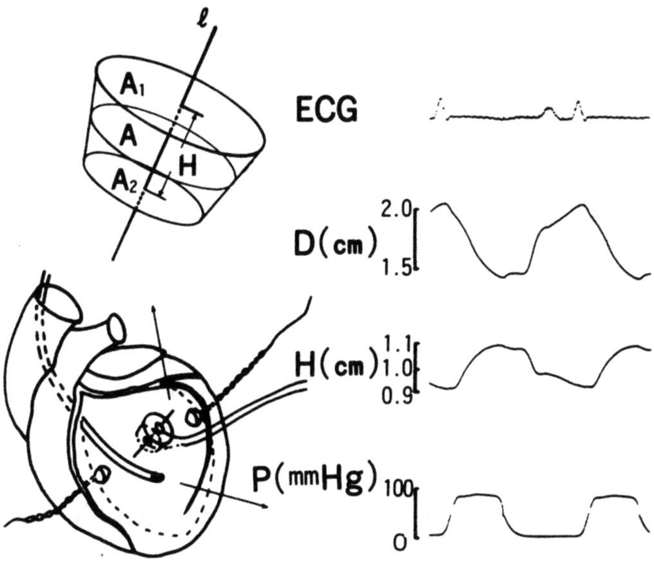

Fig. 1. The imaginary frustum of the myocardium (*left panel*), and representative recordings of electrocardiogram (*ECG*), internal diameter (*D*), wall thickness (*H*), and left ventricular pressure (*P*) (*right panel*). A_1, A_2, and A are epi-and endocardial surfaces and midwall layer of the frustum, respectively. From Nakano et al. [7]

Transformation of Eq. 3 into Eq. 4 is essential to our method. Here, AH can be moved into the integral sign, because it is constant. Using the relation $dA/A = d(\ln A)$ and defining the mean wall stress σ as $\sigma = T/H$, we obtain Eq. 6. This equation means that the area surrounded by the loop described by the σ-ln$A$ relation during a cardiac cycle is equal to RWM, that is, regional work per unit volume of myocardium per stroke.

Definition of Areal Strain and Use of (1/H) in Place of A

Since the increment of the natural logarithm of A, or dA/A, expresses a relative change in area, it can be called the 'incremental areal strain.' Then, the total areal strain, ϵ, is given by the following equation:

$$\epsilon = \int_{A_o}^{A} dA/A = \ln A - \ln A_o = \ln(A/A_o).$$

To obtain a stress-strain relation, the unstressed area A_o is required. However, for most biological materials, it is preferable to analyze the data in terms of the stiffness-stress ($d\sigma/de$ - σ) relations rather than stress-strain relations, and in the analysis of stiffness-stress relation, A_o is not required. Also in the calculation of regional work, A_o is not required. Therefore, omitting A_o, we define the areal strain by

$$\epsilon = \ln A \tag{7}$$

However, the measurement of regional area is still difficult in clinical situations. To develop a clinically feasible method, we use again the incompressibility of the myocardium. From Eq. 1,

$$\ln A = \ln V + \ln(1/H).$$

Since V here is constant,

$$d(\ln A) = d(\ln(1/H)).$$

As far as the analysis of stiffness-stress relation and the calculation of regional work are concerned, we can use $\ln(1/H)$ in place of $\ln A$. Therefore from now on, we will call the $\ln(1/H)$ the areal strain.

Meaning of the Slope of the σ-ln(1/H) Relationship

The analogy of an incompressible Hookean material enables a clear understanding of the meaning of the slope of the σ-ln(1/H) relationship. Let us consider a rectangular solid body subjected to a plane stress with components σ_x and σ_y (Fig. 2). The rectangular solid delineated by a chain dotted line shows the body before deformation, and that delineated by a solid line shows the body after deformation. The elongations in the $x-$ and $y-$ directions due to σ_x and σ_y are $\epsilon_x \Delta x$ and $\epsilon_y \Delta y$, where ϵ_x and ϵ_y are the strain components. In the z-direction, contraction occurs, the amount of which is $\epsilon_z \Delta z$. Here, ϵ_z is the contracting strain. According to the simple theory of incompressible Hookean material (Poisson's ratio = 1/2), the following equation holds:

$$(\sigma_x + \sigma_y)/2 = -E\epsilon_z, \tag{8}$$

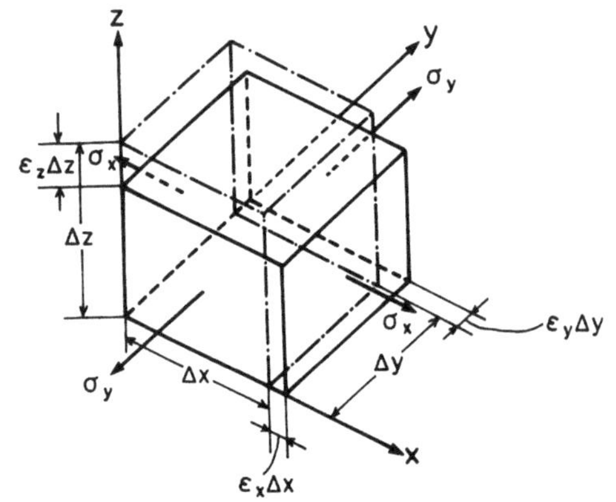

Fig. 2. Rectangular solid body subjected to a plane stress. The rectangular solid delineated by a *chain dotted line* shows the body before deformation, and that delineated by a *solid line* shows the body after deformation

which shows the relation between the stress components in the xy-plane and the strain component in the z-direction [13]. Here, E is Young's modulus. The quantity on the left-hand side of Eq. 8 is the same as our mean wall stress σ. The strain component ϵ_z is defined as $\ln(H/H_o)$, where H_o is the unstressed thickness. However, as far as the increment of ϵ_z is concerned, the unstressed thickness H_o can be omitted. Namely, the increment of ϵ_z with minus sign corresponds to that of $\ln(1/H)$, the increment of our areal strain:

$$d(\ln(1/H)) = -\, d\epsilon_z.$$

Therefore, according to Eq. 8, the gradient of the σ-$\ln(1/H)$ relation, namely, $d\sigma/d(\ln(1/H)$, corresponds to Young's modulus E.

Summary of Results of Dog Experiments

The heart was exposed according to a conventional method, and the left ventricular internal diameter (D) and the wall thickness (H) were measured wit ultrasonic dimension gauges. The left ventricular pressure (P) was measured with a catheter-tip micromanometer (Fig. 1). Since the purpose of the present study was to demonstrate our new concept, only simple measurements were made in our experiments. Therefore, a thick-walled spherical model was used to calculate the mean wall stress.

Representative changes of the σ-$\ln(1/H)$ loop during aortic constriction (pressure loading, or afterlading) in one dog are shown in Fig. 3. With the increase in σ, the end-systolic $1/H$, the end-systolic $1/H$ gradually increased, and the σ-$\ln(1/H)$ loop shifted to the right. RWM also increased from 0.42 mJ/cm^3 to 0.71 mJ/cm^3. The RWM multiplied by the total myocardial volume of the left ventricle (LVM) correlated well with the total mechanical work of the entire left ventricle (TW) calculated from the area surrounded by the pressure-volume curve delineated on the assump-

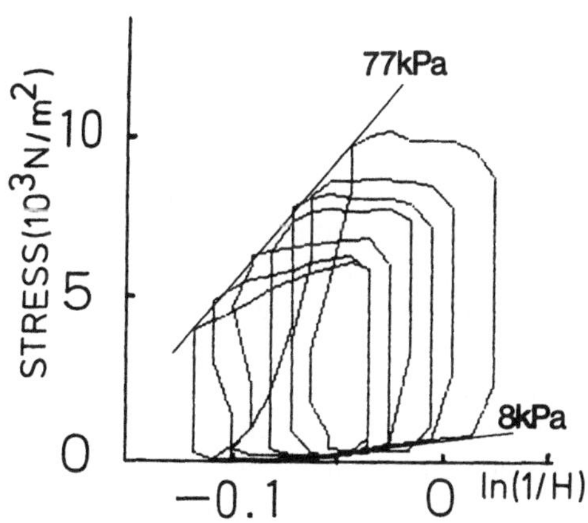

Fig. 3. Effects of acute aortic constriction on the σ-ln(l/H) loop. The end-systolic σ-ln(l/H) relation shows linearity

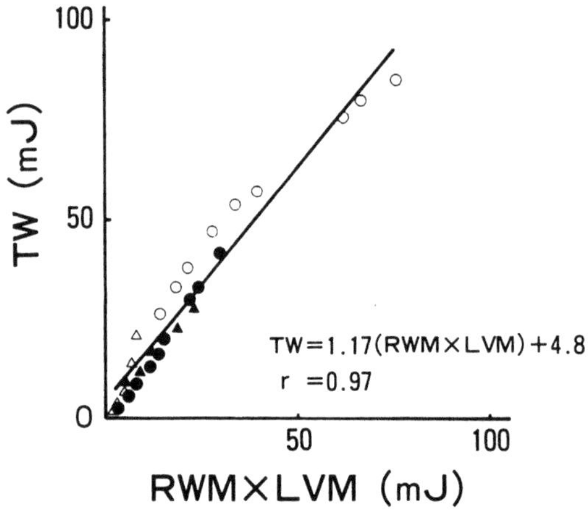

Fig. 4. Relationship between regional work per unit volume of myocardium × total myocardial volume of left ventricle (*RWM* × *LVM*) and total mechanical work of entire left ventricle (*TW*) during blood infusion. Different symbols refer to different dogs

tion that the ventricle has a spherical shape. Fig. 4 represents the relationship between the (*RWM* × *LVM*) and *TW* during blood infusion (volume loading, or preloading). The end-systolic σ-ln(1/*H*) data point moved approximately on a rectilinear line during pressure and volume loadings in each dog. Fig. 5 shows the slopes of the end-systolic σ–ln(1/*H*) relationship during aortic constriction with and without continuous administration of isoproterenol in one dog. The slope increased from 51 kPa to 144 kPa by the isoproterenol administration. Fig. 6 shows the slopes of the end-systolic σ-ln(1/*H*) relation during acute infusion of blood with and without administration of isoproterenol in one dog. The slope was increased from 36 kPa to 47 kPa by the isoproterenol administration.

Validity and Advantage of the Use of σ-ln(1/*H*) Relation

According to the law of conservation of energy, the total mechanical work accumulated over the myocardial volume of the left ventricle must correspond to the total mechanical work performed by the entire left ventricle. If a unit volume of myocardium in every part of the left ventricle performs the same amount of regional work, the regional work multiplied by the total volume of the left ventricular myocardium (*RWM* × *LVM*) must be equal to the total work calculated from the pressure-volume curve (*TW*). Our result of good correlation between (*RWM* × *LVM*) and *TW* suggests the validity of our methods of calculating regional work (Fig. 4).

The end-systolic mean wall stress vs areal strain (σ-ln(1/*H*)) data point moved approximately on a rectilinear line. The slope of this line was not affected by alterations in preload and afterload, but changed significantly due to the changes in the contractile state (Figs. 5 and 6). Therefore, the slope can be considered to correspond to the so-called E_{max}. E_{max} depends on the heart size—in other words, E_{max} is not normalized. On the other hand, the slope of the end-systolic σ-ln(1/*H*) relation

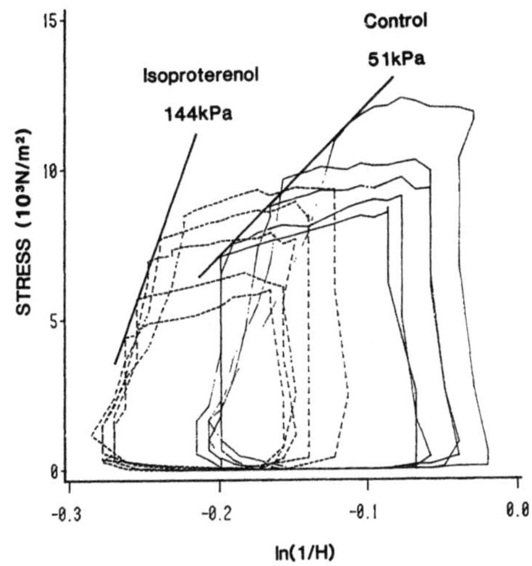

Fig. 5. Effects of isoproterenol on the slope of the end-systolic σ-ln(1/H) relation during aortic constriction

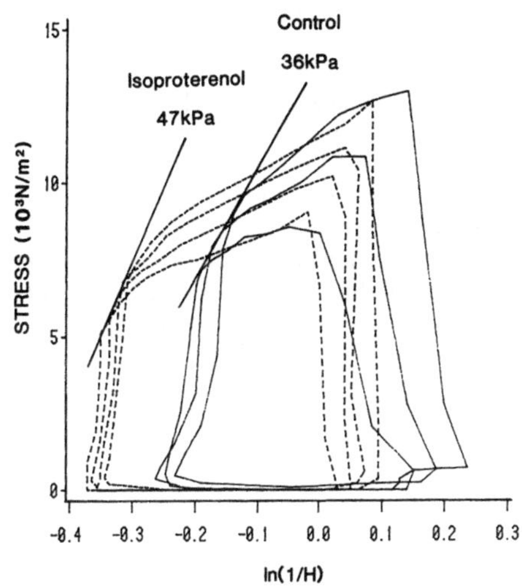

Fig. 6. Effects of isoproterenol on the slope of the end-systolic σ-ln(1/H) relation during acute infusion of blood

does not include the effects of ventricular geometry as a parameter and is expressed in kPa, which is the same unit as that for Young's modulus. In this sense, the slope of the end-systolic σ-ln$(1/H)$ relationship may be called the normalized E_{max}.

The intersection of the end-systolic σ-ln$(1/H)$ relation line and the abscissa is not important, since the 'unstressed' wall thickness, which is usually not available, was not included in our definition of areal strain, ln$(1/H)$. This is also an advantage of our method.

References

1. Huisman RM, Sipkema P, Westerhof N, Elzinga G (1980) Comparison of models used to calculate left ventricular wall force. Med Biol Eng Comput 18: 133–144
2. Yin FCP (1981) Ventricular wall stress. Circ Res 49: 829–842
3. Sugawara M, Tamiya K, Nakano K (1985) Regional work of the ventricle:Wall tension-area relation. Heart Vessels 1: 133–144
4. Goto Y, Suga H, Yamada O, Igarashi Y, Saito M, Hiramori K (1986) Left ventricular regional work from wall tension-area loop in canine heart. Am J Physiol (Heart Circ Physiol) 250(19): H151–H158
5. Goto Y, Igarashi Y, Yamada O, Hiramori K, Suga H (1988) Hyperkinesis without the Frank-Starling mechanism in a nonischemic region of acutely ischemic excised canine heart. Circulation 77: 468–477
6. Goto Y, Igarashi Y, Yasumura Y, Nozawa T, Futaki S, Hiramori K, Suga H (1988) Integrated regional work equals total left ventricular work in regionally ischemic canine heart. Am J Physiol (Heart Circ Physiol) 254(23): H894–H904
7. Nakano K, Sugawara M, Tamiya K, Satomi G, Koyanagi H (1986) A new approach to defining regional work of the ventricle and evaluating regional cardiac function: Mean wall stress-natural logarithm of reciprocal of wall thickness relationship. Heart Vessels 2: 74–80
8. Sugawara M, Nakano K (1987) A method of analyzing regional myocardial function: Mean wall stress-area strain relationship. Jpn Circ J 51: 120–124
9. Nakano K, Sugawara M, Kato T, Sasayama S, Carabello BA, Asanoi H, Umemura J, Koyanagi H (1988) Regional work of the human left ventricle calculated by wall stress and the natural logarithm of reciprocal of wall thickness. J Am Coll Cardiol 12: 1442–1448
10. Suga H, Sagawa K (1974) Instantaneous pressure-volume relationships and their ratio in excised, supported canine left ventricle. Circ Res 35: 117–126
11. Sagawa K (1981) The end-systolic pressure-volume relation of the ventricle: Definition, modifications and clinical use. Circulation 63: 1223–1227
12. Suga H, Yamada O, Goto Y, Igarashi Y, Yasumura Y, Nozawa T (1986) Reconsideration of normalization of Emax for heart size. Heart Vessels 2: 67–73
13. Sommerfeld A (1964) Mechanics of deformable bodies. (translated by Kuerti G) Academic Press, New York, pp 65–66

Regional Wall Motion in the Ischemic Heart*

HITONOBU TOMOIKE, YOSHITOSHI URABE, SHUNICHI KASEDA, and MOTOOMI NAKAMURA[1]

Summary. Determinants of regional wall motion were examined during regional coronary hypoperfusion, changes in regional coronary contractile state, and coronary occlusions. The lowest coronary perfusion pressure for maintaining the regional shortening increased when the level of ventricular pressure was elevated by pulmonary artery constriction, which suggests the importance of afterload as one of determinants of regional shortening. The relationship between regional segment length and ventricular pressure was visualized as a loop, and the end-systolic pressure-length relation was determined during progressive declines of preload following occlusion of the inferior caval vein in homogeneously or heterogeneously contracting heart in open-chest dogs. Although the changes in slope of the end-systolic pressure-length relation theoretically represent the elastance of systolic myocardium, the x-axis intercept, but not the slope of the end-systolic pressure-length relation, altered, correlating with changes in contractile state. The level of the x-axis intercept of the regional myocardium, rendered ischemic, increased following an extension of infarct size, which suggests the stretch of ischemic myocardium from the surrounding intact myocardium. Analysis of regional wall motion on the trajectories of the pressure-length (wall thickness) relation facilitates understanding of regional wall motion unique to muscle properties.

Introduction

Abrupt changes in regional wall motion after coronary ligation were first documented by Tennant and Wiggers [1]. The characteristics of regional wall motion have been studied not only from the point of variables such as heart rate, contractility, preload, afterload, and coronary perfusion pressure or coronary blood flow, but also from the point of the interactions between the intact and injured myocardium [2, 3] or the relationship between fiber orientation and the axis of the wall motion measurement [4, 5]. Since the advent of echocardiography and X-ray cineangiography, analyses of regional wall motion as a useful index of myocardial ischemia have been given increasing attention [6].

When systemic hemodynamic variables such as heart rate, preload, or afterload remain constant, regional myocardial function can be described by the end-diastolic

* Supported in part by Grants-in-Aid for Scientific Research (No. 60570396, 61232015, 61870037, 63624511) from the Japanese Ministry of Education, Science, and Culture and a grant for Research on Cardiovascular Disease (61-4) from the Japanese Ministry of Health and Welfare.

[1] Research Institute of Angiocardiology and Cardiovascular Clinic, Faculty of Medicine, Kyushu University, 3-1-1 Maidashi, Higashi-ku, Fukuoka, 812 Japan

length and the extent or percent of regional systolic shortening based on the framework of Frank-Starling's law [7, 8]. Sonnenblick et al. [9] applied the concept of muscle physiology to characterize regional wall motion instead of the overall response of the heart as a pump. They measured instantaneous intraventricular pressures and distances between right ventricular markers at 1/30-s intervals on cineradiograms and examined the force-velocity relation at the same muscle length at rest and during exercise, before and after beta-adrenergic blockade. This procedure was one of the approaches used in assessing qualitatively the contractile state on the framework of the force-velocity relation in muscle mechanics. However, such conventional indices of regional wall motion, though simple to measure and possessing sensitivity in detecting regional myocardial ischemia, have several shortcomings. For example: (1) percentage changes in shortening (percent shortening) do not necessarily indicate changes in the contractile state, (2) systolic phase indices are sensitive to changes in preload and afterload, and (3) there is a lack of quantitative information concerning global cardiac function or the stretch of hypokinetic segments due to the active contraction of the normal myocardium. In the following, we describe our results on the determinants of regional wall motion and assessment of the muscle property unique to regional contractile states.

Effects of Perfusion Pressure on Regional Wall Motion

When the region of ischemia is a moderate to large one, there is an invariable decrease in the level of arterial pressure and cardiac output along with an increase in end-diastolic pressure [10]. This phenomenon hinders the quantitative elucidation of the relationship between the perfusion pressure and regional wall motion due to decreases of afterload on the left ventricle. On the contrary, the evolution of severe regional ischemia in the right ventricular free wall did not affect cardiac output or arterial pressure, which enabled the stable coronary perfusion during the evolution of the regional myocardial ischemia in the right ventricular free wall. Then, the relationship between regional percent shortening and coronary perfusion pressure was examined in the right ventricle. Our study showed that the coronary perfusion pressure was one of the direct determinants of regional systolic shortening below the level of critical perfusion pressure [11]. Effects of afterload on the extent of regional shortening during changes in coronary perfusion pressure were evident by differences of critical perfusion pressure between conditions of normal and elevated right ventricular pressures. A positive linear relation between increases in right ventricular pressure and critical perfusion pressure suggests the importance of the level of afterload, the related oxygen consumption, and the vascular-waterfall mechanism, in determining the level of wall motion under ischemia and altered afterload.

Pressure-Chamber Dimension Relation

Suga, Sagawa, and their colleagues have shown that the load-independent property of the end-systolic volume elastance can be derived without assumption [12], and it can be represented as a slope in the trajectory of the end-systolic pressure-volume relation for a given inotropic state [12–14]. Since the end-systolic pressure-volume relation (ESPVR) on the isolated heart has been examined extensively under various conditions [15], the behavior of ESPVR in in situ heart [16–18] and its application to pressure-diameter [19], pressure-length, [20, 21] or pressure-wall thickness relations [22] has been an area of increasing interest by researchers. We applied this concept to the characterization of regional wall motion.

In a preliminary study, the end-systolic pressure-length relation (ESPLR) was so linear in control and in the presence of regional myocardial ischemia that an analysis of pressure-length relation looked promising for quantifying the regional contractile state. However, the physiological usefulness and technical limitations in an analysis of pressure-length relation have not been studied rigorously. Accordingly, we examined the trajectories of the end-systolic pressure-length relations during global changes in the contractile state [23], in different levels of regional myocardial dyskinesis [3], and before and after expansion of size of ischemia [24].

Changes in Regional Contractile State

Since regional systolic shortening in the ischemic region includes influences of contraction of normal region, shape change due to regional ischemia, and alterations of the neural control, assessment of regional contractile state in the presence of heterogenous contraction is rather complicated. Whether ESPLR is applicable for characterization of regional wall motion abnormality (hyper- and hypokinesis) remained to be tested. Accordingly, the ESPLR was examined in the heterogeneously contracting heart in open-chest dogs, in which the level of regional hypokinesis was controlled without producing coronary stenosis. Regional lengths were measured sonomicrometrically in areas of both the left circumflex coronary artery (LCX) and left anterior descending coronary artery (LAD).

In group 1 (12 dogs), saline as a control condition and isoproterenol 0.05 μg/min to augment the regional contractile state were infused into the LCX. Pressure-segment length relations were examined during decreases in preload following occlusion of the inferior vena cava. ESPLR, determined by connecting points of the left upper corner of pressure-length loops, was linear. Highly significant linearity of ESPLR was maintained even following transcient reductions in arterial pressure in either the control ($r > 0.997$) or the isoproterenol infusion ($r > 0.993$). The slope of ESPLR denoted E_{es} was increased from 57 ± 7 to 71 ± 9 mmHg/mm ($P < 0.01$) after isoproterenol without changes in the extrapolated x-axis intercept (L_0). In the LAD area, E_{es} and L_0 of ESPLR remained constant both before and after the infusion of a positive inotropic agent into the left circumflex coronary artery.

In group 2 (13 dogs), 1 mg/min and 5 mg/min of lidocaine were infused into the left circumflex coronary artery to deteriorate the systolic wall motion of that area without making changes in coronary perfusion pressure. Continuous intracoronary infusion of lidocaine resulted in regional hypokinesis, which is almost the same pattern observed in the presence of regional myocardial ischemia. Percent shortening in the LCX area was reduced from 14% $\pm$ 1% to 8% $\pm$ 2% ($P<0.01$ vs control) with 1 mg lidocaine and to 5% $\pm$ 2% ($P < 0.01$ vs control; $P <0.01$ vs 1 mg lidocaine) with 5 mg lidocaine. Linearity of ESPLR was repeatedly confirmed not only at the control state (r > 0.998), but also under a decreased contractile state (r > 0.972). The slope of the ESPLR line tended to decrease in the case of 1 mg lidocaine but increased with 5 mg lidocaine (statistically not significant). The x-axis intercept of ESPLR, denoted L_0, increased from 6.3 ± 0.3 to 7.1 ± 0.4 mm ($P < 0.5$ vs control) with 1 mg lidocaine and to 8.0 ± 0.5 mm ($P < 0.01$ vs control; $P < 0.05$ vs 1 mg lidocaine) with 5 mg lidocaine, while L_0 in the LAD area was unchanged. However, the slope of the ESPLR line decreased by 15% ($P < 0.05$) with 5 mg lidocaine, probably due to systemic effects of lidocaine infused into the left circumflex coronary artery.

The instant of end-ejection as well as end-systole may be different at various segments of the myocardium, especially under conditions of regional hyper- or hypokinesis. Even in the whole heart, there is no *a priori* reason for end-ejection to coincide with end-systole [15]. Sagawa et al. [15] have adopted the convention of deriving the ESPVR regression line from the left upper corner of pressure-volume loops obtained under various preloads and afterloads, at a constant inotropic background. The determination of end-systole of various segments of the left ventricle is more difficult than that of the whole ventricle. In the present study, we derived the ESPLR regression line from the left upper corner of pressure-segment length loops. In case of regional hypokinesis, a prematurity of end-ejection was noted [1, 25]; thus, the timing of end-systole from the beginning of systole may be a little earlier at the hypokinetic segment than the intact myocardium. Such a small difference in timing in determining the ESPLR of the hypokinetic segment may be negligible in the present study, because the ESPLR line obtained by connecting the left upper corner of the pressure-length loops was very close to the regression line in the end-systolic pressure and end-systolic length relation calculated from the 10 beats while arterial pressure was falling.

The slope of the ESPLR in the hypokinetic area did not change in parallel with perturbations applied to the region analyzed, while L_0 increased along with progression in the regional hypokinesis. Thus, changes in L_0 were sensitive as an index of regional hypokinesis, but the slope of the ESPLR could not denote a contractile state. These data suggest that the increase in L_0 incorporates the mechanical interaction between segments in the hypokinetic area and surrounding intact myocardium. In the presence of regional hypokinesis, the x-axis intercept but not the slope of ESPLR shifted rightward depending on the degree of myocardial hypokinesis or the extent of the size of the ischemia. Accordingly, the interaction between hypokinetic and intact myocardial segments during active contraction may be incorporated in changes of the x-axis intercept of ESPLR. A method for quantitative determination of the force generated between the hypokinetic and the intact segment of the ventricle remains to be developed. A simple measurement of the regional myocardial contractility is a desirable goal, but it is difficult both theoretically and practically. Further studies are needed to assess changes in regional contractile state from the pressure-length relations.

Regional Wall Motion During Extension of Infarct Size

The ESPLR at the center of the ischemic area was examined during changes in size of the ischemia in anesthetized open-chest dogs [24]. Regional shortening decreased from 17% ± 2% before coronary occlusion to − 0.2% ± 2% in the case of a small ischemia and to − 0.2% ± 1% with a large ischemia. The ESPLR was strongly linear during changes in the size of the ischemia; however, the slope of the ESPLR line showed no consistent changes. The value of the x-axis intercept progressively increased from 6 ± 1 mm before coronary occlusion to 9 ± 1 ($P < 0.01$) in small and large ischemia, respectively. Thus, expansion of the area of ischemia shifted the ESPLR rightward with no change in the degree of regional shortening or the slope of ESPLR. These observations suggest stretch of the ischemic myocardium with a consecutive expansion in size of the ischemic area due to a generation of systolic drag toward the intact myocardium.

Conclusion

Linearity of end-systolic pressure-length relation was confirmed in each perturbation; however, the behavior of this relation with regard to changes in slope and x-axis intercept differed among the three situations (positive and negative inotropic interventions, and changes in infarct size). When regional hypokinesis was produced pharmacologically by selective intracoronary infusion of lidocaine, the x-axis intercept moved rightward, depending on the degree of regional hypokinesis. Expansion in size of the ischemic area also shifted the x-axis intercept rightward. These two phenomena suggested that the hypokinetic segments were stretched by the surrounding intact myocardium. Accordingly, end-systolic pressure-length relations provide useful information concerning the functional state of regional myocardium. Systolic properties of the regional myocardium rendered hypokinetic have to be quantified so as to include total geometry of the heart.

Since regional changes in wall thickness closely parallel regional shortening of the subendocardial segments [26], the concept derived from the present study, albeit qualitative, seems applicable to the study of pressure-wall thickness relations through echocardiography. The ESPVR and ESPLR probably cannot be applied for comparisons among patients until a rational normalization procedure of ESPVR and ESPLR for different heart sizes is established.

References

1. Tennant R, Wiggers CJ (1935) The effect of coronary occlusion of myocardial contraction. Am J Physiol 112:351-361
2. Tyberg JV, Parmley WW, Sonnenblick EH (1969) in vitro studies of myocardial asynchrony and regional hypoxia. Circ Res 25: 569-579
3. Kaseda S, Tomoike H, Ogata I, Nakamura M (1984) End-systolic pressure-length relations during changes in regional contractile state. Am J Physiol 247 (Heart Circ Physiol 16): H768-H774
4. Gallagher KP, Osakada G, Hess OM, Koziol JA, Kemper WS, Ross J Jr (1982) Subepicardial segment function during coronary stenosis and the role of myocardial fiber orientation. Circ Res 50: 352-359
5. Lew WYW, LeWinter MM (1986) Regional comparison of midwall segmental and area shortening in the canine left ventricle. Circ Res 58: 678-691
6. Tomoike H (1982) Analysis of in situ heart mechanics. Jpn Circ J 46: 1108-1111

7. Tomoike H, Franklin D, McKown D, Kemper M, Guberek M, Ross J Jr (1978) Regional myocardial dysfunction and hemodynamic abnormalities during strenuous exercise in dogs with limited coronary flow. Circ Res 42: 487–496
8. Tomoike H, Franklin D, Ross J Jr (1978) Detection of myocardial ischemia by regional dysfunction during and after rapid pacing in conscious dogs. Circulation 58: 48–56
9. Sonnenblick EH, Braunwald E, Williams JF Jr, Glick G (1965) Effects of exercise on myocardial force-velocity relations in intact anesthetized man: Relative roles of changes in heart rate, sympathetic activity, and ventricular dimensions. J Clin Invest 44: 2051–2062
10. Pfeffer MA, Pfeffer JM, Fishbein MC, Fletcher PJ, Spadaro J, Kloner JA, Braunwald E (1979) Myocardial infarct size and ventricular function in rats. Circ Res 44: 503–512
11. Urabe Y, Tomoike H, Ohzono K, Koyanagi S, Nakamura M (1985) Role of afterload in determining regional right ventricular performance during coronary underperfusion in dogs. Circ Res 57: 96–104
12. Suga H, Sagawa K, Shoukas AA (1973) Load independence of the instantaneous pressure-volume ratio of the canine left ventricle and effects of epinephrine and heart rate on the ratio. Circ Res 32: 314–322
13. Sagawa K (1978) The ventricular pressure-volume diagram revisited. Circ Res 43: 677–687
14. Weber KT, Janicki JS, Hefner LL (1976) Left ventricular force-length relations of isovolumic and ejecting contractions. Am J Physiol 231: 337–343
15. Sagawa K, Sunagawa K, Maughan WL (1985) Ventricular end-systolic pressure-volume relations. In: Levine HJ, Gaasch WH (eds) The ventricle: basic and clinical aspects. Martinus Nijhoff, Boston, pp 79–103
16. Grossman W, Braunwald E, Mann T, McLaurin LP, Green LH (1977) Contractile state of the left ventricle in man as evaluated from end-systolic pressure-volume relations. Circulation 56: 845–852
17. Borow KM, Neumann A, Wynne T (1982) Selectivity of end-systolic pressure-dimension and pressure-volume relations to the inotropic state in humans. Circulation 65:988–996
18. Sodums MT, Badke FR, Starling MR, Little WC, O'Rourke RA (1984) Evaluation of left ventricular contractile performance utilizing end-systolic pressure-volume relationships in conscious dogs. Circ Res 54: 731–739
19. Sagawa K, Suga H, Shoukas AA, Bakalar KM (1979) End-systolic pressure-volume ratio: A new index of contractility. Am J Cardiol 40: 748–753
20. Miller WP, Liedke AJ, Nellis SH (1984) Regional end-systolic pressure-length relationships using a volume loading technique in the intact pig heart. Circ Res 55: 326–335
21. Way B, Victory J, LeWinter MM, Lew WY, Doyle R, Foëx P, Ryder WA, Jones LA (1986) Hysteresis of left ventricular end ejection pressure-dimension relations after acute pressure loading in the intact canine heart. Cardiovasc Res 20: 490–497
22. Osakada G, Hess OM, Gallagher KP, Kemper WS, Ross J Jr (1983) End-systolic dimension-wall thickness relations during myocardial ischemia in conscious dogs. A new approach for defining regional function. Am J Cardiol 51: 1750–1758
23. Kaseda S, Tomoike H, Ogata I, Nakamura M (1985) End-systolic pressure-volume, pressure-length, and stress-strain relations in canine hearts. Am J Physiol 249 (Heart Circ Physiol 18): H648–H654
24. Kaseda S, Tomoike H, Ogata I, Nakamura M (1982) Regional left ventricular end-systolic pressure-length relation in ischemic area shifts rightward with extension of ischemia. Circulation (Suppl II) 66: II-255
25. Kumada T, Karliner JS, Pouleur H, Gallagher KP, Shirato K, Ross J Jr (1979) Effects of coronary occlusion on early ventricular events in conscious dogs. Am J Physiol 237: H542-H549
26. Sasayama S, Franklin D, Ross J Jr, Kemper WS, McKown D (1976) Dynamic changes in left ventricular wall thickness and their use in analyzing cardiac function in the conscious dog. Am J Cardiol 38: 870–879

Assessment of Left Ventricular Filling Patterns in Patients with Hypertrophic Cardiomyopathy

Yoshio Ishida, Akihiro Tani, Noboru Matsubara, Takakazu Morozumi, Masatsugu Hori, Akira Kitabatake, Kazufumi Kimura, and Takenobu Kamada[1]

Summary. Although alterations in left ventricular (LV) early filling patterns have been observed in patients with hypertrophic cardiomyopathy (HCM), it is not well elucidated what factors are related to the altered filling patterns. We classified the filling patterns into 3 groups based on the normal values of early filling indices, peak filling rate (PFR, > 2.4 end-diastolic count [EDC]/s) and time to PFR (TPFR, < 180 ms), derived from high-temporal-resolution radionuclide ventriculography: group A with normal PFR and TPFR, group B with normal PFR but prolonged TPFR, and group C with reduced PFR and prolonged TPFR. Then, in 28 patients with nonobstructive HCM, 14 patients with localized septal or apical hypertrophy, 9 patients with extensive hypertrophy, and 5 patients with diffuse hypertrophy, we investigated the effects of the extent hypertrophy and the magnitude of diastolic asynchrony on the classified filling patterns. M-mode and two-dimensional echocardiograms demonstrated that patients with localized hypertrophy were frequently included in groups A and B, but patients with extensive hypertrophy were classed in group C. Quantitative analysis of regional filling function by sector analysis with radionuclide ventriculography revealed that the magnitude of diastolic asynchrony estimated by coefficient of variance of regional TPFR was the highest in group C among the 3 groups. Thus, the classified early filling patterns were closely related to the extent of hypertrophy and the magnitude of diastolic asynchrony. Since early filling is mainly determined by LV relaxation and abnormalities of relaxation may be related to the hypertrophic process, the classification of LV early filling patterns may be useful for assessing the degree of impaired LV relaxation in patients with HCM.

Introduction

Impaired left ventricular (LV) early filling has been observed in patients with hypertrophic cardiomyopathy (HCM) by analyzing sequential changes of LV volume with contrast or radionuclide ventriculographies [1-3], by determining LV internal dimension lengthening or posterior wall thinning rates from echocardiograms [4, 5], and by measuring peak mitral flow and deceleration from Doppler ultrasonograms [6]. It is important to note, however, that these indices of early filling have not been consistently abnormal in patients with HCM. In the studies with radionuclide ventriculography, using peak filling rate (PFR) and time to PFR (TPFR) as indices of early filling function, Bonow et al. [3] found reduced PFR in only 10 of 40 patients (25%) with HCM and prolonged TPFR in 21 (53%) of the patients. This report also noted a

[1] The First Department of Medicine and Division of Nuclear Medicine, Osaka University School of Medicine, 1-1-50 Fukushima, Fukushima-ku, Osaka, 553 Japan

disparity between PFR and TPFR in the patients. Thus, there is a variety of LV filling patterns, from normal to abnormal, in the peak rate and the time course of early filling in patients with HCM. However, it has not been well elucidated what factors are related to the variation of LV filling pattern.

Radionuclide ventriculography, which has the advantage of noninvasive measurement of sequential LV volume changes, is one of the popular means of analyzing LV filling function. The conventional method used mainly for assessing LV systolic function, is however, not appropriate in assessing LV diastolic function. Higher temporal and spatial resolution in the measurement of volume curve should be required for assessing LV diastolic filling characteristics. The first purpose of the current study was to develop a method for improving the accuracy of the diastolic volume curve derived from radionuclide ventriculography. Then, applying this method to patients with nonobstructive HCM, the second purpose was to classify LV filling patterns based on the early filling indices.

The impairment of LV early filling may be related to slowed LV relaxation and partially related to increased LV chamber stiffness. Although the underlying mechanisms of slowed LV relaxation in patients with HCM have not been clarified, changes in muscle loading conditions due to decreased LV volume and increased wall thickness [7] , and failure of myocardial inactivation due to a primary calcium overload [8, 9] and/or myocardial ischemia [10, 11] with the hypertrophic process have been suggested as the possible causes. Thus, the impaired early filling may be dependent upon the extent of hypertrophy. Also, since the extent of hypertrophy is variable, ranging from localized involvement of the basal septum to concentric involvement of the left ventricle [12], LV early filling may be influenced by the interaction of relaxation mechanics in the hypertrophied and non-hypertrophied regions. As the final goal of this study, we therefore investigated the significance of the extent of hypertrophy and the regional nonuniformity of diastolic performance as determinant factors of the variation of LV filling pattern in patients with nonobstructive HCM.

Materials and Methods

Patient Selection

We studied 28 patients with nonobstructive HCM aged 25 to 72 years (mean 53). There were 23 males and 5 females. The diagnosis was based on typical echocardiographic and angiographic criteria [13, 14]. These patients were without LV outflow obstruction at rest. From the findings of two-dimensional echocardiography, the patients were divided into 4 groups based on the extent of hypertrophy: (1) 3 patients with localized septal hypertrophy (SH), (2) 11 with localized apical hypertrophy (AH), (3) 9 with hypertrophy extended from septal to anterolateral and/or apical regions (extensive hypertrophy), and (4) 5 with diffuse (concentric) hypertrophy. All patients were in normal sinus rhythm. All medications were withheld 48 h before the study. The patients with hypertension were not included.

We also studied 9 normal control subjects (NL) aged 36–59 (mean 47). All were male and without evidence of cardiovascular disease.

Radionuclide Measurements

Data Acquisition

Radionuclide ventriculography was performed in patients at rest in the supine position using red blood cell labeled in vivo with 20–30 mCi of technetium-99m. Imaging was performed with an Anger camera equipped with a high-sensitivity, parallel-hole collimator oriented in a modified left anterior oblique position to isolate the left ventricle from other cardiac structures. Data were acquired in a list mode fashion as a series of scintigraphic data ($x-$, $y-$ coordinates), time markers (every 10 ms), and R-wave markers for 6 min. The use of list-mode data acquisition allowed for the elimination of ectopic beats and the reduction of anomalous patterns of diastolic filling caused by sinus arrhythmia, when the data were processed for generating cardiac image sequences (described below).

Data Processing for Determining LV Filling Patterns

As technical factors which lose validity for diastolic analysis in the conventional method, (1) the variabilities of R-R intervals due to sinus arrhythmia in the collecting beats, (2) the low framing rate to reduce temporal resolution, and (3) the inappropriate setting of LV region of interest (ROI) to reduce spatial resolution are listed. Therefore, to minimize errors by these factors, we improved data processing software as follows. First, we eliminated extrasystolic and postextrasystolic beats and selected the beats with the R-R interval within + 5% of its mean value from the acquired list-mode data by inspection of the distribution of the R-R intervals. It is because our preliminary study indicated that the diastolic filling portions of the volume curve is distorted when the selected range of the R-R intervals were widened as shown in Fig. 1. Second, we chose a framing rate of 20 ms/frame. The framing rate has great influence on the measured rate of LV filling. As shown in Fig. 2, we found that PFR tends to be underestimated with the framing rate decreasing, more strongly in normal subjects than in patients with coronary artery disease and with decreased PFR. At a framing rate of less than 30 ms/frame, it became difficult to indicate a difference between the measured PFR in the normal and patient groups. Then, using a framing rate of 20 ms/frame, we reformatted the data into a sequence of cardiac images in the averaged cardiac cycle by combined forward-gating and reverse-gating from the R wave. The usefulness of the reverse-gating method in determining atrial systolic LV filling is now widely accepted [15]. Third, to measure sequential changes in LV radioactivities, we used a variable ROI method instead of a fixed ROI method. Although a fixed ROI method, in which LV end-diastolic ROI is applied to all frames of the cardiac image sequences, has been preferably used because of the ease of data processing, in the frames around end-systole, the radioactivities originating from other cardiac structures are often inclusively counted as LV radioactivities in this method. Therefore, we used a variable ROI method in which LV ROI for each frame is determined with an appropriate cut-off level as shown in Fig. 3. From these technical improvements for diastolic analysis, we finally generated a LV volume curve as a "forward-backward" composite time-activity curve and calculated systolic and diastolic parameters (Fig. 4). Ejection fraction (EF) was automatically measured from the time-activity curve. Peak ejection rate (PER) and peak filling rate (PFR) were measured as the maximal slope (counts per

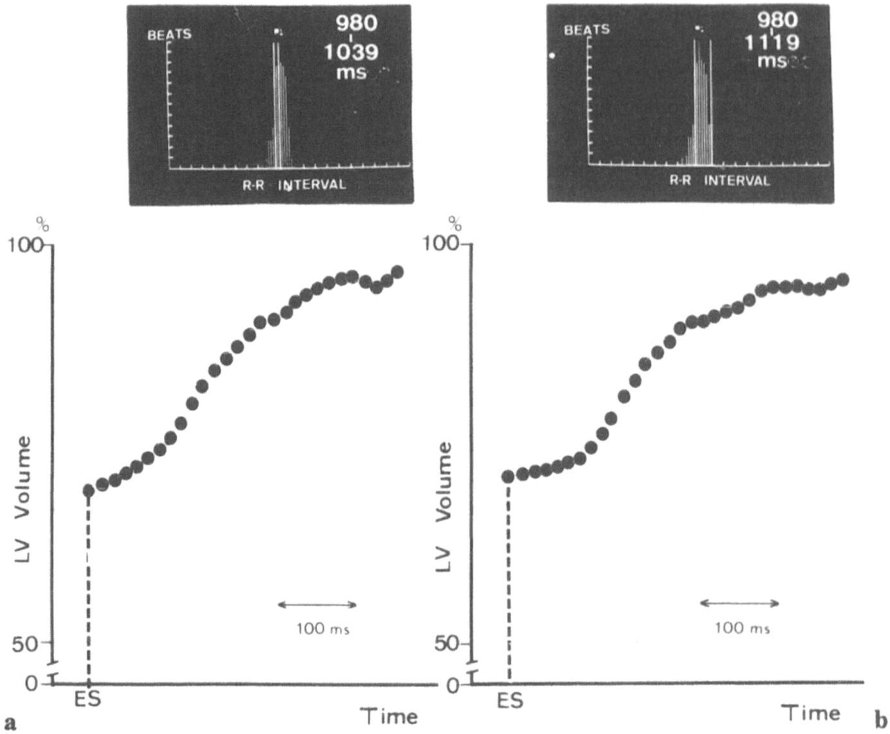

Fig. 1a, b. Effect of the selected range of R-R intervals for reformatting on the measured left ventricular (*LV*) diastolic volume curve in a patient with marked R-R variabilities due to sinus arrhythmia. **a** 980–1039 ms; **b** 980–1119 ms. *ES*, end-systole

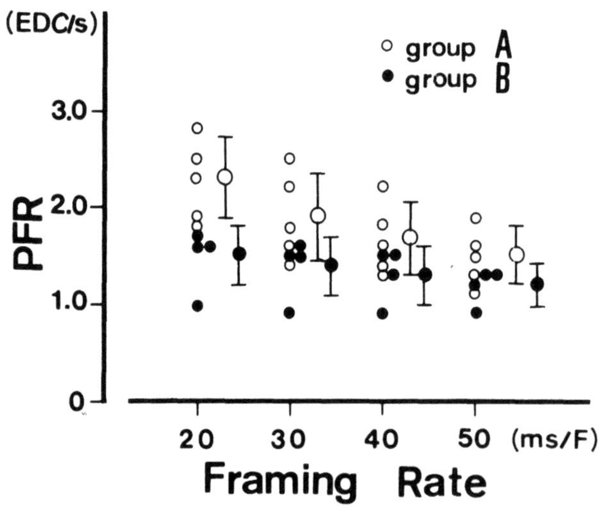

Fig. 2. Effect of framing rate in reformatting the data in a sequence of cardiac images on the measured peak filling rate (*PFR*). Group A: normal subjects (*empty circles*); group B: patients with myocardial infarction (*solid circles*). *EDC*, end-diastolic count; *F*, frame

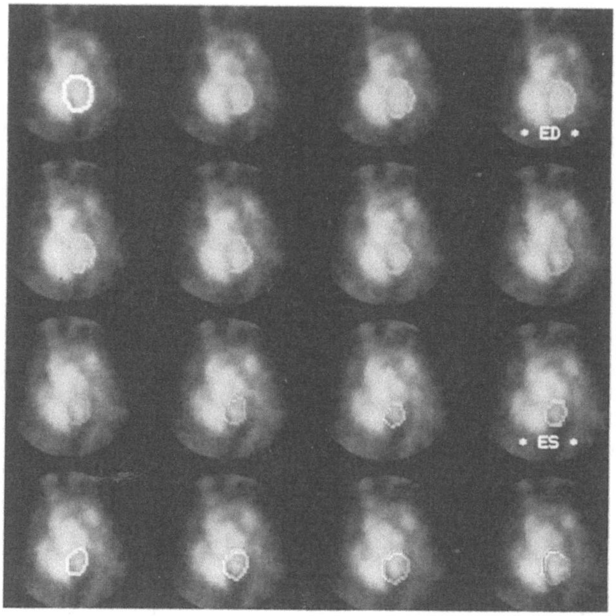

Fig. 3. A sequence of cardiac images in the averaged cardiac cycle and left ventricular regions of interest (ROIs) determined by a variable ROI method. *ED*, end-diastole; *ES*, end-systole

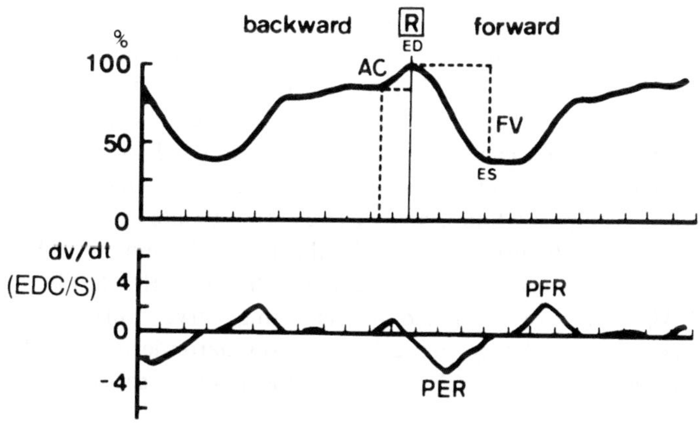

Fig. 4. A forward and backward composite left ventricular time-activity curve determined by forward-gating and backward-gating from ECG R wave and computer-derived parameters of systolic function and diastolic filling. *ED*, end-diastole; *ES*, end-systole; *AC/FC*, atrial contribution to filling; *PER*, peak ejection rate; *PFR*, peak filling rate; *EDC*, end-diastolic count

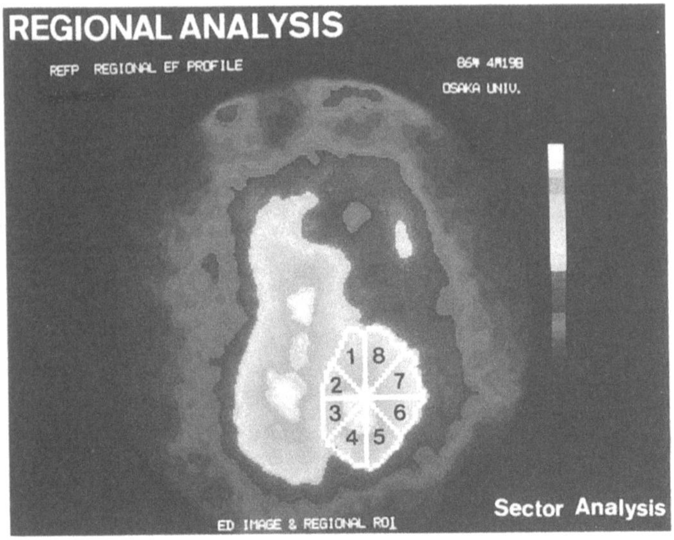

Fig. 5. Determination of 8 annular sectors of equal arc (45°), each emanating from the end-diastolic left ventricular region of interest's geometric center

second) during ejection and rapid filling phases of the curve, respectively, and were normalized for the number of end-diastolic counts (EDC/s). Time to PFR (TPFR, ms) was measured from end-systole (the nadir of the curve) to PFR. The relative contribution of atrial systole to LV filling was defined as a ratio of the filling volume during atrial systole to total filling volume (AC/FV, %).

Sector Analysis for Determining Regional Performance

To estimate LV regional performance, we used the modified "sector analysis" described by Vital et al. [16]. LV ROI in each frame was divided into 8 annular sectors of equal arc (45°), each emanating from the end-diastolic LV ROI's geometric center (Fig. 5). After setting these regional LV ROIs, sectorial time-activity curves were generated in the 8 sectors. These curves represent the changes in radioactivities within each sector and thus are related to the regional wall motion surrounding each sector. These sectorial curves were also determined as "forward-backward" composite time-activity curves. To estimate the regional performance during diastole, each curve was fitted by third-order harmonics of Fourier series for eliminating statistical errors and regional PFR (rPFR), TPFR (rTPFR), and AC/FV (rAC/FV) were calculated in the fitting curve. Furthermore, as indices of LV diastolic asynchrony, we calculated standard deviation (SD) and coefficient of variance (SD/mean, CV, %) of rTPFR among 8 sectors.

Results

LV EF in 28 patients with nonobstructive HCM was significantly greater than normal (73% ± 8% vs. 66% ± 8%, $P < 0.001$, Table 1). No patient showed

Table 1. Left ventricular systolic and diastolic performance measured by radionuclide ventriculography

	Normal ($n = 9$)	HCM ($n = 28$)	P
Heart rate (beats/min)	68 ± 13	62 ± 8	NS
LV systolic function			
LV ejection fraction (%)	66 ± 8	73 ± 8	<0.05
Peak LV ejection rate (EDC/s)	−3.4 ± 0.8	−4.0 ± 0.8	NS
LV diastolic filling			
Peak LV filling rate (EDC/s)	3.0 ± 0.5	2.7 ± 0.6	NS
Time to peak filling rate (ms)	162 ± 19	215 ± 49	<0.01
Atrial contribution to filling (%)	20 ± 6	23 ± 7	NS

Mean ± SD.
LV, left ventricular; EDC, end-diastolic count; HCM, hypertrophic cardiomyopathy

subnormal EF. PER was greater than normal (−4.0 ± 0.8 EDC/s vs −3.4 ± 0.8 EDC/s) although it was not statistically significant. Resting heart rate was not significantly different between patients with HCM (62 ± 8 beats/min) and normal subjects (68 ± 13 beats/min).

Global LV Diastolic Filling Patterns

Early Filling Patterns

We characterized LV early filling patterns by PFR and TPFR. PFR in patients with HCM was not significantly different from normal (Table 1), although 9 of 28 patients (32%) showed PFR less than 2.4 EDC/s, the lower limit of normal (mean-SD). However, TPFR and the TPFR normalized by heart rate were significantly prolonged compared with normal subjects (Table 1). Twenty-three of 28 patients (82%) showed a TPFR of more than 180 ms, the upper limit of normal (mean + SD). PFR did not correlate with TPFR in patients with HCM. From these results, the early filling patterns were classified into 3 groups based on these two early filling indices: (1) group A with normal patterns, (2) group B with prolonged TPFR but with normal PFR, and (3) group C with abnormalities in both PFR and TPFR. In the representative records illustrating the volume curves (Fig. 7), marked differences were observed in the diastolic filling pattern among three groups. In the case of group B, a delay in the onset of rapid filling was observed, but the rate of filling was preserved; while in a case of group C, a decrease in the rate of filling made the rapid filling phase prolonged and the slow filling period was consequently abbreviated. As shown in Fig. 6 indicating the patients distribution, there were 5 patients (18%) in group A, 14 (50%) in group B, and 9 (32%) in group C. PFR was 2.9 ± 0.3 EDC/s in group A, 3.0 ± 0.4 EDC/s in group B, and 2.0 ± 0.3 EDC/s in group C (Table 2). TPFR was 140 ± 25 ms in group A, 220 ± 25 ms in group B and 249 ± 44 ms in group C (Table 2). Heart rate was not significantly different among the 3 groups. Age was slightly but significantly higher in group C than in the other groups.

Atrial Systolic Filling Patterns

We characterized atrial systolic LV filling patterns by AC/FV. AC/FV in patients with HCM was not significantly greater than in normal subjects, although 9 (32%) of 28 patients showed an AC/FV of more than 26%, the upper limit of normal values (mean + SD). As related to the classification of early filling patterns, AC/FV showed a significant increase in group C, compared with normal subjects, while it did not in groups A and B (Table 2).

Table 2. Classification of LV filling patterns in patients with HCM

	HCM-A ($n = 5$)	HCM-B ($n = 14$)	HCM-C ($n = 9$)
Age	47 ± 12	50 ± 8	61 ± 8**
Heart Rate (beats/min)	64 ± 7	62 ± 7	61 ± 9
LV ejection fraction (%)	69 ± 4	73 ± 10	74 ± 5*
Peak LV ejection rate (EDC/s)	−3.6 ± 0.9	−3.9 ± 0.9	−4.3 ± 0.6*
Peak LV filling rate (EDC/s)	2.9 ± 0.3	3.0 ± 0.4	2.0 ± 0.3**
Time to peak filling rate (ms)	140 ± 25	220 ± 25**	249 ± 44**
Atrial contribution to filling (%)	20 ± 4	22 ± 6	28 ± 7*

Mean ± SD.
*P<0.05 and **P<0.001 vs normal subjects.
LV, left ventricular; EDC, end-diastolic counts; HCM, hypertrophic cardiomyopathy

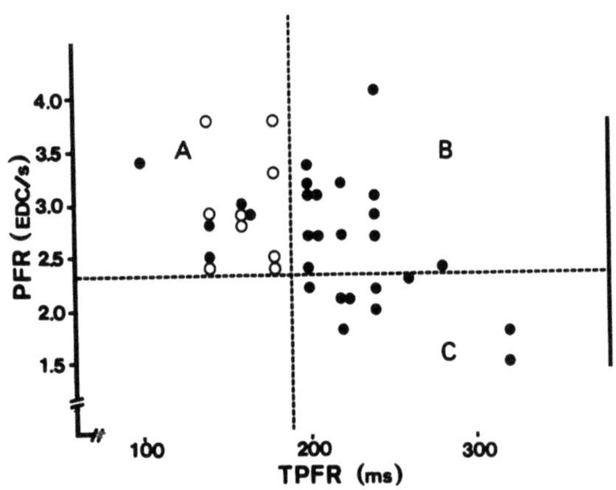

Fig. 6. Classification of left ventricular diastolic filling patterns based on early filling parameters, peak filling rate (PFR) and time to PFR (*TPFR*) and distribution of normal subjects (*solid circles, NL*) and patients with hypertrophic cardiomyopathy (*empty circles, HCM*). Group *A*, normal PFR and normal TPFR; group *B*, normal PFR but prolonged TPFR; group *C*: reduced PFR and prolonged TPFR. *EDC*, end-diastolic count

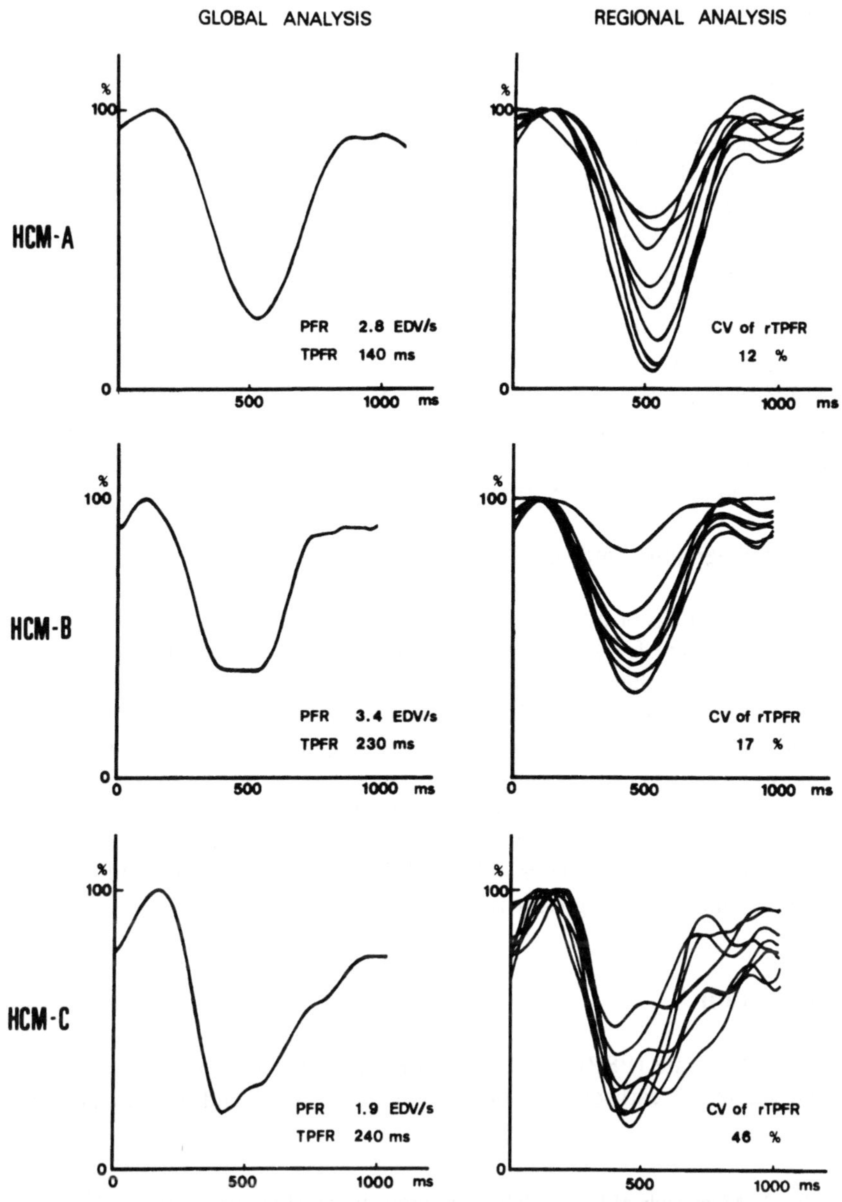

Fig. 7. Representative records illustrating global left ventricular time-activity curves and sectorial time-activity curves in HCM groups *A, B,* and *C. HCM*, hypertrophic cardiomyopathy; *PFR*, peak filling rate; *TPFR*, time to PFR; rTPFR, regional time to PFR; *CV*, coefficient of variance; *EDV*, end-diastolic volume

Regional LV Filling Patterns

Representative records illustrating global time-actively curves and sectorial time-activity curves in groups A, B, and C are shown in Fig. 7. In the case of group A (*upper panel*), homogeneous patterns in the timing of events during systole and diastole were demonstrated in the regional curves. In the case of group B (*middle panel*), a few sectorial curves showed a decrease in the slope of diastolic rapid filling (rPFR) and a delay in the timing of rPFR (rTPFR). In the case of group C (*bottom panel*), 5 sectorial curves showed irregular filling patterns (*upward* and *downward*) just after end-systole and then a decrease in rPFR and a delay of rTPFR, although the other 3 sectorial curves showed normal filling patterns. CV of rTPFR in the cases of groups A, B, and C were 12%, 17%, and 46%, respectively. SD and CV of rTPFR among 8 sectors were significantly higher in group C than in normal subjects and in groups A and B (Table 3). There was no difference in either index between groups A and B. In all cases, the SD and CV inversely correlated with global PFR (r = − 0.56 and r = − 0.48, respectively).

Echocardiographic Findings

From the findings of two-dimensional echocardiography, we examined the relationship between the LV filling patterns as classified into 3 groups and the extent of hypertrophy. As shown in Table 4, in group A, localized types of hypertrophy (SH and AH) were observed with high frequency, in 4 of 5 (80%) patients. In group B, the localized type was observed in 8 of 14 (57%) patients, the extensive type of hypertrophy in 2 patients (14%), and the concentric type of hypertrophy was in 4 patients (29%). In group C, the extensive type was observed with high frequency, in 6 of 9 patients (67%). Only 2 patients with AH were included in group C, and they were both elderly (72 and 70 years old). Both interventricular septal and posterior wall thickness increased progressively from groups A to C.

Discussion

Peak filling rate (PFR) and time to PFR (TPFR) have been commonly used to characterize LV early diastolic filling patterns in radionuclide studies, and abnorma-

Table 3. Magnitude of LV diastolic asynchrony in HCM groups A, B, and C

	Normal ($n = 9$)	HCM-A ($n = 5$)	HCM-B ($n = 14$)	HCM-C ($n = 9$)
Standard deviation of rTPFR (ms)	28 ± 6	45 ± 14*	53 ± 22*	120 ± 32**
Coefficient of variance of rTPFR (%)	22 ± 5	32 ± 9	31 ± 12	53 ± 9**

Mean ± SD.
*P < 0.05 and **P < 0.001 vs normal subjects.
LV, left ventricular; HCM, hypertrophic cardiomyopathy; rTPFR, regional time to peak filling rate

Table 4. Echocardiographic findings of LV hypertrophy

	Distribution of LV hypertrophy				Interventricular septal and posterior wall thickness	
	SH	AH	Extensive	Diffuse	IVSTh (mm)	LVPWTh (mm)
HCM-A ($n = 5$)	1 (20%)	3 (60%)	1 (20%)	0	15.2 ± 4.4	11.6 ± 0.9
HCM-B ($n = 14$)	2 (14%)	6 (43%)	2 (14%)	4 (29%)	16.5 ± 5.0	12.8 ± 3.0
HCM-C ($n = 9$)	0	2 (22%)	6 (67%)	1 (11%)	18.4 ± 3.7	13.3 ± 2.3

Mean ± SD.
LV, left ventricular; HCM, hypertrophic cardiomyopathy; SH, localized septal hypertrophy; AH, localized apical hypertrophy; Extensive, hypertrophy extended from anteroseptal to apical regions; IVSTh, interventricular septal thickness; LVPWTh, left ventricular posterior wall thickness

lities of these indices have been observed in patients with HCM [3, 17]. The present study demonstrated that there is a variation of early filling patterns in patients with HCM and the variation can be classified into 3 groups based on PFR and TPFR: (1) normal TPFR and PFR (group A), (2) prolonged TPFR but normal PFR (group B), and (3) prolonged TPFR and reduced PFR (group C). In our patients, 5 (18%) were included in group A, 14 (50%) in group B, and 9 (32%) in group C. Interestingly, 23 (82%) of 28 patients had a prolongation of TPFR, but only 9 (32%) had a decrease in PFR. Thus, PFR is not consistently abnormal in HCM, although prolonged TPFR is more often present than reduced PFR.

According to the previous studies indicating that PFR and TPFR correlated well with the time constant of LV isovolumic pressure fall [19, 20], both indices have been related to LV relaxation. The present study indicated, however, that PFR is not necessarily associated with TPFR in patients with HCM. Recently, determinant factors of PFR were investigated by using experimental models. In isolated cardiac muscles, Goethals et al. [21] and Zile et al. [22] indicated that the load applied to the muscle during isotonic lengthening (late load) affects the maximum rate of isotonic lengthening and suggested that loading conditions during early diastolic filling modulate the rate of LV filling. In the intact heart, measuring phasic mitral flow with an electromagnetic flow probe equipped at mitral annulus, Ishida et al. [23] observed that PFR is determined by the atrioventricular pressure difference at PFR and is therefore influenced by not only LV relaxation rate but also left atrial (LA) pressure at mitral valve opening. They also found that PFR is maintained at a normal level because of an increase in the LA pressure when the LV isovolumic relaxation rate is mildly impaired. These studies suggest that elevated LA pressure increases the load on the LV wall and promotes filling, counteracting the effect of slowed isovolumic relaxation. The disparity between PFR and TPFR in patients with HCM may be explained by the more sensitive property of PFR to ventricular loading conditions than TPFR. Therefore, group C (with a reduced PFR) may be in a more severely impaired state of LV relaxation than group B (without a reduced PFR). Our classification of LV early filling patterns based on PFR and TPFR may be clinically useful in estimating the degree of impaired LV relaxation in patients with HCM.

The augmented contribution of atrial systole to LV filling has been also investigated in patients with HCM [24]. Using a "reverse" gating method, we determined atrial systolic LV filling volume relative to total LV filling volume (AC/FV) and investigated it as related to the classification of LV early filling patterns. Among the 3 groups, group C (with a reduced PFR) showed a significant increase in AC/FV (28%) compared with normal subjects (20%). In groups A and B (without a reduced PFR), AC/FV was maintained at normal values. Thus, LV late diastolic filling characterized by AC/FV also showed, in group C, a different pattern from the other groups. Since absolute LV volumes cannot be estimated due to the limitation of our noninvasive technique, it is difficult to determine whether the increase in AC/FV in group C reflects only a decrease in early filling volume or associates with an increase in atrial systolic filling volume. However, it is clear that the dependence on atrial systole increases to maintain adequate LV filling volumes in compensation for the reduced rate and volume of early filling.

The impairment of early filling function, prolonged TPFR, and reduced PFR may be mainly caused by slowed relaxation. In HCM, based on evidence of load-dependent properties of relaxation [7], a primary calcium overload [8, 9], and myocardial ischemia [10, 11] (which induces failure of myocardial inactivation), the slowed relaxation has been explained by changes in muscle loading conditions due to increased wall thickness and intrinsic myocardial relaxing properties. Since these factors may be associated with the hypertrophic process, we investigated the relationship between the classified early filling patterns and the extent of hypertrophy. In result, in groups A and B, the localized type of hypertrophy was observed with high frequency. In group B, the incidence of the concentric type of hypertrophy was also high. On the other hand, in group C, with reduced PFR, the extensive type of hypertrophy was observed with high frequency, but the incidence of the localized type and the concentric type of hypertrophy was low. Although there were no significant changes in septal and posterior wall thicknesses among the 3 groups, both tended to increase progressively from group A to group C. Thus, the magnitude and distribution of hypertrophy appear to determine the variation of early filling patterns. However, the concentric type of hypertrophy was an exception to the rule, suggesting the additional role of regional nonuniformity on the filling patterns. Then, we investigated the effect of diastolic asynchrony on the filling patterns. When the degree of diastolic asynchrony was evaluated by standard deviation (SD) or coefficient of variance (CV) of LV 8 sectorial rTPFRs derived from sector analysis, we found good correlations between PFR and SD and CV in patients with HCM. Particularly, in group C with reduced PFR, the values of both indices were much higher than those in groups A and B. This may correspond with the high incidence of the extensive type of hypertrophy in this group. Group B (including 4 patients with concentric hypertrophy), however, did not show higher values than group A in either index. Thus, these results suggest that a regional difference of the time course of relaxation modulates early filling and is particularly related to reduced PFR, and so, in addition to the extent of hypertrophy, plays a significant role in determining the filling pattern.

In summary, compared with pressure overload hypertrophy, the hypertrophic lesions in HCM commonly show localized characteristics. Therefore, differences of relaxation mechanics in the hypertrophied and non-hypertrophied regions are present and demonstrate diastolic asynchronous wall motions. The interaction of regional relaxation mechanics may result in a variety of LV diastolic filling pattern.

Elevated LA pressure is often seen in this disease. Because of the load dependency of muscle lengthening, the non-hypertrophied myocardium may therefore effectively correct for the abnormal relaxation in the hypertrophied regions. The hypertrophied regions could also act to enhance relaxation in response to increased wall stress. These compensatory mechanisms would act to maintain a normal early filling pattern and to maintain a normal peak rate of rapid filling despite a prolongation of the relaxation process. These compensations, however, may be limited by an increase in the extent of hypertrophy and a decreased capacity for enhancing relaxation in the non-hypertrophied regions with aging and other unknown factors [26, 27]. Consequently, early filling would be severely impaired. Thus, the radionuclide ventriculographic observation of LV global filling patterns as classified based on PFR and TPFR may provide useful information about the degree of LV relaxation abnormalities which closely relate to the extent of hypertrophy and the magnitude of diastolic asynchrony.

References

1. Hammermeister KE, Wabasse JR (1974) The rate of change of left ventricular volume in man. II. Diastolic events in health and disease. Circulation 49: 739-
2. Sanderson JE, Gibson DG, Brown DJ, Goodwin JF (1977) Left ventricular filling in hypertrophic cardiomyopathy: An angiographic study. Br Heart J 39: 661-670
3. Bonow RO, Rosing DR, Bacharach SL, Green MV, Kent KM, Lipson LC, Maron BJ, Leon MB, Epstein SE (1981) Effects of verapamil on left ventricular systolic function and diastolic filling in patients with hypertrophic cardiomyopathy. Circulation 64(4): 787-796
4. Hanrath P, Mathey DG, Siegert R, Bleifeld W (1980) Left ventricular relaxation and filling pattern in different forms of left ventricular hypertrophy: An echocardiographic study. Am J Cardiol 45: 15-23
5. Inouye I, Massie B, Loge D, Topic N, Silverstein D, Simpson P, Tubau J (1984) Abnormal left ventricular filling: An early finding in mild to moderate systemic hypertension. Am J Cardiol 53: 120-126
6. Takenaka K, Dabestani A, Gardin JM, Russel D, Clark S, Allfie A, Henry WL (1986) Left ventricular filling in hypertrophic cardiomyopathy: A pulsed Doppler echocardiographic study. J Am Coll Cardiol 7: 1263-1271
7. Brutsaert DL, Rademakers FE, Sys SU (1984) Triple control of relaxation: Implications in cardiac disease. Circulation 69: 190-196
8. Paulus WJ, Lorel BH, Craig WE, Wynne J, Murgo JP, Grossman W (1983) Comparison of the effects of nitroprusside and nifedipine on diastolic properties in patients with hypertrophic cardiomyopathy: Altered left ventricular loading or improved muscle inactivation? J Am Coll Cardiol 2: 879-886
9. Gwathmey JK, Morgan JP (1985) Altered calcium handling in experimental pressure overload hypertrophy in the ferret. Circ Res 57: 836-843
10. Cannon RO, Rosing DR, Maron BJ, Leon MB, Bonow RO, Watson RM, Epstein SE (1985) Myocardial ischemia in hypertrophic cardiomyopathy: Contribution of inadequate vasodialtor reserve the elevated left ventricular filling pressures. Circulation 71- 234-243
11. Bertrand ME, Tilmant PY, Lablance JM, (1983) Apical hypertrophic cardiomyopathy: Clinical and metabolic studies. Eur Heart J 4 (suppl F): 127-133
12. Wnigle BJ, Epstein SE (1980) Hypertrophic cardiomyopathy: recent observations regarding the specificity of three hallmarks of the disease: asymmetric septal hypertrophy, septal disorganization and systolic anterior motion of the anterior mitral leaflet. Am J Cardiol 45: 141-154

13. Maron BJ, Epstein SE (1980) Hypertrophic cardiomyopathy: recent observations regarding the specificity of three hallmarks of the disease: asymmetric septal hypertrophy, septal disorganization and systolic anterior motion of the anterior mitral leaflet. Am J Cardiol 45: 141-154
14. Yamaguchi H, Ishimura T, Nishiyama S, Nagasaki F, Nakanishi S, Takatsu F, Nishijo T, Umeda T, Machii K (1977) Hypertrophic nonobstructive cardiomyopathy with giant negative T waves (apical hypertrophy): Ventriculographic and echocardiographic features in 30 patients. Am J Cardiol 44(3): 401-412
15. Bacharach SL, Green MV, Borer JS (1979) Instrumentation and data processing in cardiovascular nuclear medicine: evaluation of ventricular function. Semin Nucl Med 9: 257-274
16. Vital DF, Green MV, Bacharach SL, Bonow RO, Watson RM, Findley SL, Jones AE (1983) Assessment of regional left ventricular function by sector analysis: A method for objective evaluation of radionuclide blood pool studies. Am J Cardiol 52: 1112-1119
17. Sugrue DD, Dickie S, Myers MJ, Lavender JP, Mckenna WJ (1984) Effects of amiodarone on left ventricular ejection and filling in hypertrophic cardiomyopathy. 54: 1054-1058
18. Betocchi S, Bonow RO, Bacharach SL, Rosing DR, Maron BJ, Gree MV (1986) Isovolumic relaxation period in hypertrohic cardiomyopathy: Assessment by radionuclide angiography. J Am Coll Cardiol 7: 74-81
19. Fioretti P, Brower RW, Meester GT, Serruys PW (1980) Interaction of left ventricular relaxation and filling during early diastole in human subjects. Am J Cardiol 46: 197-203
20. Magorien DJ, Shaffer P, Bush C, Magorien RD, Kolibash AJ, Unverferth DV, Bashore TM (1984) Hemodynamic correlates for timing intervals, ejection rate and filling rate derived from the radionuclide angiographic volume curve. Am J Cardiol 53: 567-571
21. Goethals MA, Housmans PR, Brutsaert DL (1982) Loading determinants of relaxation in cat papillary muscle. Am J Physiol 242: H303-H309
22. Zile MR, Gaasch WH, Wiegner AW, Robinson KG, Bing OHL (1987) Mechanical determinants of maximum isotonic lengthening rate in rat left ventricular myocardium. Circ Res 60: 815-823
23. Ishida Y, Meisner JS, Tsujioka K, Gallo JI, Yoran C, Frater RWM, Yellin EL (1986) Left ventricular filling dynamics: Influence of left ventricular relaxation and left atrial pressure. Circulation 74(1): 187-196
24. Bonow RO, Frederick TM, Bacharach SL, Gree MV, Goose PW, Maron BJ, Rosing DR, (1983) Atrial systole and left ventricular filling in hypertrophic cardiomyopathy: Effect of verapamil. Am J Cardiol 51: 1386-1391
25. Bonow RO, Vitale DF, Bacharach SL, Rosing DR, Frederick TM, Green MV (1984) Regional left ventricular asynchrony and impaired global left ventricular diastolic filling in hypertrophic cardiomyopathy. Effect of verapamil. (abstract) Circulation 70 (suppl 2): 303
26. Spirito P, Maron BJ, Chiarella F, Bellotti P, Tramarin R, Pozzoli M, Vecchio C (1985) Diastolic abnormalities in patients with hypertropic cardiomyopathy: relation to magnitude of left ventricular hypertrophy. Circulation 72(2): 310-316
27. Ishida Y, Kim BH, Matsubara N, Takeda H, Tsuneoka Y, Hiraoka T, Kimura K, Kamada T (1986) LV early filling function is diffusely impaired in hypertrophic cardiomyopathy with localized hypertrophy. (abstract) Circulation 74(4) (suppl 2): 451

Assessment of Diastolic Function of Hypertrophic Heart by Doppler Echocardiography

Akira Kitabatake[1], Tohru Masuyama[2], Jun Tanouchi[1], Masatsugu Hori[1], and Takenobu Kamada[1]

Summary. Abnormalities of left ventricular diastrolic filling were studied in hypertensive patients and patients with hypertropic cardiomyopathy by the measurement of transmitral flow velocity patterns using pulsed Doppler echocardiography. Peak velocity in the early diastolic filling phase (peak E) and the ratio of peak velocity in the atrial contraction phase (peak A) to peak E were used as parameters of left ventricular diastolic filling. A decrease in peak E and an increase in the ratio of peak A to peak E were observed in hypertensive patients without apparent left ventricular hypertrophy, and a further decrease in peak E and a further increase in the ratio of peak A to peak E were observed in hypertensive patients with left ventricular hypertrophy. In patients with hypertrophic cardiomyopathy, a decrease in peak E and an increase in the ratio of peak E to peak A were observed only in patients with a decreased percentage of systolic wall thickening of the left ventricular posterior wall, and both parameters were normal in patients without a decrease in the percentage of systolic wall thickening of the left ventricular posterior wall. Thus, left ventricular diastolic filling in patients with hypertrophied hearts was noninvasively assessed with pulsed Doppler echocardiography.

Introduction

Abnormal left ventricular diastolic function and/or filling has been documented in hypertensive patients as well as in patients with hypertrophic cardiomyopathy [1-4]. These abnormalities may be detectable with cineangiography, radionuclide angiography, or pulsed Doppler echocardiography. Using pulsed Doppler echocardiography, left ventricular diastolic filling is assessed by measurement of transmitral flow velocity pattern at the level of the mitral anulus or tip. Various Doppler parameters determined on the transmitral flow velocity pattern are proposed as parameters of left ventricular diastolic filling. Peak velocity in the early diastolic filling phase (peak E) and the ratio of peak velocity in the atrial contraction phase (peak A) to peak E may be included in the most popular parameters (Fig. 1).

Peak E is shown to correlate well with a pressure-derived variable of left ventricular relaxation, i.e., the time constant of isovolumic left ventricular pressure fall [5, 6]. Also, a positive correlation has been shown between the ratio of peak E to peak A and the cineangiographic peak filling rate [7]. A good correlation was found

[1] The First Department of Medicine, Osaka University School of Medicine, 1-1-50 Fukushima, Fukushima-ku, Osaka, 553 Japan
[2] Cardiology Division, Stanford University School of Medicine, Stanford, CA 94305, USA

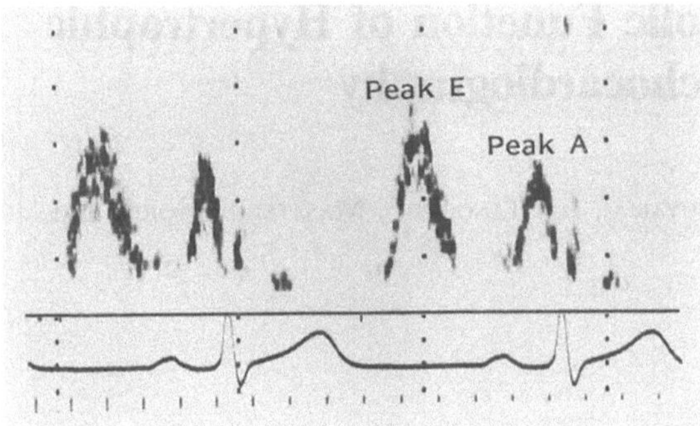

Fig. 1. Determination of Doppler parameters of left ventricular diastolic filling at peak velocity in the early diastolic filling phase (*Peak E*) and the ratio of peak velocity in the atrial contraction phase (*Peak A*) to Peak E

between the ratio of peak E to peak A and the ratio between the percent of left ventricular filling during early diastolic filling and the percent of left ventricular filling during atrial contraction as assessed with radionuclide angiography [8]. Thus, a decrease in peak E and an increase in the ratio of peak A to peak E are most likely to indicate impaired left ventricular diastolic filling or function. However, peak filling velocities are determined by the pressure differences between the atrium and ventricle at the corresponding instants. So, it is necessary to be aware of the limitations, placed by the fact that both peak E and the ratio of peak A to peak E can be influenced by left atrial pressure as well [9-11].

On the basis of these findings, pulsed Doppler echocardiography is used to assess the abnormalities of left ventricular diastolic filling in patients with hypertrophied heart, i.e., hypertensive patients and patients with hypertrophic cardiomyopathy.

Findings in Hypertensive Patients

Philips et al. [2] provided data on left ventricular diastolic filling obtained using pulsed Doppler echocardiography in treated hypertensive patients. Their data revealed early diastolic filling is normal, but late diastolic filling is increased, in treated hypertensive patients. They also suggested that the increased filling velocity in the atrial contraction phase may be independent of left ventricular hypertrophy and persist despite effective blood pressure control.

The relation between the impairment of left ventricular diastolic filling and the presence of left ventricular hypertrophy was studied in our 27 patients with essential hypertension who were not being administered antihypertensive drugs. They were divided into two groups on the basis of echocardiographically determined left ventricular wall thickness. The first group (HT-1) consisted of 17 patients without apparent left ventricular hypertrophy, i.e., mean wall thickness of the interventricular septum and left ventricular posterior wall of less than 13 mm (10 males and 7 females; ages 34-67 years, mean 54). The other group (HT-2) consisted of 10 patients with left ventricular hypertrophy, i.e., mean wall thickness of the interven-

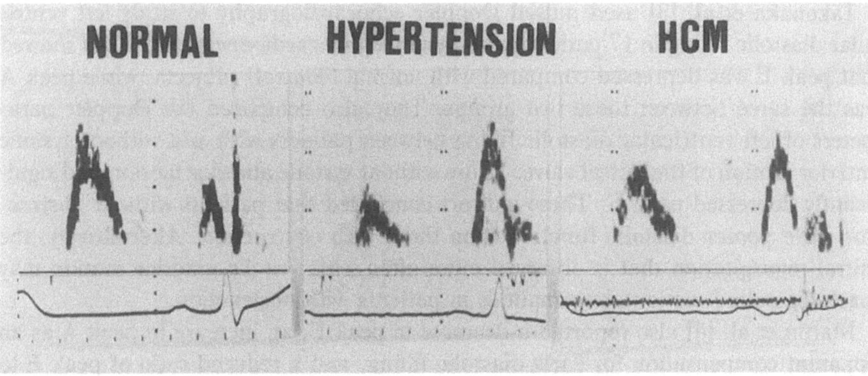

a, b c

Fig. 2a-c. Representative recordings of transmitral flow velocity pattern in **a** a normal subject (M, 58 years), **b** a patient with systemic hypertension (M, 61 years), and **c** a patient with hypertrophic cardiomyopathy (*HCM*) (M, 52 years). It is noted that in the hypertensive patient, peak velocity in the early diastolic filling phase decreased with an apparently compensatory augmentation in the flow during the atrial contraction phase. In the patient with HCM, peak velocity in the early diastolic filling phase is also decreased

tricular septum and left ventricular posterior wall of greater than or equal to 13 mm (9 males and 3 females; ages 44–74 years, mean 57).

Peak E significantly decreased in group HT-1 compared with that determined in 24 normal subjects (58 ± 8 versus 65 ± 10 cm/s, $P < 0.01$) (Fig. 2). Further decrease was observed in group HT-2 (48 ± 6 cm/s $P < 0.01$ versus normal subjects and $P < 0.01$ versus HT-1 group). The ratio of peak A to peak E was higher in group HT-1 than in normal subjects (1.03 ± 0.13 versus 0.79 ± 0.21, $P < 0.01$), and it was even higher in group HT-2 than in group HT-1 (1.37 ± 0.28, $P < 0.01$ versus group HT-1). Thus, early diastolic filling is reduced even in hypertensives without apparent left ventricular hypertrophy and further impaired in patients with left ventricular hypertrophy. These findings are comparable with previous reports obtained using radionuclide angiography and M-mode echocardiography [12–14]. However, Dianzumba et al. [15] reported that peak E in hypertensives was not different from normal value. The discrepancy between our study and Dianzumba's study may be explained by the difference in the age of the study population; their patients were younger than ours, and aging has been demonstrated to affect peak E in normal subjects as well as in hypertensives [16, 17].

Findings in Patients with Hypertrophic Cardiomyopathy

The rate of left ventricular diastolic filling can be derived from the volume curve obtained from M-mode echocardiography, from left ventricular cineangiography, or from radionuclide angiography. In contrast to the fact that this derivation has been limited by irregular and or small cavity size of the left ventricle, pulsed Doppler echocardiography provides a measurement of velocity across the mitral valve which is a direct flow assessment independent of cavity size or shape. This is particularly important in patients with hypertrophic cardiomyopathy.

Takenaka et al. [3] used pulsed Doppler echocardiography to study left ventricular diastolic filling in 17 patients with hypertrophic cardiomyopathy. They showed that peak E was depressed compared with normal (control) subjects, while peak A was the same between these two groups. They also compared the Doppler parameters of left ventricular diastolic filling between patients with and without systolic anterior motion of the mitral valve. Those without systolic anterior motion had significantly depressed peak E. These authors concluded that patients without obstruction have poorer diastolic function than those with obstruction. Alternatively, the mitral regurgitation that is observed more often with systolic anterior motion may mask diastolic function abnormalities in patients with obstruction.

Maron et al. [4] also reported a decrease in peak E, an increase in peak A as an apparent compensation for early diastolic filling, and a reduced ratio of peak E to peak A. They also observed more severe alterations in the Doppler parameters of left ventricular diastolic filling in patients with nonobstructive hypertrophic cardiomyopathy than in patients with obstruction. Further, these abnormalities were apparent in the vast majority of the study patients with hypertrophic cardiomyopathy, independent of the presence or absence of cardiac symptoms or an intracavitary pressure gradient. Thus, they suggest that impaired diastolic performance may be present at a time when functional limitation is not yet evident.

The relation between left ventricular diastolic filling abnormalities and regional contractile performance of the left ventricular posterior wall was studied in our patients with hypertrophic cardiomyopathy. Nineteen patients with hypertrophic cardiomyopathy were divided into three groups on the basis of percentage of wall thickening of the left ventricular posterior wall as follows: (1) percentage of wall thickening of greater than or equal to 70% as a group of hyperkinetic posterior wall motion (HCM-I), (2) percentage of wall thickening of less than 70% and greater than or equal to 40% as a group of normokinetic posterior wall motion (HCM-II), and (3) percentage of wall thickening of less than 40% as a group of hypokinetic posterior wall motion (HCM-III).

There was no difference in peak E between a group of 24 normal subjects and group HCM-I (65 ± 10 versus 60 ± 7 cm/s, P = NS); however, peak E was lower in group HCM-II (53 ± 10 cm/s, P = NS) and in group HCM-III (39 ± 10 cm/s, $P < 0.01$) than in the normal subjects (Fig. 3). The ratio of peak A to peak E became higher, although not significantly, in this order: normal group, group HCM-I, group HCM-II, and group HCM-III. Thus, abnormalities of left ventricular diastolic filling are minimal in patients in whom regional systolic function of the left ventricular posterior wall is maintained. On the other hand, abnormalities of left ventricular diastolic filling were remarkable in patients with depressed regional function of the left ventricular posterior wall. These findings suggest that left ventricular diastolic filling is concomitantly affected by the regional function of the interventricular septum and left ventricular posterior wall and that pulsed Doppler echocardiographic measurement of transmitral flow in patients with hypertrophic cardiomyopathy provides information about "global" or "total" rather than "regional" diastolic function of the left ventricle.

End-diastolic wall thickness of the interventricular septum was higher in group HCM-I (13 ± 2 mm, $P < 0.01$), group HCM-II (16 ± 5 mm, $P < 0.01$), and group HCM-III (28 ± 8 mm, $P < 0.01$) than in the normal subjects (8 ± 2 mm). End-diastolic wall thickness of the left ventricular posterior wall was 10 ± 2 mm in group

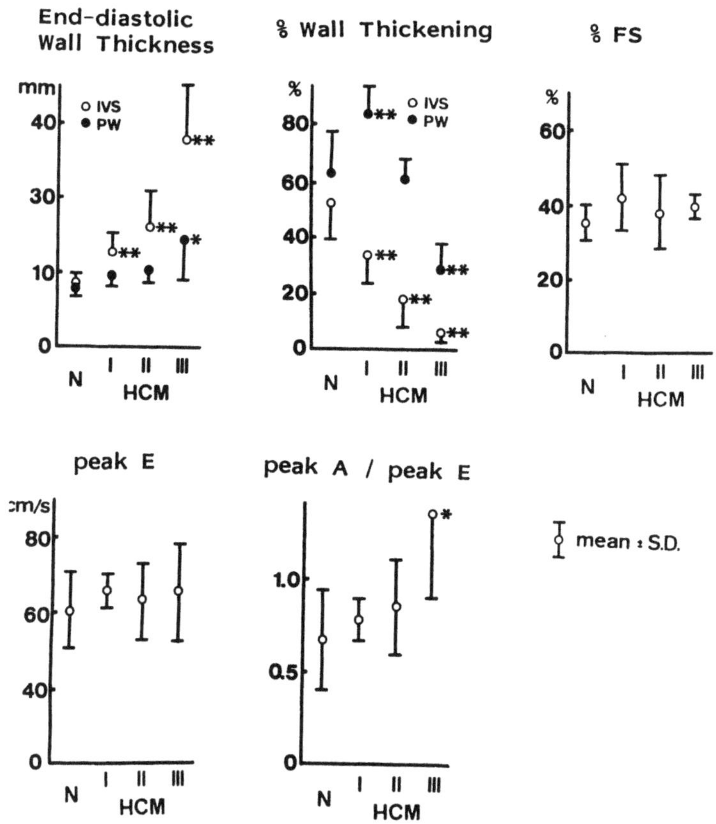

Fig. 3. M-mode echocardiographic parameters and Doppler parameters of left ventricular function. *N*, normal subjects; *HCM*, hypertrophic cardiomyopathy (*HCM-I*, patients with wall thickening of the posterior wall of greater than or equal to 70%; *HCM-II*, patients with wall thickening of the posterior wall of less than 70% and greater than or equal to 40%: *HCM-II*, patients with wall thickening of the posterior wall of less than 70% and greater than or equal to 40%; *HCM-III*, patients with wall thickening of the posterior wall of less than 40%); *IVS*, interventricular septum; *FS*, δ fractional shortening; *Perak A/Peak E*, the ratio of velocity in the atrial contraction phase to peak velocity in the early diastolic filling phase; *Peak E*, peak velocity in the early diastolic filling phase; *PW*, left ventricular posterior wall. *P<0.05, **P<0.01 versus normal subjects

HCM-I, 10 ± 2 mm in group HCM-II, and 15 ± 5 mm in group HCM-III and was significantly ($P < 0.01$) higher than in the normal subjects (9 ± 1 mm) only in group HCM-111. These findings may suggest that our classification of patients on the basis of percentage of thickening of the posterior wall partially reflects the severity of hypertrophic cardiomyopathy. If so, the degree of abnormalities of left ventricular diastolic filling may reflect the severity of hypertrophic cardiomyopathy.

Conclusion

We showed pulsed Doppler echocardiography provides useful information about left ventricular diastolic filling abnormalities in hypertrophied hearts. As noted above, however, physiological and pathological determinants of left ventricular diastolic filling have not yet been clarified, and it is necessary to be aware that there must be many factors that mimic or mask the Doppler estimates of left ventricular diastolic filling even in hypertrophied hearts.

References

1. Kitabatake A, Inoue M, Asao M, Tanouchi J, Masuyama T, Abe H, Morita H, Senda S, Matsuo H (1982) Transmitral blood flow reflecting diastolic behavior of the left ventricle in health and disease. A study by pulsed Doppler technique. Jpn Circ J 46: 92-102
2. Phillips RA, Coplan NL, Krakoff LR, Yeager K, Ross RS, Gorlin R, Goldman ME (1987) Doppler echocardiographic analysis of left ventricular filling in treated hypertensive patients. J Am Coll Cardiol 9: 317-322
3. Takenaka K, Dabestani A, Gardin JM, Russell D, Clark S, Allfie A, Henry WL (1986) Left ventricular filling in hypertrophic cardiomyopathy. A pulsed Doppler echocardiographic study. J Am Coll Cardiol 7: 1263-1271
4. Maron BJ, Spirito P, Green KJ, Wesley YE, Bonow RO, Arce J (1987) Noninvasive assessment of left ventricular diastolic function by pulsed Doppler echocardiography in patients with hypertrophic cardiomyopathy. J Am Coll Cardiol 10: 733-742
5. Kitabatake A, Tanouchi J, Inoue M, Asao M, Morita T, Masuyama T, Ito H, Yasui K, Shimazu T, Hori M, Abe H (1983) Relations between transmitral flow and ventricular relaxation. A study by pulsed Doppler flowmetry. In: Spencer MP (ed) Cardiac Doppler diagnosis. Martinus Nijhoff, Boston, pp 111-121
6. Tanouchi J, Kitabatake A, Asao M, Morita T, Masuyama T, Hori M, Inoue M, Abe H (1983) Role of left ventricular relaxation on transmitral flow dynamics during early diastole. A study with pulsed Doppler flowmetry. J Cardiogr 13: 301-307
7. Rokey R, Kuo LC, Zoghbi WA, Limacher MC, Quinone MA (1985) Determination of parameters of left ventricular diastolic filling with pulsed Doppler echocardiography. Comparison cineangiography. Circulation 71: 543-550
8. Spirito P, Maron BJ, Bonow RO (1986) Noninvasive assessment of left ventricular diastolic function. Comparative analysis of Doppler echocardiographic and radionuclide angiographic techniques. J Am Coll Cardiol 7: 518-526
9. Ishida Y, Meisner JS, Tsujioka K, Gallo JI, Yoran C, Frater RW, Yellin EL (1986) Left ventricular filling dynamics. Influence of left ventricular relaxation and left atrial pressure. Circulation 74: 187-196
10. Choong CY, Herrmann HC, Weyman AE, Fifer MA (1987) Preload dependence of Doppler-derived indexes of left ventricular diastolic function in humans. J Am Coll Cardiol 10: 800-808
11. Appleton CP, Hatle LK, Popp RL (1988) Demonstration of restrictive ventricular physiology by Doppler echocardiography. J Am Coll Cardiol 11: 757-68
12. Drelinski GR, Frohlich ED, Dunn FG, Messerli FH, Suarez DH, Reisin E (1981) Echocardiographic diastolic ventricular abnormality in hypertensive heart disease: atrial emptying index. Am J Cardiol 47: 1087-90
13. Inouye I, Massie B, Loge D, Topic N, Silverstein D, Simpson P, Tubau J (1984) Abnormal left ventricular filling: an early finding in mild to moderate systemic hypertension. Am J Cardiol 53: 120-126
14. Shapiro L, McKenna W (1984) Left ventricular hypertrophy. Relation of structure to diastolic function in hypertension. Br Heart J 51: 637-42
15. Dianzumba SB, DiPette DJ, Cornman C, Weber E, Joyner CR (1986) Left ventricular filling characteristics in mild untreated hypertension. 8 (Suppl I): I-156-160
16. Miyatake K, Okamoto M, Kinoshita N, Owa M, Nakasone I, Sakakibara H, Nimura Y (1984) Augmentation of atrial contribution to left ventricular inflow with aging as assessed by intracardiac Doppler flowmetry. Am J Cardiol 53: 586-589
17. Sartori MP, Quinones MA, Kuo LC (1987) Relation of Doppler-derived left ventricular filling parameters to age and radius/thickness ratio in normal and pathologic state. Am J Cardiol 59: 1179-82

Assessment of Cardiac Function in Diseased Heart by Catheterization Study

Yuzo Hirota, Gen Shimizu, Yoshio Kita, and Keishiro Kawamura[1]

Summary. The assessment of cardiac function by catheterization began with the application of Frank-Starling's mechanism in clinical practice. Although numerous indexes of contractility have been proposed in the past 20 years, the most reliable methods currently used are: (1) E_{max} (the slope of end-systolic pressure volume line obtained by changing afterload) by means of pressure and volume recordings, (2) the force-length (end-systolic stress–end-systolic volume) relationship either in the two-dimensional framework or these ratio, and (3) the stress-shortening (afterload-shortening) relationship in the two-dimensional framework for the analysis of systolic function.

Left ventricular relaxation is usually assessed by obtaining the time constant of the pressure fall during the isovolumic relaxation period. The assessment of chamber compliance can be obtained by analysis of the diastolic pressure-volume relationship, and that of muscle compliance can be done by analysis of the stress-strain relationship.

For these evaluations of systolic and diastolic function, the accurate measurement of pressure, volume, and wall thickness are mandatory. Every index proposed in the past, and in current use, has theoritical as well as methodological limitations; there is no golden standard for the assessment of cardiac function.

Introduction

The evaluation of ventricular function by cardiac catheterization started with the application of Frank-Starling's mechanism in the two-dimensional framework of preload and performance [1]. Measurements of left ventricular (LV) volume and ejection fraction (EF) have become routine work in the catheterization laboratory since cineangiography was introduced in clinical practice. The indexes of isovolumic contraction period, i.e., maximum velocity of myocardial fiber shortening at zero stress (V_{max}) and peak value of instantaneous dP/dt/p (Vpm), were received great attention in the late 1960s as ideal indexes of contractility, but these indexes were proven to have relatively little sensitivity in the detection of depressed contractility in patients with heart disease.

It is well known that overall pump function is maintained by the beautifully integrated workings of preload, afterload, contractility, heart rate, relaxation, and compliance. Therefore, the understanding or assessment of cardiac function in normal

[1] The Third Division, Department of Internal Medicine, Osaka Medical College, 2-7 Daigakucho, Takatsuki, Osaka, 569 Japan

and diseased hearts should be based on precise analysis and reconstruction of these six important determinants.

In this review article, currently used indexes of these determinants are explained, and their theoretical and methodological limitations are discussed.

Preload

The actual preload to the ventricle is defined as the sarcomere length just before the beginning of contraction. As it is impossible to obtain sarcomere length in vivo heart, end-diastolic pressure (EDP), end-diastolic volume (EDV), or end-diastolic stress (σED) have been used as indexes. At present, σED is considered to be the most reliable index [2]. This is based on the assumption that sarcomere length is increased with the elevation of passive stress. There is no evidence, however, that sarcomere length increases to the same level in different hearts with the same level of σED. Furthermore, there are some other problems with σED as the index of preload: (1) wall thickness is not uniform in the ventricle, (2) ellipsoid revolution cannot be applied in some heart diseases such as hypertrophic cardiomyopathy, and (3) EDP measured in the ventricle does not always represent the transmyocardial pressure when pericardial pressure is taken into account [3] (Fig. 1).

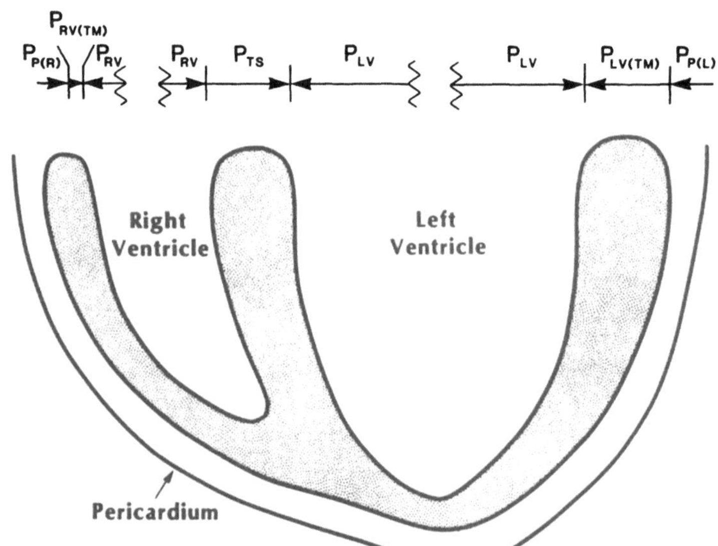

Fig. 1. The concept of transmyocardial pressure. The 4 chambers of the heart are surrounded by the relatively stiff pericardium. As the atrium is both a thin-walled chamber and a compliant cavity in comparison with the ventricle, intrapericardial pressure should be elevated in cases of elevated atrial pressures. Thus, the transmyocardial pressure in congestive heart failure should be far less than intraventricular diastolic pressure. $P_{P(R)}$, right-sided pericardial pressure; $P_{RV(TM)}$, right ventricular transmyocardial pressure; P_{RV}, right ventricular pressure; P_{TS}, transmyocardial pressure of the interventricular septum; P_{LV}, left ventricular pressure; $P_{LV(TM)}$, left ventricular transmyocardial pressure; $P_{P(L)}$, left-sided pericardial pressure. From [3] with the author's permission

Afterload

Afterload to the ventricle is defined as the resistance against the ejection of blood from the ventricle. Considering this resistance from the outside of the ventricle, it is the aortic input impedance which consists of systemic vascular resistance, elastance of the aorta and large arteries, and blood viscosity [4]. This aortic input impedance, however, is meaningless in some diseased hearts, e.g., those with aortic stenosis or mitral regurgitation. Alternatively, this resistance against systolic ejection can be defined as systolic force against contracting myocardium, and the force is able to be obtained as mean systolic stress (σSMEAN) [2].

By measuring instantaneous pressure, radius and wall thickness, instantaneous stress can be obtained, and σSMEAN can be calculated by averaging instantaneous stress during systole. Although it is possible to obtain instantaneous stress by the simultaneous recordings of LV pressure and cineangiogram or echocardiogram, this procedure is time consuming and impractical. Therefore, mid-systolic stress (σMS) [5] and end-systolic stress (σES) [6] have been proposed as the conventional indexes of afterload.

Relaxation

During the isovolumic contraction period, intracellular calcium bound to the troponin is released and sequestrated into the sarcoplasmic reticulum. As a result, the actomyosin cross-bridge is disconnected, and the sarcomere increases its length to the unstressed state. It is well accepted that this process, relaxation, is disturbed from the very early stages of any heart disease. To evaluate relaxation, many invasive and noninvasive indexes have been proposed: peak velocity of LV pressure fall (peak (−) dP/dt), time constant of LV pressure fall (T), time between the aortic valve closure to the mitral valve opening (isovolumic relaxation time), peak velocity of transmitral blood flow (peak filling rate or velocity), and time between the aortic valve closure to the peak filling rate (time to peak filling) among them. These indexes, however, are strongly affected by other factors, especially the level of left atrial pressure and heart rate. Nobody knows the meaning and significance of the measurements of these indexes in diseased hearts which have no isovolumic relaxation period, such as those with mitral and aortic regurgitations.

Weiss et al. [7] have proposed T as an index of relaxation based on the observation in canine hearts that LV pressure decay during isovolumic relaxation is exponential, and T is relatively independent of other cardiac determinants. Hori et al. [8], however, have proven that T is markedly changed by the timing of the afterload of preceeding systole, and Kumada and his co-workers [9] have shown that LV pressure decay in diseased heart is not exponential. Even though these theoretical limitations do exist, T seems to be the most reliable index of relaxation at present. The clinical significance of this relaxation disturbance has not been elucidated yet.

Compliance

Clinical evaluation of LV compliance is based on the experimental observation that there is an exponential relationship between LV diastolic pressure and volume [10].

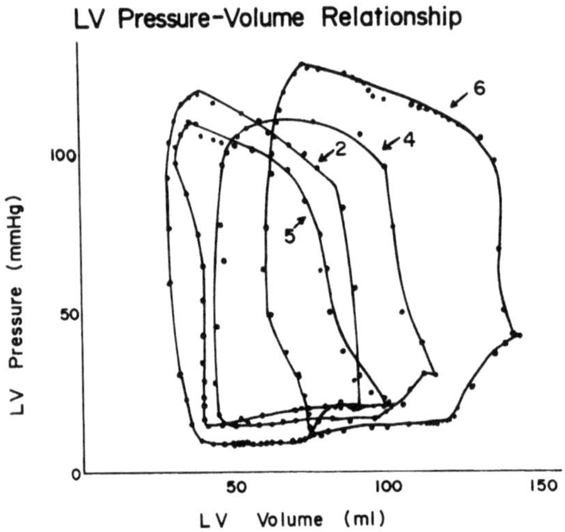

Fig. 2. Pressure-volume loops of patients with hypertrophic cardiomyopathy. Diastolic pressure-volume curves are apparently not exponential. In cases of normal sinus rhythm (*2, 4, 6*), the curves shift abruptly upward with the increment of diastolic filling by atrial contraction. This phenomenon is considered to be due to myocardial viscous component. The ventricle becomes stiffer when filling rate is increased. Looking at these loops, it is self-evident that the assessment of compliance by the fitting of the exponential curve is impossible in the human left ventricle (*LV*), line 5 is taken from a patient with atrial fibrillation

LV chamber compliance is estimated by analysis of the pressure-volume relation, and muscle compliance by stress-strain relation [11]. There are 2 major limitations to this approach. As myocardium is not elastic but visco-elastic properties, diastolic pressure-volume curve shifts with the changes of filling rate. For example, the diastolic pressure-volume curve does not fit on the exponential line when filling is suddenly increased by atrial contraction (Fig. 2). Another important limitation is that the diastolic pressure-volume curve is not steep, but flat, in cases of restrictive cardiomyopathy or constrictive pericarditis (Fig. 3), both of which are considered to have the least compliant left ventricles among diseased hearts. The reason why the line in these diseases is flat is unclear. It is postulated that LV diastolic pressure does not represent transmyocardial pressure in these conditions. The left ventricle is encapsulated by the stiff pericardium in constrictive pericarditis, and pericardial pressure may be elevated via high right atrial pressure in restrictive cardiomyopathy [3].

It must be recognized that we do not currently have an appropriate method to assess human LV compliance accurately.

Evaluation of Cardiac Performance

Frank-Starling's Curve Revived

The classical method for the analysis of LV function by means of Frank-Starling's mechanism has been revived since the introduction of the Swan-Ganz catheter at bedside. The therapeutic value of Forrester et al.'s subset [12] in acute myocardial

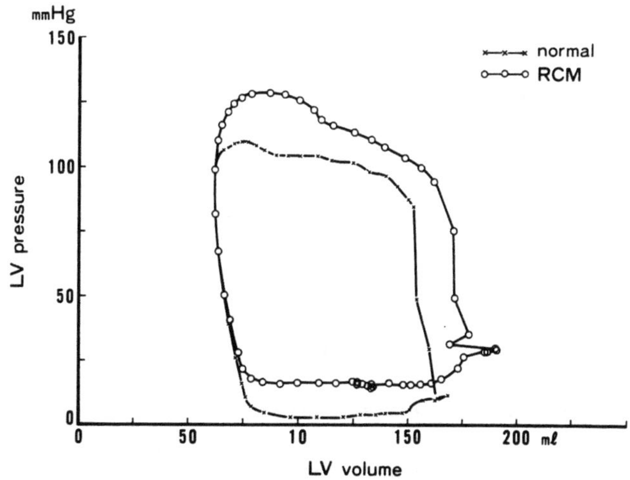

Fig. 3. Pressure-volume loops of normal heart and restrictive cardiomyopathy (*RCM*). Restrictive cardiomyopathy is defined as the disease of stiff left ventricle. If the chamber or the myocardium is stiff, the diastolic curve should be steeper than that of a normal heart. The diastolic pressure-volume curve in a patient with restrictive cardiomyopathy is not steep at all, and only shifted upward during whole diastole. The same phenomenon was observed in 3 other patients with this disease and in 4 cases of constrictive pericarditis

infarction is well known. It must be recognized that the 2 major life-threatening factors are the reduction of cardiac output, which leads to cardiogenic shock, and the elevation of pulmonary capillary pressure, which results in acute pulmonary edema. The measurements and attenuations of these 2 important factors are really useful for the management of not only patients with acute myocardial infarction but any critically ill patient. The cardiac output itself does not represent overall cardiac performance, and the pulmonary capillary pressure differs from LV EDP or preload in many occasions. Therefore, one must be aware that Forrester et al.'s classification is a therapeutic guideline derived by the modification of Frank-Starling's mechanism.

End-Systolic Pressure Volume Relation (E_{max})

The concept of E_{max} is discussed extensively by Sagawa (p 81, this volume) and is omitted in this article. The clinical application was relatively difficult until recently, as LV volume and pressure must be measured repeatedly by changing loading conditions. Grossman and his associates [13] reported E_{max} obtained in the catheterization laboratory for the first time. The usefulness of this concept will be validated in the near future as instantaneous measurements of LV pressure and volume have become relatively easy with the clinical application of conductance catheters.

Force-Length (End-Systolic Stress-Volume) Relation

The evaluation of contractility by the σES-ESV relation is based on the concept of the force-length relation in the excised papillary muscle study by Sonnenblick [14]. The clinical usefulness of this analysis has been emphasized by both Spann et al.

[15] by observation in patients with aortic stenosis, and Carabello et al. [16] in patients with mitral regurgitation. Hirota and his co-workers [17], however, have reported that it is impossible to distinguish between patients with volume overload and dilated cardiomyopathy, as both of these groups of patients have an elevated σES and an enlarged end-systolic volume (ESV). The limitation of this approach for the assessment of contractility could be explained as follows. In hearts with depressed contractility, ESV is increased as the result of the elongation of end-systolic sarcomere length due to inadequate contraction. On the other hand, the increase in ESV in hearts with volume overload is mediated by the longitudinally hypertrophied myocyte with normal sarcomere length at the end-systole. It is impossible to differentiate these two conditions by the measurement of ESV.

Stress-Shortening (Force-Velocity) Relation

The evaluation of contractility in the two-dimensional framework of afterload and shortening rate or velocity is also based on the concept of the force-velocity relationship of the papillary muscle study [14]. The clinical application of this approach began with the introduction of the concept of "afterload mismatch and preload reserve" by Ross [18]. Since Gunther and Grossman [19] reported the importance of inappropriately elevated afterload or afterload mismatch in the genesis of congestive heart failure (CHF) in aortic stenosis, many investigators have tried this approach for the evaluation of cardiac function in the diseased heart. The major limitation of this approach is the absence of the preload contribution in this framework, since systolic shortening is not only dependent on afterload but also on preload via the Frank-Starling mechanism. The other problem is the evaluation of afterload as mentioned above. This approach, however, seems to be most reliable for the assessment of systolic function at present.

Current Concepts and Future Studies in Systolic and Diastolic Function in the Diseased Heart

Concentric Hypertrophy or Pressure Overload Heart

Systemic hypertension and aortic stenosis are the two representative diseases of pressure overload. In both conditions, systolic pressure is elevated, but afterload or systolic stress and systolic shortening are maintained in the normal range during the early stage by the mechanism of compensatory hypertrophy.

Disturbance or prolongation of relaxation does exist from the early stage. The genesis of this disturbance is not clear, but abnormalities of the calcium-handling process in the myocardium via hypertrophy itself or subendocardial ischemia has been postulated. The appearance of atrial fibrillation and CHF with normal systolic shortening is occasionally seen in moderately or severely hypertrophied hearts. This has been attributed to "diastolic failure," i.e., prolonged relaxation and reduced compliance. The stiff heart with hypertrophy seems to be self-evident, but there is no method of proving stiff heart as a pathogenetic factor in CHF with our current methods as discussed above.

Whether a severely hypertrophied heart with a normal systolic shortening has a normal or depressed contractility is a matter of controversy. Both myocardial failure due to depressed contractility and inappropriately elevated afterload or afterload mismatch due to inadequate hypertrophy are important for CHF with reduced systolic shortening. The separation between depressed contractility and afterload mismatch is reported to be possible with the analysis of the stress-shortening relation [20]. The pathogenesis and turning point of deterioration of contractility in the hypertrophied heart are not clear and will need to be elucidated.

Eccentric Hypertrophy or Volume Overload Heart

Mitral (MR) and aortic regurgitations (AR) are the two representative diseases of volume overload. Preload is reported to be elevated from early stages, while afterload and contractility remain in the normal range, according to Grossman et al. [21]. Our observation suggests that preload and afterload remain in the normal range during the compensated stage, and both of these loads begin to elevate with the appearance of CHF in association with a mild decline in EF [17]. Detailed analysis suggests that the compensatory mechanism and genesis of CHF differ between these two diseases. While MR can be classified as pure volume overload, AR is both pressure and volume overload. The genesis of CHF in MR is mainly related to the elevation of left atrial pressure due to the regurgitation itself, and that in AR is the sequela of depressed contractility due to the failure of compensatory hypertrophy. Considering these factors, mitral valve surgery is recommended in symptomatic patients when CHF cannot be controlled well with medical therapy, and aortic valve surgery should be performed before the appearance of CHF or myocardial failure (unpublished data).

References

1. Sarnoff SJ, Berglund E (1954) Ventricular function. I. Starling's law of the heart studied by means of simultaneous right and left ventricular function curve in the dog. Circulation 9: 706–718
2. Braunwald E, Ross J Jr (1979) Control of cardiac performance. In: Berne RM (ed): Cardiovascular system. American Physiological Society, Bethesda, Maryland, pp 533–580 (Handbook of physiology, vol 1/2)
3. Tyberg JV (1989) Ventricular interaction and the pericardium. In: Levine HJ, Gaasch WH (eds) The ventricle. Basic and clinical aspects. Martinus Nijhoff, Boston, pp 171–207
4. Milnor, MR (1975) Arterial impedance as ventricular afterload. Circ Res 36: 565–570
5. Hirota Y, Suwa M, Takatsu T (1978) Mid-systolic endocardial stress as a noninvasive index of afterload. J Cardiogr 8: 99–106
6. Reicheck N, Wilson J, St John Sutton M, Plappert TA, Goldberg S, Hirshfeld JW (1982) Noninvasive determination of left ventricular end-systolic stress: Variation of the method and initial application. Circulation 65: 99–108
7. Weiss JL, Frederiksen JW, Weisfeldt ML (1976) Hemodynamic determinants of the time-course of fall in canine left ventricular pressure. J Clin Invest 58: 751–760
8. Hori M, Inoue M, Kitakaze M, Tsujioka K, Ishida Y, Fukunami M, Nakajima S, Kitabatake A, Abe H (1984) Ejection timing as a major determinant of left ventricular ralaxation rate in isolated perfused canine heart. Circ Res 55: 31–38

9. Kumada T, Takayama K, Matsuzaki M, Matsuda Y, Kusukawa R (1982) Assessment of left ventricular relaxation in the diseased heart in man. Jpn Circ J 46: 58-63
10. Diamond G, Forrester JS, Hergis J, Parmley WW, Danzig R, Swan HJC (1971) Diastolic pressure-volume relationship in the canine left ventricle. Circ Res 29: 267-275.
11. Gaasch WH, Apstein CS, Levine HJ (1985) Diastolic properties of the left ventricle. In: Levine HJ, Gaasch WH (eds) The left ventricle. Basic and clinical aspects. Martinus Nijhoff, Boston, pp 143-170
12. Forrester JS, Diamond G, Chatterjee K, Swan HJC (1976) Medical therapy of acute myocardial infarction by application of hemodynamic subsets. N Engl J Med 295: 1356-1362, 1404-1413
13. Grossman W, Braunwald E, Mann T, McLaurin LP, Green LH (1977) Contractile state of the left ventricle in man as evaluated from end-systolic pressure-volume relations. Circulation 56: 845-852
14. Sonnenblick EH (1962) Force-velocity relations in mammalian heart muscle. Am J Physiol 202: 931-939
15. Spann JF, Bove AA, Natarajan G, Kreulen T (1980) Ventricular performance, pump function, and compensatory mechanisms in patients with aortic stenosis. Circulation 62: 576-582
16. Carabello BA, Nolan SP, McGuire LB (1981) Assessment of preoperative left ventricular function in patients with mitral regurgitation: Value of the end-systolic stress end-systolic volume ratio. Circulation 64: 1212-1217
17. Hirota Y, Shimizu G, Kaku K, Saito T, Kawamura K (1983) Two dimensional analysis of left ventricular function by afterload ejection fraction and end-systolic stress volume relationships. (abstract) Circulation (Suppl III) 68: 238
18. Ross J Jr (1976) Afterload mismatch and preload reverve. A conceptual framework for the analysis of ventricular function. Prog Cardiovasc Dis 18: 255-264
19. Gunther S, Grossman W (1979) Determinants of ventricular function in pressure-overload hypertrophy in man. Circulation 59: 679-688
20. Carabello BA, Green LH, Grossman W, Cohn LH, Koster JK, Collins JJ Jr (1980) Hemodynamic determinants of prognosis of aortic valve replacement in critical aortic stenois and advanced congestive heart failure. Circulation 62: 42-48
21. Grossman W, Jones D, McLaurin LP (1975) Wall stress and patterns of hypertrophy in the human left ventricle. J Clin Invest 56: 56-64

Clinical Diagnosis and Management of Left Ventricular Diastolic Dysfunction

William H. Gaasch[1] and Gerald Izzi[2]

Summary. Diastolic dysfunction is a relatively common problem that may be mild and asymptomatic or it may present with severe disabling symptoms. It is most frequently due to coronary artery disease and/or left ventricular (LV) hypertrophy, and it is especially common in the older populations. Patients with diastolic dysfunction and normal-systolic function are best treated with calcium-channel blocking agents or beta-blocking agents (drugs that are generally avoided in patients with significant systolic dysfunction). These drugs are used in the same dosage as used in patients with angina or hypertension. Most treatment is empirical or based on symptomatic relief and, therefore, periods of cautious trial and error are the rule. When diastolic dysfunction is associated with systolic dysfunction, it may be necessary to treat both conditions, but in general, positive inotropic agents and arterial vasodilators are not useful in patients with diastolic dysfunction.

Introduction

In recent years, clinicians and physiologists alike have re-examined their concepts about the pathophysiology of heart failure. A major focus of this deliberation has been to clarify the important distinction between systolic and diastolic dysfunction of the left ventricle (Fig. 1). Simply stated, systolic dysfunction can be considered a defect in the ability of the myofibrils to shorten against a load; thus, the left ventricle loses its ability to *eject* blood into a high pressure aorta. The term diastolic dysfunction implies that the ventricle cannot *accept* blood at low pressures; ventricular filling is slow, delayed, or incomplete unless atrial pressure increases. Consequently, pulmonary and/or systemic venous congestion develop. Thus, the signs and symptoms of pulmonary and/or systemic venous congestion are not necessarily the result of systolic dysfunction, but rather they are commonly related to alterations in diastolic properties of the left ventricular (LV) chamber. These may consist of changes in the passive "elastic" properties of the left ventricle and/or changes in the "active" processes of relaxation and filling.

[1] Department of Medicine (Cardiology), The Med Center/Memorial, 119 Belmont Street, Worcester, MA 01605, USA
[2] The Medical Center of Boston, 125 Parker Hill Avenue, Boston, MA 02120, USA

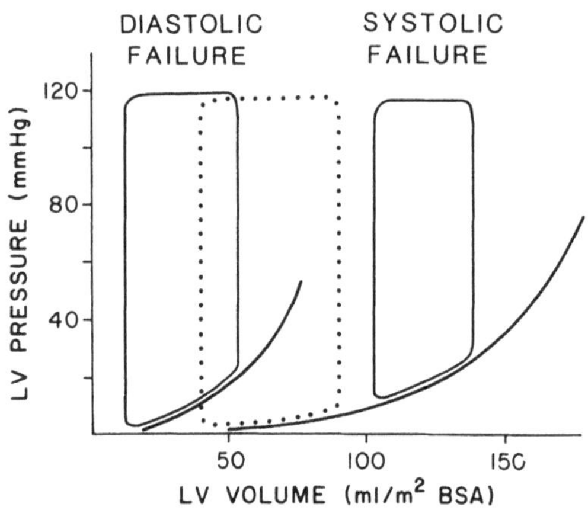

Fig. 1. Schematic diagram of left ventricular (*LV*) pressure-volume loops in diastolic and systolic failure. The *dotted line* represents a normal loop. In diastolic failure the chamber volume is normal or decreased, the systolic ejection fraction is normal or above normal, and the chamber stiffness is increased; as a consequence of the increased stiffness, end-diastolic pressure is elevated. In systolic failure the chamber is dilated and the systolic ejection fraction is low; the end-diastolic pressure may or may not be increased. *BSA*, body surface area

Definitions

A detailed discussion of terminology is beyond the scope of this clinical article; the reader is referred to more extensive reviews of this subject [1–10].

Chamber stiffness is defined by analyzing the characteristics of the curvilinear diastolic pressure-volume relations (Fig. 2). The slope of a tangent (dP/dV) to this curvilinear relation defines chamber stiffness at any level of filling pressure. Since diastolic pressure-volume data can be fit by an exponential relation, a modulus of chamber stiffness (k_c) can be derived as the slope of the linear relation between dP/dV and pressure. Thus, chamber stiffness (dP/dV) may vary if there is a change in filling pressure or if there is a change in the pressure-volume curve.

Myocardial stiffness refers to the intrinsic properties of a unit of myocardium. The myocardial stiffness constant (k_m) is derived as the slope of the linear relation between myocardial elastic stiffness and wall stress; elastic stiffness is defined as the slope of a tangent to the curvilinear myocardial stress-strain relation. *Stress* is defined as the force per unit cross-sectional area of myocardium. *Strain* is the deformation of a material produced by the application of a force.

Myocardial relaxation refers to the processes by which the myocardium returns to its initial or presystolic (end-diastolic) length and tension. The energy dependent process of deactivation interacts with the prevailing loading conditions to control relaxation, and these interrelated factors are modulated by neurohumoral, metabolic, and pharmacologic influences. Attempts to identify relaxation abnormalities in the intact heart have developed from measurements made during two periods of the cardiac cycle: (1) the period between the aortic valve closure and mitral valve opening (the isovolumic-relaxation period), and (2) the period after mitral valve opening (the diastolic-filling period).

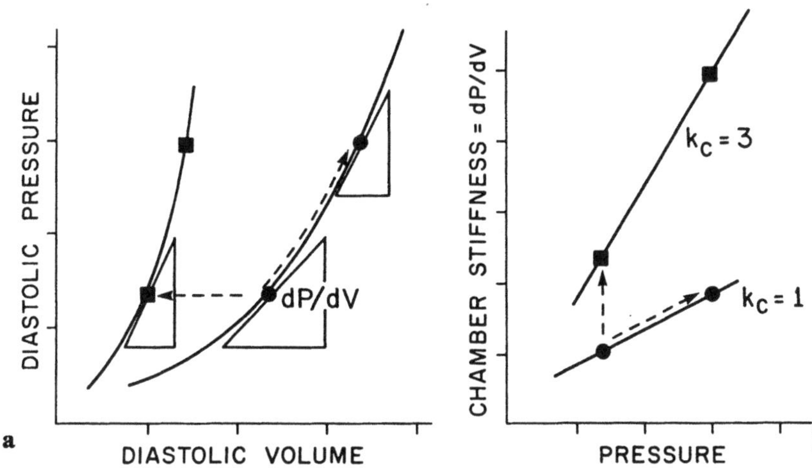

a b

Fig. 2a, b. Schematic diagram of left ventricular diastolic pressure-volume relations (**a**) and chamber stiffness-pressure relations (**b**). A progressive increase in volume will increase chamber stiffness (*dP/dV*). A leftward shift of the pressure-volume relation will also increase chamber stiffness. As the diastolic pressure-volume relation is exponential, the relation between *dP/dV* and pressure is linear; the slope of this relation represents the chamber stiffness constant (k_c)

Methods

Cardiac catheterization with angiography has been the standard method for assessing the diastolic properties of the left ventricle. Simultaneous measurements of LV pressure, volume, and geometry are made throughout the cardiac cycle. Thus, the rate of LV pressure decline (isovolumic relaxation rate), the rate and timing of diastolic filling (auxotonic relaxation), and diastolic compliance curves (chamber stiffness and myocardial stiffness) may be described. Cardiac catheterization not only provides a complete description of the diastolic properties of the heart, but this technique also allows an accurate assessment of the multiple factors known to influence diastolic events (i.e., right ventricular and pericardial effects, left atrial pressure, coronary perfusion pressure, reflected aortic pressure waves, and ventricular asynergy).

The M-mode *echocardiogram* can provide regional lengthening and wall thinning rates as well as diastolic time intervals and diastolic mitral valve motion. This method is especially suited for serial measurements over time in a single patient and for studies of diastolic heart sounds. The two-dimensional echocardiogram has a slow acquisition rate and limited resolution, but despite these limitations, it has been used to measure a first-half filling fraction as an index of diastolic function.

Doppler echocardiography allows an assessment of LV inflow velocity at the level of the mitral valve and in this fashion LV "filling rates" are evaluated. Blood velocity is closely related to flow only if the mitral annulus size remains constant; because this is not generally true, it must be emphasized that Doppler echocardiography provides a measure of velocity (meters per second) not flow (milliliters per second). Under steady-state conditions, early and late diastolic velocity measurements correlate well with data derived from other methods. It should be recognized, however, that alterations in left atrial and LV diastolic pressure can cause substantial changes in the Doppler velocity pattern—independent of changes in the intrinsic cardiac relaxing system (i.e., the sarcoplasmic reticulum).

Radionuclide angiography provides a measure of changes in the radionuclide emitted counts from the LV blood pool throughout the cardiac cycle, and thus, it provides ejection and filling parameters that are not dependent on LV geometry. Peak filling rate, the time to peak filling rate, and filling fractions can be obtained from the radionuclide angiocardiogram. This technique has the same disadvantages of any "non-invasive method": intracardic pressures are not measured, and the multitude of factors that affect diastolic function cannot be evaluated.

Diagnosis

A complete evaluation of the diastolic properties of the left ventricle requires calculation of LV chamber and myocardial stiffness constants as well as operating (end-diastolic) pressure, stress, and stiffness; LV pressure and volume transients (i.e., dP/dt and dV/dt) during diastole must also be described. If these parameters and indices of diastolic function are to provide meaningful clinical information however, the hemodynamic and other factors that influence them should also be considered. For example, left atrial pressure is a major determinant of the LV filling rate; thus, changes in filling rate are difficult to interpret unless simultaneous measurements of left atrial pressure are made. This is a formidable task requiring invasive catheterization techniques, and therefore, a complete assessment of cardiac diastole is rarely made in clinical practice. For these and other reasons, a clinical assessment of diastole requires less invasive methods or tests that are tailored to answer the question at hand.

Population Studies

If the purpose of the study is to estimate the incidence or prevalence of diastolic dysfunction in a large population, it is appropriate to use non-invasive radionuclide angiocardiography or echocardiography. The rate of change of LV volume and other volume transients are easily determined with the radionuclide technique while diastolic time intervals, wall thinning rates, and transmitral flow velocities can be determined with echocardiographic and Doppler techniques. This purely descriptive non-invasive approach to the assessment of diastolic function might well be used in cardiovascular studies of aging, in patients with hypertensive cardiovascular disease, as well as in other populations with or without symptoms referable to the heart.

Patients with Dyspnea

This category includes patients with the symptoms of congestive heart failure; such patients are sometimes found to have a normal LV ejection fraction and the diagnosis of diastolic dysfunction is therefore considered. Although many, if not most, of these patients have at least a component of diastolic dysfunction, the diagnosis

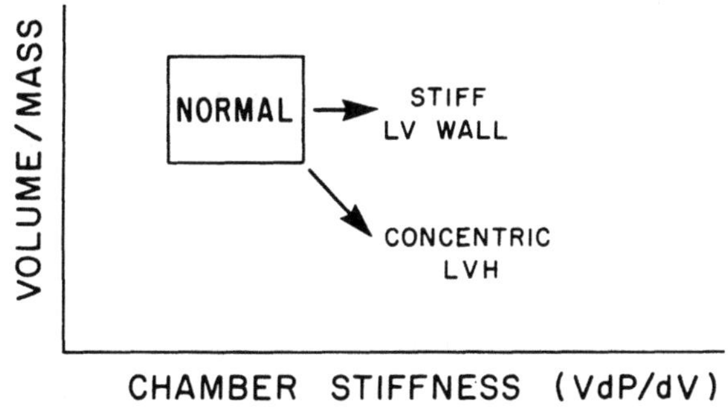

Fig. 3. Left ventricular volume-to-mass ratio (V/M) is plotted against normalized chamber stiffness (*VdP/dV*). Chamber stiffness may be increased in the presence of a normal volume-to-mass ratio, as may be seen in restrictive cardiomyopathies (*stiff LV wall*). In concentric left ventricular hypertrophy (*LVH*), the increased chamber stiffness is secondary, at least in part, to the decreased volume-to-mass ratio

requires exclusion of pulmonary disorders and mitral valve disease as well as other less common causes of impaired LV filling (i.e., cor triatriatum, pulmonary venous occlusive disease). Ideally, the pulmonary artery wedge or left atrial pressure should be measured at the same time as the ejection fraction. In clinical practice, however, the presence of pulmonary rales and radiographic evidence of pulmonary venous hypertension in combination with a normal ejection fraction is usually sufficient to suggest the diagnosis of diastolic dysfunction.

Having made the diagnosis of diastolic dysfunction, it may be possible to dissect out the cause by considering the LV end-diastolic volume/mass or radius/thickness ratio (Fig. 3); this is especially true if fixed abnormalities of the passive pressure-volume relation are present. If the LV wall thickness is increased, myocardial hypertrophy or infiltrative disease are likely culprits; if not, stiff (fibrotic) myocardium or intermittent myocardial ischemia is probably responsible for the diastolic dysfunction.

Patients with Congestive Heart Failure

Patients in this category include those admitted to the intensive care unit with acute pulmonary edema and others that develop pulmonary congestion in the perioperative setting; some of these patients have a normal systolic ejection fraction and abnormal diastolic function. Under these circumstances it is necessary to confirm an elevated pulmonary venous pressure (pulmonary artery wedge pressure) and obtain echocardiographic or radionuclide data to define the extent and cause of diastolic dysfunction. As discussed above, mitral valve disease, pulmonary disease, and non-cardiac causes of pulmonary edema must be excluded.

Clinical Research

Another consideration involves the acquisition of detailed functional and structural data as might be necessary in clinical investigation designed to answer basic physiologic questions or in those designed to assess mechanisms of drug actions. Thus, it is necessary to calculate the chamber and myocardial stiffness constants, to measure chamber relaxation and filling rates, and the factors that influence these parameters (i.e., systolic load, length, shortening, and left atrial pressure); myocardial biopsies may also be obtained if, for example, it is necessary to assess myocardial fibrosis and collagen content.

Factors Responsible for Diastolic Dysfunction

Attempts to design a rational therapeutic approach require an understanding of the mechanisms underlying diastolic dysfunction. A detailed discussion of these factors, however, is well beyond the scope of this article. Thus, the general overview shown in Fig. 4 illustrates the complexity of the problem. Multiple factors alone or in combination can be responsible for elevated LV diastolic pressures, left atrial and pulmonary venous hypertension, and the signs and symptoms of congestive heart failure.

Increased passive stiffness of the left ventricle may be seen in patients with myocardial scar or aneurysm after transmural myocardial infarction. Chronic coronary artery disease and dilated cardiomyopathy also result in scattered myocardial fibrosis and increased LV stiffness. Hypertrophy, particularly the concentric distribution found in response to a pressure overload and the inappropriate hypertrophy seen in hypertrophic cardiomyopathy, is variably associated with an increase in passive chamber stiffness. Although increased LV mass per se can influence chamber stiffness through simple geometric mechanisms, myocardial fibrosis may contribute to increased chamber stiffness through an influence on intrinsic myocardial stiffness. In hypertrophic hearts, ischemia and delayed relaxation may cause an additional stiffening effect.

Abnormal myocardial relaxation may also play a role in producing elevated diastolic pressures in a variety of conditions including coronary disease and hypertrophy. Prolonged and incomplete relaxation in these disorders is due in part to sub-

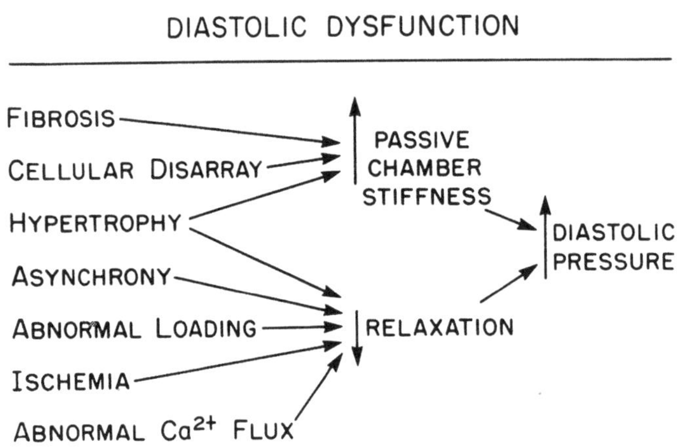

Fig. 4. Factors responsible for diastolic dysfunction and increased left ventricular diastolic pressure

endocardial ischemia, but it may also be related to primary alterations in cytosolic calcium fluxes or to asynchronous contraction-relaxation. Conduction delays in areas of ischemia or fibrosis within the myocardium might likewise produce abnormal isovolumic relaxation and a slow early diastolic filling rate. In patients with acute or chronic ischemia, alterations in clinical indices of relaxation likely represent the participation of several dynamic factors, including altered ventricular loads, changes in cellular calcium sequestration, and nonuniformity of LV contraction-relaxation.

Treatment

General Approach to Therapy

Traditional therapy directed at improving systolic performance has little place in the management of heart failure due to diastolic dysfunction. Indeed, arterial vasodilators and perhaps even digitalis may be harmful. In patients with hypertrophic or coronary heart disease, diastolic dysfunction is best managed with beta adrenergic receptor blocking agents, calcium channel blocking drugs, or both. These drugs slow the heart rate, improve the balance between myocardial oxygen supply and demand, and thus, they may increase myocardial relaxation rate. A general approach to therapy is outlined in Table 1.

A major goal of therapy should be to reduce pulmonary and systemic venous pressures and, in this way, lessen the congestive symptoms. Although diuretics and salt restriction are effective, care must be taken to avoid excessive preload reduction. Because of the steep (stiff) diastolic pressure-volume curve, large changes in pres-pressure can be achieved with only modest changes in volume. Venodilators, such as nitrates, are useful preloading-reducing agents, and they have the additional benefit of providing anti-ischemic therapy.

Maintenance of a properly timed atrial contraction is important in preserving LV filling in late diastole. Thus, efforts should be made to preserve normal sinus rhythm and to employ cardioversion when atrial arrhythmia is present; when a pacemaker is indicated for heart block, an atrioventricular sequential pacemaker should be considered.

Table 1. A general approach to treatment of congestive heart failure caused by diastolic dysfunction

Goal of therapy	Suggested method
Reduce venous pressure	Diuretics, nitrates
Maintain atrial contraction	Cardioversion for atrial fibrillation, sequential A-V pacemaker
Prevent or reduce LVH	Antihypertensive drugs, surgery (AVR for AS)
Prevent/treat ischemia	Beta-blocker, Ca^{2+} entry blocker, Nitrates
Reduce hr/increase filling time	Beta-blocker, Ca^{2+} entry blocker
Avoid arterial vasodilators	

LVH, left ventricular hypertrophy; A-V, aortic valve; AVR, aortic valve replacement; AS, aortic stenosis

Systemic arterial hypertension should be treated early, in an attempt to prevent progressive LV hypertrophy. If hypertrophy is already established, medications that reduce both arterial pressure and LV mass should probably be used; alpha-methyldopa and beta-blocking agents accomplish both these aims, whereas hydralazine and minoxidil do not. Aortic valve replacement when performed in properly selected patients with aortic valve disease usually results in a substantial reduction in LV mass and filling pressures.

The acute increase in LV diastolic pressure that occurs during myocardinal ischemia can be treated with nitroglycerin, but a more rational goal should be to prevent myocardial ischemia and its consequences, both acute (impaired LV relaxation) and chronic (fibrosis and scar). Such consequences are seen in patients with coronary disease and in those with cardiac hypertrophy alone. Thus, beta-adrenergic receptor blockers, calcium-channel blocking drugs, and nitroglycerin are essential ingredients in the treatment and prevention of diastolic dysfunction.

Beneficial slowing of the heart rate can be achieved with the use of beta-adrenergic receptor-blocking agents or calcium-channel blocking drugs. Whereas a reduction in heart rate may promote LV filling by increasing the filling time, a most important effect of these drugs is energy sparing (improved balance of myocardial oxygen supply and demand) and, in some instances, a decrease in myocardial ischemia. The optimal heart rate in patients with diastolic dysfunction is not known.

Positive inotropic drugs and arterial vasodilators appear to have no place in the management of congestive heart failure that is due to diastolic dysfunction. Indeed, some vasodilators may cause substantial hypotension, especially if used with diuretics and nitrates; therefore, this class of drugs should be avoided in the treatment of patients with diastolic dysfunction.

References

1. Gaasch WH, Levine HJ, Quinones MA, Alexander JK (1976) Left ventricular compliance: Mechanisms and clinical implications. Am J Cardiol 38: 645–653
2. Grossman W, McLaurin LP (1976) Diastolic properties of the left ventricle. Ann Intern Med 84: 316–326
3. Mirsky I (1976) Assessment of passive elastic stiffness of cardiac muscle: Mathematical concepts, physiologic and clinical considerations, directions of future research. Prog Cardiovasc Dis 18: 277–308
4. Glantz SA, Parmley WW (1978) Factors which affect the diastolic pressure-volume curve. Circ Res 42: 171–180
5. Brutsaert DL, Housmans PR, Goethals MA (1980) Dual control of relaxation: Its role in the ventricular function in the mammalian heart. Circ Res 47: 637–652
6. Mirsky I (1984) Assessment of diastolic function: suggested methods and future considerations. Circulation 69: 836–841
7. Gaasch WH, Apstein CS, Levine HJ (1986) Diastolic properties of the left ventricle. In: Levine HJ, Gaasch WH (eds) The ventricle. Martinus Nijhoff, Boston, 143–170
8. Grossman W, Lorell BH (1988) Diastolic relaxation of the heart. Martinus Nijhoff, Boston
9. Harizi RC, Bianco JA, Alpert JS (1988) Diastolic function of the heart in clinical cardiology. Arch Intern Med 148: 99–109
10. Katz AM (1988) Influence of altered inotropy and lusitropy on ventricular pressure-volume loops. J Am Coll Cardiol 11: 438–445

Chapter E.
Cardiac Function in Heart Failure

Masatsugu Hori

Heart failure is a clinical syndrome that may be a manifestation of inadequate myocardial contraction (myocardial failure), volume overload, or inadequate ventricular filling. To compensate for the impaired cardiac function, the Frank-Starling mechanism works initially when the end-diastolic volume is increased; however, this mechanism's primary service is for beat-to-beat regulation of cardiac performance. For a longer-term activation of the heart, e.g., during exercise, sympathetic activity is enhanced to increase the contractile state of the heart. This regulation is mediated by neural reflex through the sympathetic efferent activity mediated by the cardiovascular center in the hypothalamus. The heart responds to the sympathetic activation through adrenergic receptors on the cell surface. In the failing heart, however, the response of the heart to sympathetic activity is attenuated partly due to the down-regulation of beta-adrenergic receptors.

Hori et al. review the neurohumoral regulation of the heart in chronic heart failure, particularly focusing on the role of beta-adrenergic receptors during compensated and decompensated periods and the clinical significance of the enhanced renin-angiotensin-aldosterone system in chronic heart failure. They also review the long-term effect of beta-adrenergic antagonists in chronic heart failure. More detailed mechanisms of signal transduction through adrenergic receptors are reviewed by Karliner. He describes the possible role of alpha-adrenergic receptors coupled with phosphatidyl inositol response in cardiac hypertrophy which is often associated with chronic heart failure. Recently, the abnormality of G proteins which transfer the signal from beta-receptors to adenylate cyclase has been a point of interest. An enhanced coupling between Gs and beta-receptors may partially account for the beneficial effect of beta-blocker therapy in chronic heart failure.

Sympathetic stimulation is prominent during exercise. Sato et al. describe the role of somatic reflex in activation of sympathetic nerves. A close correlation between the mixed venous oxygen tension and plasma norepinephrine levels indicates that the augmented sympathetic response to exercise in heart failure is attributed to an increase in hypoxic stimulus to muscle afferents caused by an inadequate perfusion of active muscles. Although sympathetic stimulation is a major mechanism of augmented cardaic output during exercise, the contribution of the Frank-Starling mechanism is still unclear. Sasayama and Asanoi analyze circulatory adjustments during exercise stress using the coupling framework of the ventricle and the arterial load. They describe that in advanced heart failure the regulation of stroke volume during exercise shifts from a catecholamine-mediated reduction in end-systolic volume to a greater reliance on the Frank-Starling mechanism. In the last contribution, Hashimoto reviews new inotropic agents developed in Japan which are under preclinical or early clincical trials. They include several compounds with various mechanisms of action. Some of them seem promising for treatment of chronic heart failure, although further clinical tests are needed to establish a new inotropic agent.

Neurohumoral Abnormalities and Adrenoceptor Changes in Chronic Heart Failure

MASATSUGU HORI, KATSUOMI IWAKURA, AKIRA KITABATAKE, and TAKENOBU KAMADA[1]

Summary. In congestive heart failure, sympathetic activity is excessively enhanced, partly due to blunted baroreceptor sensitivity. Renin-angiotensin and arginine-vasopressin levels are also elevated. These neurohumoral factors accelerate systemic vasoconstriction which impedes the cardiac output, and may exert a direct toxic effect on the myocardium. Sustained sympathetic stimulation also induces the down-regulation of myocardial β_1-adrenergic receptors which may cause a manifestation of overt heart failure, whereas β-adrenergic receptors in compensatory periods are increased in the experimental animal models. Thus, down-regulation of β-receptors indicates a loss of a compensatory mechanism for cardiac dysfunction. A long-term treatment with β-blocking drugs is effective for some patients with advanced dilated cardiomyopathy. Beneficial effects of β-blocker therapy may be attributed in part to the up-regulation of β-adrenergic receptors, although improvement of symptoms and cardiac function are not directly related to β-receptor changes.

Autonomic Nervous System in Congestive Heart Failure

Cardiopulmonary and arterial baroreceptors are located in the heart and lungs, and in the great vessels. They are stimulated by local wall stretch in response to increases in pressure and volume, and discharge inhibitory signals to the central nervous system via the glossopharyngeal nerve and vagus nerve. There are several lines of evidence both in animal preparations and in humans with heart failure which showed blunted baroreceptor responses to changes in pressure and volume [1]. Animals with low-output heart failure exhibit impaired responses to bilateral occlusion of the carotid artery: increases in heart rate, arterial pressure, and peripheral vascular resistance [1]. It is also reported that the heart rate response to administration of a pressor dose of a vasoconstrictive agent (phenylephrine) is markedly blunted in dogs with heart failure [2]. In patients with congestive heart failure, also, increases in peripheral vascular resistance and heart rate in response to head-up tilt or lower body negative pressure are considerably impaired [3, 4]. Under normal conditions, these postural maneuvers induce venous pooling and hence a decrease in arterial pressure which attenuates the inhibitory signals from the cardiopulmonary and arterial baroreceptors to the central nervous system, leading to increased sympathetic nervous activity, activation of the renin-angiotensin system,

[1] The First Department of Medicine, Osaka University School of Medicine, 1-1-50 Fukushima, Fukushima-ku, Osaka, 553 Japan

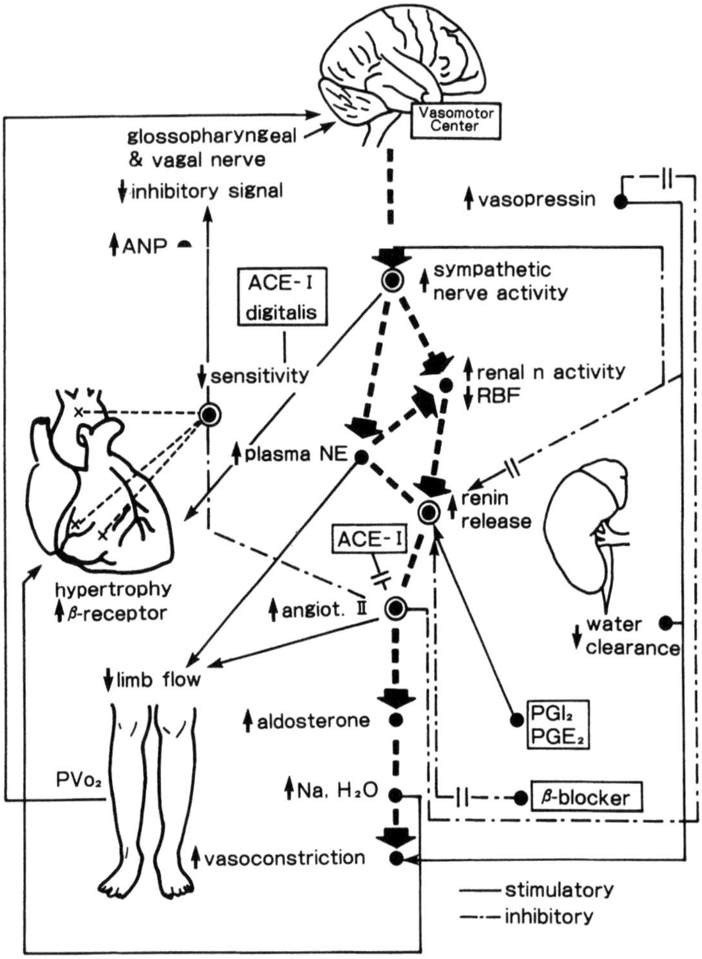

Fig. 1. Schema of the neurohumoral abnormalities in congestive heart failure. In congestive heart failure, the sensitivity of baroreceptors in the cardiac chambers and the great vessels is decreased and thus, baroreflex inhibition of central sympathetic activity is attenuated. Consequent activation of sympathetic, renin-angiotensin, and arginine-vasopressin systems exerts vasoconstriction, and induces renal retention of sodium and water. Digitalis, angiotensin-converting enzyme inhibitor (ACE-I) and β-adrenergic receptor antagonists may block this vicious circle in heart failure. (↑) indicates stimulation process and (↓) indicates inhibitory process. *angiot II*, angiotensin II; *ANP*, atrial natriuretic peptide; *n*, nerve; *NE*, norepinephrine; PV_{O_2}, venous oxygen pressure; *RBF*, renal blood flow

and the release of vasopressin [5]. These neurohumoral activations serve to ensure cardiovascular homeostasis; maintenance of systemic blood pressure through vasoconstriction and preservation of blood volume through retention of sodium and water (Fig. 1).

In heart failure, the sensitivity of cardiopulmonary and aortic baroreceptors is attenuated and inhibitory signals to the medullary cardiovascular control region are depressed. The consequent decrease in parasympathetic activities augments sympathetic outflow and hence, renin and vasopressin release. Although the mechanisms of an attenuated sensitivity in congestive heart failure are still unknown, the degenerative changes in the neuronal end-plates of the aortic arch and carotid sinuses may be involved [6]. A parasympathetic defect is also involved as evidenced by impaired heart rate response to atropine or baroreceptor stimulation in heart failure [7]. However, it is clinically important that the blunted sensitivity of the baroreceptor is functional in part since it reverses with administration of digitalis [8], chronic treatment with angiotensin-converting enzyme (ACE) inhibitor, or heart transplantation [10]. Improvement in baroreceptor function can be achieved even by a decrease in cardiac loading and an increase in myocardial contractility [11]. The mechanism of the effect of digitalis on restoration of baroreceptor function may be a combination of direct stimulation of receptors, improvement in contractility, and inhibition of the sodium-potassium ATPase pump [12]. It should also be noted that administration of digitalis decreases the plasma renin and angiotensin levels, which may improve the reflex response since angiotensin II attenuates baroreceptor sensitivity [13].

Plasma norepinephrine levels are often increased in patients with congestive heart failure [14] and the degree of elevation correlates with severity of cardiac dysfunction and the exercise capacity. It is widely accepted that the plasma norepinephrine level reflects the sympathetic activation, although it represents only a small percentage of the total neurotransmitter pool [15]. Therefore, the high plasma norepinephrine level in heart failure indicates excessive sympathetic nervous activity. However, it is not established whether the excessive sympathetic activity is a cause of acceleration of cardiac dysfunction or an outcome of the vicious circle of heart failure. There is no doubt that enhanced sympathetic activities induce excessive vasoconstriction which, in turn, impedes the cardiac output. Increased plasma norepinephrine may be deleterious for myocardium through a direct insult to myocardial cells and an indirect insult by increasing myocardial oxygen consumption. At the same time, however, the excessive sympathetic activity may serve as a compensatory support in the depressed heart. Since the cardiac response to sympathetic stimulation during exercise is markedly depressed in patients with severe heart failure, enhanced sympathetic activation in heat failure may play a compensatory role in maintenance of cardiac pump function in response to the oxygen demand of peripheral tissues [16].

It is noteworthy that a large discrepancy is observed between the effects of sympathetic stimulation in the acute hemodynamic effect and in the long-term prognosis; sympathetic stimulation which could compensate the hemodynamic depression may be deleterious for the prognosis of the patients. Cohn et al. [16] reported that plasma norepinephrine is the only independent multivariate predictor for 5-year mortality. The unfavorable response to long-term therapy with β-adrenergic agonists observed in the previous studies [17] also supports the hypothesis that the chronic excessive sympathetic outflow is harmful for cardiac function and long-term prognosis though it is effective in short-term treatment.

It is believed that the renin-angiotensin system also contributes to the hemodynamic and clinical abnormalities of patients with chronic heart failure. This is supported by two lines of evidence; plasma renin activity is elevated in relation to the clinical severity of the heart [18], and long-term converting-enzyme inhibition improves cardiac performance, exercise duration and prognosis in a large proportion of patients with congestive heart failure of various etiologies [19, 20]. Although renin and angiotensin are potent vasoconstrictors, the peripheral vasoconstriction alone is unlikely to be a cause of cardiac deterioration, since elevated plasma renin or angiotensin activity is not always accompanied by stable heart failure but they may be transient in the early stage of heart [21]. It should also be noted that renin and angiotensin II may act as local tissue hormones, so that circulating levels may not accurately reflect the concentrations at active receptor sites [22]. It is more conceivable that the renin-angiotensin system is important in the pathogenesis of hyponatremic heart failure [23] which may be directly related to the prognosis of patients. A long-term follow-up study supports this hypothesis, since treatment with converting-enzyme inhibitor improves the prognosis of the hyponatremic patients [24].

Down-regulation of Adrenergic Receptors in Chronic Heart Failure

It has been well known that long-term treatment with a β-adrenergic agonist results in desensitization of the inotropic response of the heart both in vitro and in vivo [25]. It is also known that inotropic and chronotropic responses of the heart to β-adrenergic stimulation are markedly depressed in congestive heart failure [26]. However, direct evidence that desensitization of the heart in chronic heart failure is attributable to the down-regulation of β-adrenergic receptors was first demonstrated in 1982 by Bristow and his co-workers [27]. Since their original work, several reports have confirmed that the density of ventricular myocardial β-adrenergic receptors is reduced in patients with chronic heart failure, regardless of etiology [28]. Fig. 2

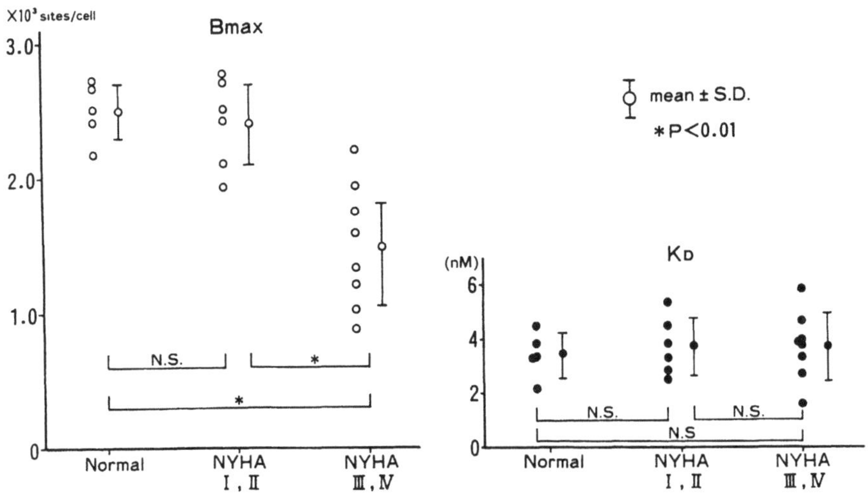

Fig. 2. The β-adrenergic receptor number (*Bmax*) and dissociation constant (K_d) of lymphocytes obtained from the patients with congestive heart failure. Down-regulation of β_2-adrenergic receptor is observed in severe heart failure

shows the down-regulation of β_2-adrenergic receptors of lymphocytes in severe heart failure (NYHA III, IV).

The significance of β-adrenergic receptor changes in the inotropic response of the heart is further validated by a close correlation between β-adrenergic receptor density and contractile response of the heart to β-adrenergic stimulation. Bristow et al. reported that subcellular signal transduction via β-adrenergic receptors is not impaired since the maximal increase in cyclic AMP production with sodium fluoride (NaF) or forskolin is not reduced in chronic heart failure [27]. A recent report by Feldmann et al., however, suggested that in the failing heart an inhibitory G protein (Gi) activity is increased as evidenced by a decrease in basal and Gpp(NH)p- or forskolin-stimulated adenylate cyclase activity [29]. A tight coupling of muscarinic receptors to adenylate cyclase inhibition has been reported in human heart obtained from patients with dilated cardiomyopathy. Although these findings imply the existence of abnormalities in intracellular signal transduction in chronic heart failure, there is no doubt that alterations in β-adrenergic receptors play an important role in regulation of the contractile response.

It is noteworthy that β-adrenergic receptors are often increased in the pre-failure stage, such as in a compensatory period of cardiac hypertrophy [30] and ischemic cardiomyopathy [34] in animal models. In pressure-overloaded heart failure in guinea pigs, β-adrenergic receptors are increased 3 weeks after aortic banding in accordance with development of cardiac hypertrophy [30, 31] (Fig. 3). We also

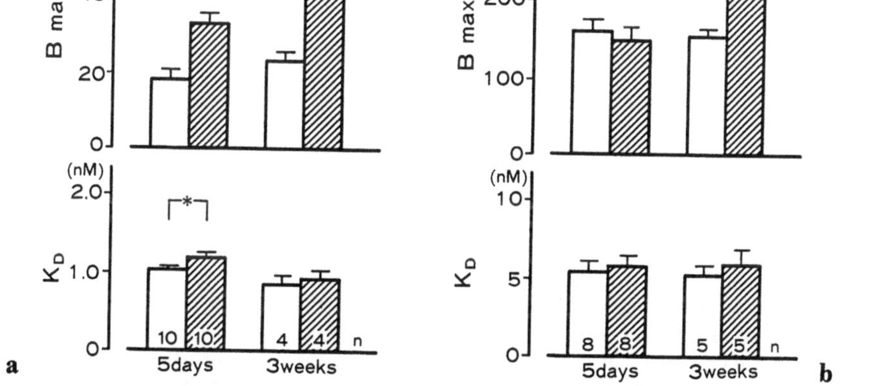

Fig. 3a, b. The maximal binding (*Bmax*) and dissociation constant (*K_d*) *of* **a** [^{3}H] prazosin (α_1-adrenergic receptor) and **b** [^{3}H] dihydroalprenolol (β-receptor) 5 days and 3 weeks after aortic banding in guinea pigs. Both α- and β-receptors are significantly increased 3 weeks after pressure overload. * $P < 0.05$; ** $P < 0.01$; *shaded bars*, aortic banding (mean ± SE); *white bars*, sham operation observed an increase in β-adrenergic receptor density in dogs 7 days after recovery from acute myocardial ischemia produced by intracoronary microembolization [32]. This increase in β-adrenergic receptors was associated with norepinephrine depletion in myocardium but with normal cardiac function. It is of interest that the inotropic response to intravenous administration of forskolin was significantly depressed whereas the reponse to isoproterenol was apparently normal (Fig. 4). This may indicate that an increase in β-adrenergic receptors in this model of ischemic

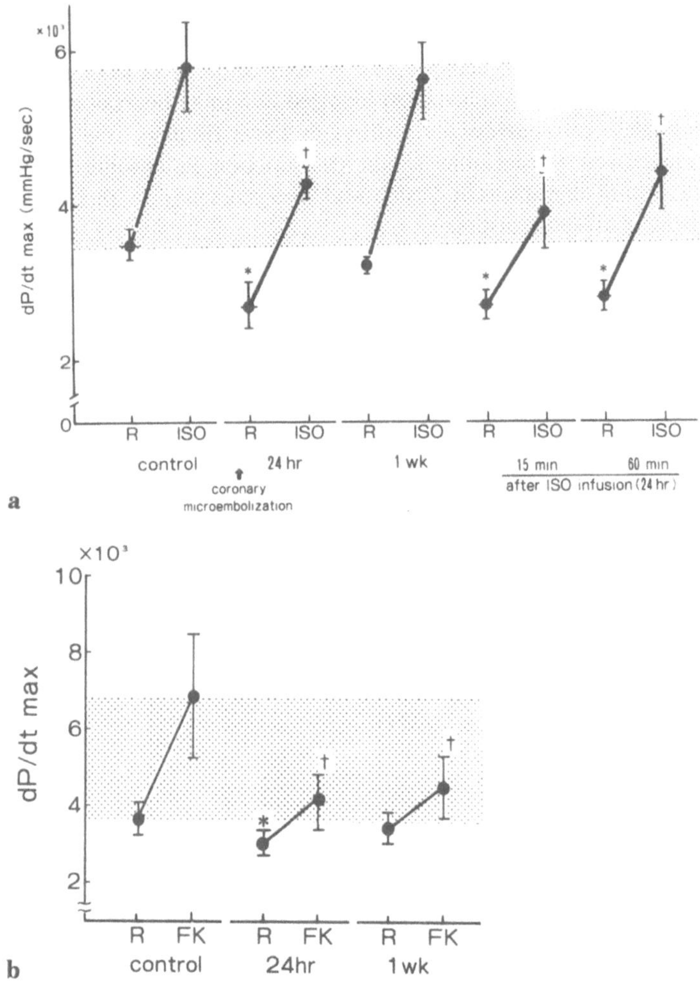

Fig. 4a, b. Inotropic response to administration of **a** isoproterenol (0.05) μg/kg bolus i.v., and **b** in dogs with coronary microembolization. Left ventricular (LV)dP/dt max at rest and its response to isoproterenol had recovered by 7 days after embolization, whereas the response to forskolin was still impaired 7 days after embolization. The apparently normal response to isoproterenol during this chronic period was achieved by a compensatory increase in β-adrenergic receptors since LVdP/dt max was depressed when the β-adrenergic receptors were down-regulated by infusion of isoproterenol for 24 hours (panel (**a**). I mean ± SE (**a**) or ± SD (**b**); *R*, at rest; *ISO*, isoproterenol 0.05 μg/kg i.v.; *FK*, forskolin; * $P < 0.01$ vs control (R); † $P < 0.01$ vs control (ISO in **a**, FK in **b**)

cardiomyopathy could compensate for the impaired subcellular signal transduction of β-receptors. This hypothesis is further supported by a depressed inotropic response to β-adrenergic stimulation after β-receptors were down-regulated by a sustained infusion of isoproterenol. In cardiomyopathic Syrian hamsters (Bio 14.6) also β-adrenergic receptors were increased in the early stages while they were decreased with the progression of myocardial lesions [33]. The increase in β-adrenergic receptors in the compensatory period of heart failure may suggest that the myocardial β-adrenergic receptors are increased to compensate for the underlying myocardial deterioration although direct evidence has not been reported in humans. If this is the case, however, down-regulation of β-adrenergic receptors could cause manifestation of overt heart failure (Fig. 5).

The mechanism of down-regulation of β-adrenergic receptors in chronic heart failure is not fully understood. It is currently believed that a decrease in β-adrenergic receptor density in chronic heart failure is related to elevated sympathetic and systemic levels of norepinephrine, which are primarily correlated with the degree of

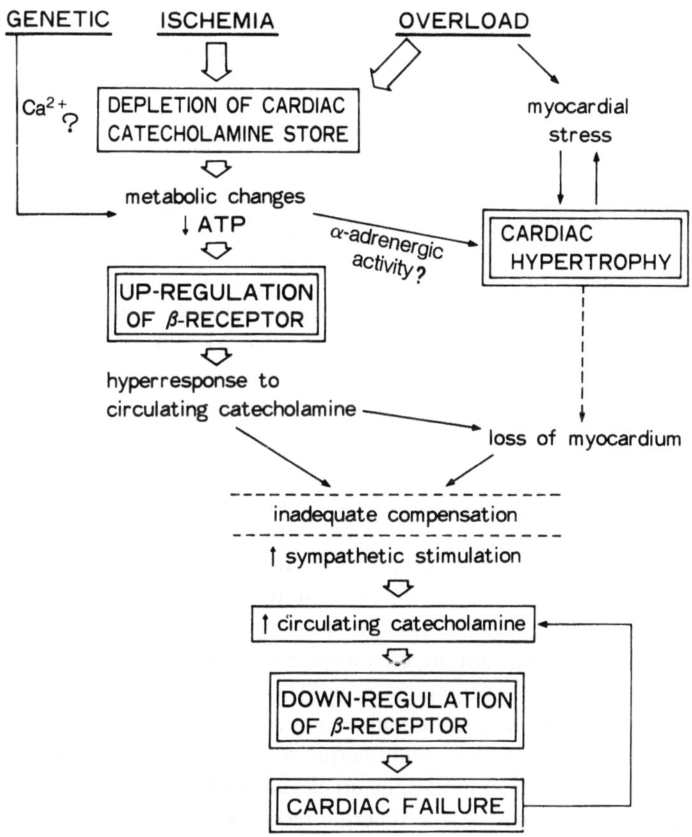

Fig. 5. Hypothesis of the role of β-adrenergic receptor changes in congestive heart failure: β-receptors are increased in the compensatory phase, and down-regulation of the receptors, probably induced by excessive circulatory catecholamines, may lead to overt heart failure

ventricular dysfunction [34]. This hypothesis is supported by the experimental finding that long-term exposure of myocardial tissue to receptor agonist, either in vitro or in vivo, reduces the density of β-adrenergic receptors, adenylate cyclase activity, and contractile response [35, 36].

In relation to the subcellular mechanism of the down-regulation, it should be noted that the reduction in β-adrenergic receptor number is "chamber-specific." In patients with right ventricular failure, a greater reduction in β-adrenergic receptors in the right ventricle was observed whereas in biventricular failure the down-regulation was observed in both ventricles [37]. This suggested that the circulating catecholamine is not a sole regulator of β-adrenergic receptors but unknown local factors are also involved. Recently Strasser et al. isolated the cyclic AMP-independent β-adrenergic receptor kinase which phosphorylates the β-adrenergic receptor protein leading to the desensitization of receptors [38]. Local alterations in receptor kinase or microtubules responsible for down-regulation [39] may be involved as a local factor.

Another interesting observation is that in chronic heart failure β_1-adrenergic receptors are decreased whereas β_2-adrenergic receptors remain unchanged [40]. In normal human myocardium, β_1-adrenergic receptor is dominant; 70%–80% of myocardial β-adrenergic receptors are β_1-specific [41]. In chronic heart failure, however, the relative occupancy of β_2-adrenergic receptors is increased with selective reduction in β_1-receptors [40]. This suggests that β_2-adrenergic receptor agonists may be more effective in advanced heart failure. Selective down-regulation in β_1-adrenergic receptors may be attributed to the fact that β_1-adrenergic receptors are innervated and hence, exposed to high synaptic concentrations of neuronally released norepinephrine [42]. Although this hypothesis seems to be logical, however, it cannot fully explain the selective down-regulation, since myocardial norepinephrine stores in the sympathetic nerve terminals are depleted in failing heart and thus, the norepinephrine released may be much less in chronic heart failure [43].

The role of α-adrenergic receptors in chronic heart failure has not been extensively studied. Although postsynaptic α-adrenergic receptors are found in many species including human myocardium, their roles in the pathophysiology in chronic heart failure are still unclear. Recent studies on the subcellular mechanism of α-adrenergic receptors have revealed that phosphatidylinositol turnover is activated by α_1-adrenergic receptor stimulation, and that diacylglycerol (DG) enhances Ca^{2+} mobilization and protein kinase C activity [44]. Although there are several lines of evidence that α_1-adrenergic receptor stimulation enhances inotropic action, the contribution of α_1-adrenergic receptor to inotropic activity may be minor at least in normal hearts. α-Adrenergic receptor activity may be subservient to the longer-term cellular effects of catecholamine, e.g., cardiac hypertrophy and related metabolic changes. In contrast to β-adrenergic receptors, α_1-receptors are not down-regulated in chronic heart failure despite an elevation of plasma catecholamine [45]. Thus, the relative occupancy of α-adrenergic receptors may be increased in heart failure.

Karliner et al. reported that α-adrenergic receptors are increased in guinea pigs with heart failure [30]. We also observed that in cardiomyopathic Syrian hamsters (Bio 14.6 strain) α_1-adrenergic receptor density was higher even in heart failure stage than that of the age- and sex-matched control animals, and that a long-term adminstration of an α_1-blocker (bunazosin) attenuated cardiac hypertrophy and myocardial calcium overload [46]. If calcium overload is deleterious for the progression of heart failure, an α_1-adrenergic receptor antagonist may prevent, at least in part, the progression of heart failure. α_2-Adrenergic receptor agonists including clonidine and methyldopa reduce plasma norepinephrine levels, heart rate, and arterial blood pressure, and improve exercise performance in patients with chronic heart failure by decreasing overall sympathetic tone and renin-angiotensin activities [47]. Taken together, a long-term inhibition of α_1-adrenergic receptors may have a salutary effect on the progression of heart failure.

β-Adrenergic Blocker Therapy for Chronic Heart Failure

Conventional treatment of patients with chronic heart failure is based on the use of diuretics, digitalis, and vasodilator drugs. β-Adrenergic agonists are also useful for treatment of severe cardiac dysfunction, although a sustained treatment with β-adrenergic agonists may induce tachyphylaxis or even deteriorate the cardiac function [25]. A provocative challenge to such a conventional therapy for chronic heart failure was first reported in 1975 by Waagstein et al [48]. They treated severe patients of advanced dilated cardiomyopathy with β-adrenergic antagonists for 2–12 months, and observed that all of them experienced improvement of symptoms associated with increases in exercise capacity and circumferential fiber shortening velocity of the left ventricle. Furthermore, the same Swedish group reported that the survival of 24 patients with dilated cardiomyopathy treated with β-blocker agents was prolonged compared with that of a control group [49]. In another study, they also observed that after withdrawal of metoprolol, patients showed exacerbation of heart failure [50].

The beneficial effect of a long-term treatment with β-blocking agents was confirmed by other studies mainly conducted at Stanford; 9 of 10 patients who tolerated therapy showed clinical improvement with a considerable increase in ejection fraction during exercise [51]. Consequently, four of these patients were removed from the heart transplantation waiting list. Although several studies failed to show the beneficial effect of β-blocking therapy, it is generally accepted that there are some patients with chronic heart failure who respond to a long-term blocker treatment. It is noteworthy that the patients who did not respond to β-blocker therapy were either in a relatively better initial function class (NYHA II or III) or in extremely advanced heart failure with marked elevations in plasma norepinephrine and angiotensin II concentrations. Absence of beneficial effects was reported with treatments lasting only for short-terms, mostly of less than one month [52]. This indicates that β-blocker therapy is effective only when it is tolerated for a long period.

Potential mechanisms for the beneficial effects of β-blocker therapy have been suggested [52]: (1) inhibition of sympathetically mediated vasoconstriction, (2) protection from catecholamine-induced myocardial damage, (3) improvement of diastolic relaxation and ventricular compliance, (4) up-regulation of β-adrenergic receptors, (5) decreased myocardial oxygen consumption, and (6) inhibition of tachyarrythmias. Waagstein et al. postulated the mechanism that a reduction in heart rate results in a decrease in myocardial oxygen demand [48]. This effect is particularly beneficial for ischemic cardiomyopathy. It is believed that sympathetically mediated vasoconstriction exacerbates cardiac dysfunction due to an increase in ventricular afterload. However, an increase in arterial pressure is not often observed in advanced heart failure, and thus, this mechanism may not be important. The enhanced renin-angiotensin system secondary to renal hypertension may play a primary role in aggravation of heart failure through its potent vasoconstriction, sodium retention, and attenuation of baroreceptor sensitivity [2]. It should be noted that the beneficial effect of β-blocker therapy is accompanied by a decrease in plasma renin activity, but plasma norepinephrine is not decreased [53]. We observed no sigificant changes in resting plasma norepinephrine levels during withdrawal and after readministration of metoprolol in a long-term treatment (mean 8.8 months) of patients with dilated cardiomyopathy, whereas plasma renin activity was significantly increased during withdrawal of the β-blocker [54]. Beneficial effects of converting-enzyme inhibitor may also strengthen the importance of renin-angiotensin system in chronic heart failure.

Several previous studies indicated the deleterious effects of tachycardia on diastolic function in patients with dilated cardiomyopathy [55]. Fleckenstein et al. [56] also demonstrated that a high concentration of isoproterenol led to excessive calcium entry into cells. This may cause a decrease in ventricular compliance during exercise. To test this hypothesis, we obtained the left ventricular pressure-volume relationship during exercise in patients with dilated cardiomyopathy, from the left ventriculograms obtained by digital subtraction angiography (DSA) and concomitant measurement of left ventricular pressure. Of interest, the diastolic pressure-volume relationship was shifted upward during exercise and this shift was prevented by administration of propranolol [57].

If a reduction of ventricular distensibility by β-adrenergic stimulation could be attributed to excessive calcium entry, this mechanism could lead to energy depletion, mitochondrial damage, and cell necrosis. Even though irreversible damage is not produced, calcium overload could mimic myocardial stunning [58].

Up-regulation of myocardial β-adrenergic receptors observed during long term β-blocker therapy may also lead to hemodynamic improvement. Although the direct measurement of myocardial β-adrenergic receptors in biopsy samples is rather hard in a clinical setting, the β_2-adrenergic receptor activity of lymphocytes reflects the myocardial β-receptor activity. We measured lymphocyte β-adrenergic receptors before and after withdrawal of administration of β-blocker during a long-term treatment of dilated cardiomyopathy [54]. β-Receptors, assessed by radioligand assay with [^{3}H]CGP 12177, were down-regulated 7 days after withdrawal of metoprolol (60 mg/day) and were restored to the pre-withdrawal level 3 days after readministration (40 mg/day). These results indicate that β-adrenergic receptor density could alter promptly in response to adminstration and withdrawal of this β-blocking drug. However, the improvements of symptoms and hemodynamics are usually observed only after a couple of months of treatment. Therefore, up-regulation of β-adrenergic receptors may not primarily contribute to the improvement of cardiac function, although a correlation between β-receptor density and cardiac function still supports the importance of β-adrenergic receptor regulation in the pathophysiology of chronic heart failure.

Acknowledgments. This study was supported in part by a grant-in-aid for Scientific Research (No. 62570392) and Scientific Reserach on Priority Areas (No. 62624005) from the Ministry of Education, Science, and Culture, Japan and a grant from the Japan Cardiovascular Research Foundation.

References

1. Hirsch AT, Dzau VJ, Creager MA (1987) Baroreceptor function in congestive heart failure: effect on neurohumoral activation and regional vascular resistance. Circulation 75 (Suppl IV): IV36-IV48
2. Higgins CB, Vatner SF, Eckberg DL, Braunwald EG (1972) Alterations in the baroreceptor reflex in conscious dogs with heart failure. J Clin Invest 51: 715-724
3. Goldsmith SR, Francis GS, Levine TB, Cohn JN (1983) Regional blood flow response to orthostasis in patients with congestive heart failure. J Am Coll Cardiol 1: 1391-1395
4. Creager MA, Faxon DP, Cutler SS, Kohlmann O, Ryan TJ, Gavras H (1986) Contribution of vasopressin to vasoconstriction in patients with congestive heart failure: comparison with the renin-angiotensin system and the sympathetic nervous system, J Am Coll Cardiol 7: 758-765
5. Goldsmith SR, Francis GS, Cowley AW, Cohn JN (1981) Reflex control of vasopressin and norepinephrine release in normal man. Circulation 64 (Suppl IV): IV-155
6. Abraham D (1967) The structure of baroreceptors in pathological conditions in man. In: Kezdi P (ed) Baroreceptors and hypertension. Pergamon Oxford, pp 273-291
7 Eckberg DL, Drabinski M, Braunwald E (1971) Defective cardiac parasympathetic control in patients with heart disease, N Engl J Med 285: 877-883
8. Ferguson DW, Abboud FM, Mark AL (1984) Selective impairment of baroreflex-mediated vasoconstrictor responses in patients with ventricular dysfunction. Circulation 69: 451-460
9. Cody RJ, Franklin KW, Kluger J, Laragh JH (1982) Mechanism governing the postural response and baroreceptor abnormalities in chronic congestive heart failure: effects of acute and long-term converting-enzyme inhibition. Circulation 66: 135-142
10. Levine TB, Olivari MT, Carlyle P, Francis GS, Cohn JN (1982) Reversibility by heart transplantation of abnormal neurohumoral control mechanism in chronic congestive heart failure. Circulation 66 (SuppL II): 192
11. Zucker IH, Peterson TV, Gilmore JP (1980) Ouabain increases left atrial stretch receptor discharge in the dog. J Pharmacol Exp Ther 212: 320-324
12. Heesch CM, Abboud FM, Thames MD (1984) Acute resetting of carotid sinus baroreceptors: II. Possible involvement of electrogenic Na^+ pump. Am J Physiol 247: H833-H839
13. Stein RD, Stephenson RB, Weaver LC (1984) Central actions of angiotensin II oppose baroreceptor-induced sympathoinhibition. Am J Physiol 246: R13-19
14. Thomas JA, Marks BH (1978) Plasma norepinephrine in congestive heart failure. Am J Cardiol 41: 233-243
15. Goldstein DS, McCarty R, Polinsky RJ, Kopin IJ (1983) Relationship between plasma norepinephrine and sympathetic neural activity. Hypertension 5: 552-559
16. Cohn JN, Levine TB, Olivari MT, Garberg V, Lura D, Francis GS, Simon AB, Rector T: (1984) Plasma norepinephrine as a guide to prognosis in patients with chronic congestive heart failure. N Engl J Med 311: 819-823
17. Kenakiñ TP, Ferris RM (1983) Effects of in vivo β-adrenoceptor down-regulation on cardiac response to prenalterol and pirbuterol. J Cardiovasc Pharmacol 5: 90-97
18. Dzau VJ, Colucci WS, Hollenberg NK, Williams GH (1981) Relation of the renin-angiotensin-aldosterone systems to clinical state in congestive heart failure. Circulation 63: 645-651
19. Franciosa JA, Wilen MM, Jordan RA (1981) Effects of enalapril, a new angiotensin-converting enzyme inhibitor, in a controlled trial in heart failure. J Am Coll Cardiol 5: 101-107

20. Creager MA, Massie BM, Faxon DP, Friedman SD, Kramer BL, Weiner DA, Ryan TJ, Topic N, Melidossian CD (1985) Acute and long-term effects of enalapril on the cardiovascular response to exercise and exercise tolerance in patients with congestive heart failure. J Am Coll Cardiol 6: 163-170
21. Watkins L, Burton JA, Haber E, Cant JR, Smith FW, Barger AC (1976) The renin-angiotensin-aldosterone system in congestive failure in conscious dogs. J Clin Invest 57: 1606-1617
22. Swales JD: (1979) Arterial wall or plasma renin in hypertension? Clin Sci 56: 293-298
23. Packer M, Medina A, Yushak M (1984) Correction of dilutional hyponatremia in severe chronic heart failure by converting-enzyme inhibition. Ann Intern Med 100: 782-789
24. Parker M, Lee WH, Kesser PD, Gottlieb SS, Bernstein JL, Kukin ML (1987) Role of neurohumoral mechanism in determining survival in patients with severe chronic heart failure. Circulation 75 (Suppl IV): IV-80
25. Chang HY, Klein RM, Kunos, G (1982) Selective desensitization of cardiac beta receptors by prolonged *in vivo* infusion of catecholamines in rats. J Pharmacol Exp Ther 221: 784-789
26. Newman WH (1977) A depressed response of left ventricular contractile force to isoproterenol and norepinephrine in dogs with congestive heart failure. Am Heart J 93: 216-221
27. Bristow MR, Ginsburg R, Minobe W, et al. (1982) Decreased catecholamine sensitivity and beta-adrenergic receptor density in failing human hearts, N Engl J Med 307: 205-211
28. Fowler MB, Bristow MR, Hopkins DG, et al. (1984) Impaired beta-adrenergic inotropic response in severe heart failure. Circulation 70 (Suppl II): II-191
29. Feldman AM, Cates AE, Veazey WB, Hershberger RE, Bristow MR, Baughman KL, Baumgartner WA, Cornelis VD (1988) Increase of the 40,000-mol wt toxin substrate (G protein) in the failing human heart. J Clin Invest 82: 189-197
30. Karliner S, Barnes P, Brown M, Dollery C (1980) Chronic heart failure in the guinea pig increases cardiac alpha$_1$- and beta-adrenoceptors. Eur J Pharmacol 67: 115-118
31. Tamai J, Hori M, Kagiya T, Iwakura K, Iwai K, Kitabatake A, Watanabe Y, Yoshida H, Inoue M, Kamada T (1989) Role of α_1-adrenoceptor activity in progression of cardiac hypertrophy in pressure overloaded guinea pig hearts. Cardiovasc Res 23: 315-322
32. Hori M, Koretsune Y, Kagiya T, Watanabe Y, Iwakura K, Iwai K, Kitabatake A, Yoshida H, Inoue M, Kamada T (1989) A compensatory increase in beta-adrenoceptors for postischemic dysfunction following coronary microembolization in dogs. Cardiovasc Res 23: 424-431
33. Karliner JS, Alabaster C, Stephens H, Barnes P, Dollery C (1981) Enhanced noradrenaline response in cardiomyopathic hamsters: possible relation to changes in adrenoceptor studies by radioligand binding. Cardiovasc Res 15: 296-304
34. Bristow MR (1984) Myocardial beta-adrenergic receptor down regulation in heart failure. Int J Cardiol 5:648-652
35. Mickey JV, Tate R, Mullikin D, Lefkowitz RJ (1976) Regulation of adenylate cyclase-coupled beta-adrenergic receptor binding sites by beta adrenergic catecholamines in vitro. Mol Pharmacol 12: 409-419
36. Hayes JS, Pollock GD, Fuller RW (1984) In vivo cardiovascular responses to isoproterenol, dopamine and tyramine after prolonged infusion of isoproterenol. J Pharmacol Exp Ther 231:633-639
37. Bristow MR, Ginsburg R, Minobe WA, Harrison DC, Reitz BA, Stinson EB (1982) Beta-adrenergic receptor measurement in normal and failing human right and left ventricle. Circulation 66 (Suppl II): II-207
38. Strasser RH, Sibley DR, Lefkowitz RJ (1986) A novel catecholamine-activated adenosine cyclic 3', 5'-phosphate independent pathway for β-adrenergic receptor phosphorylation in wild-type and mutant S_{49} lymphoma cells: mechanism of homologous desensitization of adenylate cyclase. Biochemistry 25: 1371-1377
39. Limas CJ, Limas C (1984) Rapid recovery of cardiac β-adrenergic receptors after isoproterenol-induced "down"-regulation. Circ Res 55: 524-531

40. Bristow MR, Ginsburg R, Umans V, Fowler M, Minobe W, Rasmussen R, Zera P, Menlove R, Shah P, Jamieson S, Stinson EB (1986) β_1- and β_2-adrenergic receptor subtypes to muscle contraction and selective β_1-receptor down-regulation in heart failure. Circ Res 59: 297–309
41. Ginsburg R, Bristow MR, Zera P (1984) Beta$_2$-receptors are coupled to muscle contraction in human ventricular myocardium. Circulation 70 (Suppl II) II-67
42. Ariens EJ (1981) The classification of beta-adrenoceptor. Trends Pharmacol Sci 2: 170–172
43. Chidsey CA, Braunwald E, Morrow AG (1965) Catecholamine excretion and cardiac stores of norepinephrine in congestive heart failure. Am J Med 39: 442–451
44. Brown JH, Buxton IL, Brunton LL (1985) α_1-adrenergic and muscarinic cholinergic stimulation of phosphoinositide hydrolysis in adult rat cardiomyocytes. Circ Res 57: 532–537
45. Hayes JS, Bowling N (1987) Role of the alpha agonist activity of dobutamine in mediating cardiac output: effects of prolonged isoproterenol infusion. J Pharmacol Exp Ther 241: 861–869
46. Iwakura K, Hori M, Kagiya T, Watanabe Y, Yoshida H (1988) Increased α_1-adrenergic receptors coupled with phosphatidylinositol response augument Ca^{2+} overload in cardiomyopathic hamster. Circulation 78 (Suppl II): II-334
47. Giles TD, Thomas MG, Sander GE, Quiroz AC (1985) Central alpha-adrenergic agonists in chronic heart failure and ischemic heart disease. J Cardiovasc Pharmacol 7 (Suppl 8): S51–S55
48. Waagstein F, Hjalmarson Å, Varnauskas E, Wallentin I (1975) Effect of chronic beta-adrenergic receptor blockade in congestive cardiomyopathy. Br Heart J: 1022–1036
49. Swedberg K, Hjalmarson Å, Waagstein F, Wallentin I (1975) Prolongation of survival in congestive cardiomyopathy by beta-receptor blockade. Lancet 1: 1374–1376
50. Swedberg K, Hjalmarson Å, Waagstein F, Wallentin I (1980) Adverse effects of beta-blockade withdrawal in patients with congestive cardiomyopathy. Br Heart J 44: 134–142
51. Fowler MB, Bristow MR, Laser JA, Ginsburg R, Scott LB, Schroelder JS (1984) Beta blocker therapy in severe heart failure: improvement related to beta$_1$-adrenergic receptor up-regulation? Circulation 70 (Suppl II): II-112
52. Alderman J. Grossman W (1985) Are β-adrenergic-blocking drugs useful in the treatment of dilated cardiomyopathy? Circulation 71: 854–857
53. Binkley PF, Lewe R, Lima J, Unverferth DV, Leier CV (1988) Neurohumoral profile in congestive heart failure: response to β-blockade. J Lab Clin Med III: 393–398
54. Kagiya T, Hori M, Fukunami M, Hoki N, Iwakura K, Kurihara T, Watanabe Y, Kamada T (1987) Role of beta-adrenergic receptors in beta-blocker therapy for chronic heart failure in patients with dilated cardiomyopathy. Circulation 76 (Suppl IV): IV-308
55. Mitchell JH, Linden RJ, Sarnoff SJ (1960) Influence of cardiac sympathetic and vagal nerve stimulation on the relation between left ventricular diastolic pressure and myocardial segment length. Circ Res 8: 1100–1107
56. Fleckenstein A, Janke J, Doring HJ, Pachinger O (1973) Calcium overload as the determinant factor in the production of catecholamine-induced myocardial lesions. Recent Adv Study Cardiac Structure Metab 2: 455–466
57. Ozaki H, Sato H, Matsuyama T, Ishida Y, Takeda H, Inoue Y (1988) β-blocker prevents failing heart from myocardial stiffening during dynamic exercise. Circulation 78 (Suppl II): II-589
58. Kitakaze M, Weisman HF, Marban E (1988) Contractile dysfunction and ATP depletion after transient calcium overload in perfused ferret hearts. Circulation 77: 685–695

Adrenergic Receptors and Signal Transduction in Myocardial Hypertrophy, Ischemia, and Failure

JOEL S. KARLINER[1]

Summary. Signal transduction mediated by both α_1- and β-adrenergic receptors may be important in the pathogenesis of myocardial hypertrophy, ischemia, and failure. In neonatal rat ventricular myocytes, α_1-adrenergic agonism produces cardiac cell enlargement by mechanisms that ultimately involve regulation of genes controlling myocardial protein synthesis. In this system α_1-adrenergic receptors do not down-regulate, and second messenger production (inositol trisphosphate) does not undergo desensitization. On the gene level, the rates of transcription of several genes are regulated in a highly complex manner which is currently under intense investigation. Both in animal models and in patients with congestive cardiac failure, down-regulation of β-adrenergic receptors appears to be a constant finding. This at least in part accounts for the reduction of adenylate cyclase activity and myocardial function that is observed. Recent attention has focused on the role of G proteins in heart failure. There appears to be an abnormality in G protein function but whether this results from a reduction in the guanine nucleotide stimulatory protein, Gs, or an increase in the guanine nucleotide inhibitory protein, Gi, or both, has not been established. In an animal model of acute myocardial ischemia, Gs function is diminished, but in the presence of chronic β-blockade there appears to be enhanced coupling between Gs and the β-adrenergic receptor resulting in relative preservation of agonist-stimulated adenylate cyclase activity, particularly in the subepicardium. This observation could account, at least in part, for the beneficial effects of chronic β-adrenergic blockade in acute myocardial ischemia.

α_1-Adrenergic Receptors in Experimental Myocardial Hypertrophy

Myocardial hypertrophy is an important component in the pathophysiology of systemic arterial hypertension and valvular heart disease. Physiologic hypertrophy also occurs in the well-trained athlete. Such hypertrophy is usually considered to be a compensatory mechanism which is the result of hemodynamic overload. Exceptions may be certain cases of coronary artery disease and idiopathic hypertrophic states, whether localized or generalized.

Since all of these entities involve multiple unknown humoral, neural, and hemodynamic variables, it is difficult to identify molecular signals that mediate the pathogenesis of myocardial hypertrophy in the intact organism. Accordingly, we have used a heart cell culture preparation derived from neonatal rat ventricles which

[1] Department of Medicine, University of California, San Francisco; Cardiology Section, Veterans Administration Medical Center, San Francisco, CA 94121, USA

has the characteristic of increasing in size (hypertrophy) in response to stimuli without exhibiting proliferation. The cells are prepared in such a way as to eliminate mesenchymal cell proliferation and therefore the culture consists of 85%-90% pure cardiac myocytes [1, 2]. Another advantage of this preparation is that cells can be grown in serum-free medium which permits a variety of interventions to study the regulation of growth and the properties of adrenergic receptors and adrenergically-mediated signal transduction [3].

Studies using this model system have revealed a possibly unique role for the α_1-adrenergic receptor in the promotion of cardiac growth. In serum-free medium, cardiac myocytes in control cultures do not undergo hypertrophy over time. However, incubation with norepinephrine or epinephrine at 2 μM increased myocardial cell size by 1.5 to 2-fold [4]. Both DNA synthesis and cell number were unchanged. Increased protein content was detected at 6 h, the half maximum was at 18 h and the plateau at 36 h. The EC_{50} for norepinephrine or epinephrine was 200 nM. This is within the physiologic range for an active sympathetic synapse. The response was specific for ℓ-isomers and was not produced by catecholamine metabolites. This hypertrophy was an α_1-mediated response [4]. Only 20% or less of the maximum effect was produced by the β-agonist isoproterenol or the α_2-antagonist clonidine, while dopamine was inactive. The naturally occurring neurotransmitter norepinephrine and the natural circulating catecholamine epinephrine were more potent than the synthetic agonists methoxamine and phenylephrine. The growth-promoting effect of norepinephrine or epinephrine was inhibited by the α_1-antagonists prazosin and terazosin, but not only by the α_2-antagonist yohimbine, the β-adrenergic antagonist propranolol or the selective β_1-antagonist betaxolol or the selective β_2-antagonist ICI 118551.

In serum-free cultures less than 5% of cardiac myocytes beat. Incubation with norepinephrine or epinephrine induces spontaneous contractile activity in over 90% of the cells. The time course of beating induction was slightly more rapid than for growth and the EC_{50} was 20 nM. In contrast with the hypertrophic response, which requires only α_1 activation, the development of beating requires both α_1 and β_1 stimulation [5]. The beating response to norepinephrine or epinephrine was eliminated equally by the antagonists propranolol, metaxolol, terazosin, or prazosin, but not by yohimbine or ICI 118551. The agonists isoproterenol and clonidine did not induce beating (25% or less). In further contrast with the growth response, the beating response was not blocked by inhibition of protein or RNA synthesis as discussed later.

Thus there are two trophic effects of α_1-adrenoceptor activation: stimulation of hypertrophy and in conjunction with β_1-agonism, induction of spontaneous contractile activity. Data also indicate that contractile activity can be induced without hypertrophy while α_1 plus β_1 stimulation in the presence of cycloheximide to block protein synthesis produces beating cells that are not enlarged [5]. The independent development of hypertrophy and contractile activity suggests that these effects may be regulated by different intracellular signals. Further, it appears that the trophic effects of norepinephrine require continuous receptor occupancy by the agonist, since regression occurs if the agonist is removed or an antagonist is added. It should also be noted that these effects of α_1-agonism are not apparent in adult myocardial cells but are confined to the neonatal rat ventricular myocyte.

In order to investigate the potential signalling mechanisms in the induction of myocardial hypertrophy via an α_1-receptor mediated mechanism, we have undertaken both radioligand binding studies and characterization of molecular signalling after α_1-agonism by measuring the products of phosphoinositide turnover. It is well known that the heart is richly endowed with α_1-receptors [6] and we have confirmed the existence of these receptors in our neonatal preparation utilizing ^{125}I-HEAT (^{125}I-(2-[β-(4-hydroxyphenyl)-ethyl-aminomethyl] tetralone) as a radioligand and measurement of inositol phosphate production after agonist exposure as an indicator of α_1-adrenergic receptor-mediated signal transduction. In the heart and in many other tissues, α_1-adrenoceptors are coupled via a phospholipase C to stimulation of phosphatidylinositide turnover that initiates many calcium-dependent responses [7–10]. Exposure of cells to agonists causes activation of phospholipase C that selectively hydrolyzes phosphatidylinositol (4, 5)-bisphosphate, yielding inositol (1, 4 , 5)-trisphosphate (IP_3) and *s, nl,* 2-diacylglycerol [11]. These products of phosphatidylinositol turnover act as intracellular second messengers; IP_3 mobilizes calcium from sarcoplasmic reticulum in cardiac and other tissues and diacylglycerol activates protein kinase C which mediates protein phosphorylation.

In contrast to the β-adrenergic system and in contrast to some studies in smooth muscle cells, we found that the α_1-adrenergic receptor does not desensitize after prolonged exposure to norepinephrine (NE) and the induction of cardiac myocyte hypertrophy (1μM NE for 48 h) [12]. Similarly, signal transduction is not affected and phosphoinositide turnover remains responsive to α_1-agonism after 48 h of exposure to norepinephrine [13]. This is consistent with the observation that continuous occupancy of the receptor is required for the effective maintenance of the production of hypertrophy in this model system.

In order to assess the effects of potential mediation of signal transduction by the inhibitory guanine nucleotide regulatory protein (Gi), we used pertussis toxin, which ADP-ribosylates this moiety [14]. We found that incubation of myocardial cells overnight with 0.5 μg/ml of pertussis toxin did not inhibit the inositol phosphate response to norepinephrine and that there was no effect on myocardial cell growth over 48 h after exposure to pertussis toxin in the presence of norepinephrine [15]. These data suggest that a pertussis toxin-sensitive guanine nucleotide regulatory protein is not an important mechanism of signal transduction for growth induced by α_1-agonism. Additional data also indicate that exposure to norepinephrine in these cells is accompanied by a rise in the activity of protein kinase C [16]. Thus, exposure of these cells to norepinephrine with the production of hypertrophy is accompanied by increases in the two putative arms of the phosphoinositide cascade, i.e., increase in inositol phosphates and induction of protein kinase C activity, presumably via an increase in diacylglycerol.

Additional observations indicate that the α_1-hypertrophic response is blocked by inhibition of protein synthesis by cycloheximide or of RNA synthesis by actinomycin D. Additional studies indicate that norepinephrine increases the cell content of rRNA, α_1-actin mRNA, and *c-myc*-encoded mRNA [17, 18]. The increase of the latter was detectable at 30 min, peaked at 2 h, and returned to baseline by 6 h after stimulation. This response was abolished by the α_1-antagonist terazosin but was not affected by the β-adrenergic antagonist propranolol and was only slightly attenuated by the α_2-adrenergic antagonist yohimbine. Exposure of cardiac myocytes to serum and the phorbol ester tumor promoter TPA also enhances *c-myc* expression in cardiac myocyte cultures [17]. Further work will be required to determine whether the stimulation of *c-myc* expression in cardiac myocytes occurs by increasing transcription or by altering message stability. Since the induction of *c-myc* expression is rapid and shortlived, while the hypertrophic response occurs much later and requires continuous exposure to the catecholamine, it is likely that the hypertrophic response involves other intracellular pathways in addition to *c-myc* expression. Thus *c-myc* expression may be necessary but not sufficient for cellular hypertrophy.

Since these studies have been confined to cultured cells from the neonatal rat ventricle, it is unknown whether the α_1-receptor is also involved in the stimulation of growth in vivo. Since it appears that this mechanism does not occur in the normal adult myocyte, it is possible that an α_1 growth mechanism may be important in developmental hypertrophy. Further, certain types of hypertrophy in the adults of some species are associated with reexpression of early developmental genes, such as V3 isomyosin. It may be possible that under some abnormal circumstances an α_1 growth mechanism may be reexpressed in adults when compensatory hypertrophy is required [19]. The α_1-receptor hypothesis also could be important in other cells since α_1-mediated growth of other types of cultured cells have been shown, including adult rat hepatocytes, bovine endothelial cells, and adult male aortic smooth muscle cells. We believe these studies provide the basis for future investigations, including the trophic role of the α_1-receptor; the mechanism of α_1-mediated signal transduction; the possible synergistic role of calcium mobilization and diacylglycerol in molecular signalling; and the effects of activation of proto-oncogenes in regulation of both mRNA and rRNA in processess involving myocardial cell growth and beating [19, 20].

There is considerable evidence that myocardial α_1-adrenoceptors mediate a variety of other effects, including positive inotropic effects both in animals and in man, and electrophysiologic abnormalities in acute myocardial ischemia [9, 21, 22]. Positive inotropic effects mediated by α_1-agonists are of slower onset but longer lasting than those mediated by β-adrenergic agonists. As noted above, it has been observed in neonatal rat ventricular myocytes that prolonged exposure (48 h) to norepinephrine does not result in down-regulation of α_1-adrenergic receptors [12]. Similarly, in failing human left ventricles, when α_1-adrenergic receptor density was measured by ^{125}I-IBE 2254, a selective α_1-adrenergic antagonist, there was no change either in the maximum number of binding sites or in the affinity of the antagonist for the receptor [23]. These observations indicate that α_1-mediated responses might provide a potential therapeutic approach to the treatment of congestive cardiac failure. However, whether second messenger formation and positive inotropic responses would occur in failing animal or human myocardium in response to α_1-agonists has not yet been ascertained and is a topic for future investigation.

Adenylate Cyclase Activity, G-Proteins and β-Adrenergic Receptors in Heart Failure and Ischemia

Both in animal models and in humans with congestive cardiac failure, there is considerable evidence that adenylate cyclase activity in response to isoproterenol stimulation is depressed. In patients with congestive heart failure due to idiopathic dilated cardiomyopathy or "ischemic" cardiomyopathy, β-adrenergic receptor density is reduced and the inotropic response to β-adrenergic agonists is impaired [24]. In an animal model of left ventricular failure due to chronic pressure overload, β-adrenergic receptor density was increased, but uncoupling from the adenylate cyclase system with loss of high-affinity receptors was noted [25]. In experimental models of acute myocardial ischemia or hypoxia, increases, no change, or a decline in β-adrenergic receptor density have been observed [26–36]. In most recent studies, however, it has been found that adenylate cyclase activity stimulated by isoproterenol is diminished [34–37].

In addition to abnormalities in the β-adrenergic receptor, alterations in post-receptor signal transduction may help to explain many of the above results. Specifically, the abnormalities appear to lie in changes in the quantity or the function of the guanine-nucleotide stimulatory protein(s). Thus, Longabaugh et al. recently reported a decrease in the stimulatory guanosine triphosphate (GTP)-binding protein Gs in dogs with left ventricular failure due to chronic pressure overload [38]. Using the aortic-banded puppy they obtained membrane fractions from the failing left ventricle and employed cholera toxin and ^{32}P-nicotinamide adenine dinucleotide (NAD) to label the α-subunit of Gs. They found a 59% reduction in the level of this protein. Further, they noted a 50% reduction in Gs activity in a reconstitution assay utilizing membranes from mouse S49 lymphoma *cyc*$^-$cells, which are deficient in Gs. Further evidence favoring an abnormality in Gs has been noted in human subjects. Using cholera toxin-stimulated labelling, Horn et al. noted that Gs was markedly reduced in 14 patients with advanced heart failure [39]. In a small series of patients with myocardial dysfunction undergoing electrophysiologic study for evaluation of ventricular tachycardia, we noted a marked decrease in both NaF- and Gpp(NH)p-stimulated adenylate cyclase activity along with the expected reduction in isoproterenol-stimulated adenylate cyclase activity [40]. In 20 patients with cardiac failure, Deniss et al. [41] made similar observations and, consistent with the report of Bristow et al. [42], noted that forskolin-stimulated adenylate cyclase activity remained within the normal range. These data in dogs with pressure-overload left-ventricular hypertrophy, and congestive cardiac failure and myocardial dysfunction in patients, suggest that a common abnormality in the heart failure state is an alteration in coupling proteins, particularly in Gs.

In this connection, Vatner et al. have recently noted in the same aortic-banded puppy model with heart failure that the inhibitory pathway involving the muscarinic cholinergic receptor is impaired in congestive heart failure [43]. These conclusions were based on a decrease in muscarinic receptor density and in particular in a specific loss of the high-affinity agonist-binding component of this receptor population. Levels of the GTP inhibitory protein known to couple the muscarinic receptor to adenylate cyclase as measured by pertussis toxin labelling were not depressed in this model of left ventricular failure. By contrast, Feldman et al., recently, suggested that there may be an increase in Gi without an alteration in Gs in humans with congestive cardiac failure [44]. Horn et al. reported a decrease in Gi but, as already indicated, in their series Gs was also reduced [39]. These apparent discrepancies will require further studies for resolution.

Previous animal and human studies of β-receptor density and affinity after chronic β-adrenergic blockade have yielded conflicting results. Glaubinger and Lefkowitz demonstrated an increase in β-adrenergic receptor density after chronic propranolol administration in rats [45]; similar results were reported by Cramb et al. and Aarons and Molinoff [46, 47]. In the latter study, K_d values in rat heart were doubled, suggesting that residual propranolol was present in the particulate preparation from ventricular tissue. This may be because propranolol exhibits considerable lipophilicity and likely persists in a given tissue after the drug has been eliminated from the plasma. Long-term β-blockade in patients with chronic congestive cardiac failure results in an increase in myocardial β-adrenergic receptor density [48]. In other human studies, chronic β-adrenergic receptor blockade has resulted in increased β-receptor density on lymphocytes and in right atrial tissue obtained at the time of cardiac surgery [49, 50]. By contrast, several studies have demonstrated no increase in β-adrenergic receptor density or affinity after chronic treatment with propranolol. These include studies in the rat, in the guinea pig, in the cat, and in the dog [51-55]. In human studies Golf and Hanson showed no changes in β-adrenergic receptor density after propranolol, timolol, or metroprolol treatment [56]. Cooper et al. also demonstrated no change in agonist competition curves after propranolol administration in the cat [55].

In a number of these studies, adenylate cyclase activity has been measured and the results of these observations suggest that β-adrenergic blockade may result in enhanced enzyme activity following incubation with β-adrenergic agonists. Although Cooper et al. observed no changes in maximal adenylate cyclase activity after stimulation by isoproterenol, they did report a shift of the isoproterenol dose-response curve upward and to the left [55]. Others have also reported an increase or no depression of adenylate cyclase activity in response to isoproterenol after chronic β-adrenergic blockade [54, 57]. In the right ventricle of the guinea pig three days after left ventricular myocardial infarction, Baumann et al. noted that previous β-blockade prevented the loss of isoproterenol-stimulated adenylate cyclase activity [58]. Recent observations in our laboratory have also suggested that chronic β-adrenergic blockade in dogs subjected to acute myocardial ischemia may preserve signal transduction mediated by the β-adrenergic receptor [59].

As indicated above, Cooper et al. noted a shift of the isoproterenol dose-response curve upward and to the left in the absence of any alterations in the agonist competition curve for isoproterenol after chronic propranolol treatment in the cat [55]. These investigators speculated that there was enhanced coupling between the β-adrenergic receptor and the stimulatory guanine-nucleotide regulatory protein following chronic treatment with propranolol. It is now well-established in experimental studies, including our own, that there is a loss of high-affinity β-adrenergic receptors as a result of acute ischemia [33, 37]. There is also a decrease in adenylate cyclase activity stimulated both via the β-adrenergic receptor and directly by Gpp(NH)p, a non-hydrolyzable analog of GTP, which acts directly on the guanine nucleotide stimulatory protein, Gs [59]. Recent work in our laboratory also suggests that in the chronically β-blocked dog subjected to acute ischemia, propranolol may enhance coupling between the βadrenergic receptor and the guanine nucleotide stimulatory protein [59].

These observations may also be related to the potential effectiveness of long-term β-blockade in patients with chronic congestive cardiac failure. Because of β-adrenoceptor down-regulation due to chronic catecholamine elevation and decreased Gs function, the response to β-adrenergic agonists is impaired in such patients. The myocardium contains few spare receptors. In the absence of such spare receptors, Bristow et al. suggested that increased doses of agonist cannot recoup the lost response [60]. Thus, he and his colleagues suggested that treatment with β-blockers could protect the β-adrenergic receptors from chronic down-regulation and thereby effectively cause up-regulation of β-adrenergic receptors. They speculated that if adrenergic receptor density returned to normal levels, the adrenergic system could again participate in inotropic support during periodic increases in local neurotransmitter release or elevation of circulating catecholamines required by exercise or other activities, despite the presence of a competitive β-blocking agent. These observations lead to two interrelated and testable hypotheses, i.e., whether chronic β-blockade may not only lead to relative up-regulation of β-adrenergic receptors but also to enhanced coupling of a reduced number of β-adrenergic receptors to a diminished number of guanine nucleotide stimulatory units both in the presence of congestive cardiac failure and during acute ischemia. Preliminary reports in patients with advanced heart failure are encouraging [48], and a controlled clinical trial is currently in progress.

Whether withdrawal of chronic β-blockade leads to clinical manifestations associated with up-regulation of β-adrenergic receptors, resulting in hypersensitivity to catecholamines is controversial. A number of reports of adverse reactions to abrupt propranolol withdrawal were reported in the late 1970s. Despite evidence of an increased number of β-adrenergic receptors demonstrated both in human and animal studies after chronic β-blockade, the bulk of the evidence indicates that hemodynamic, metabolic, and clinical responses do not appear to be augmented after discontinuation of β-blockers under most circumstances [61-65]. Although such a rebound phenomenon may occur in some patients, this does not represent a common clinical problem, possibly because of the long tissue half-life of β-blockers so that despite abrupt withdrawal and reduced plasma levels, tissue levels decline much more slowly, permitting adjustment of adrenergic influences to a new equilibrium.

References

1. Simpson P, McGrath A, Savion S (1982) Myocyte hypertrophy in neonatal rat heart cultures and its regulation by serum and by catecholamines. Circ Res 51: 787-801
2. Karliner JS, Simpson P, Honbo N, Woloszyn W (1986) Mechanism and time course of β_1-receptor desensitization in mammalian cardiac myocytes. Cardiovasc Res 20: 221-228
3. Karliner JS, Simpson P, Taylor JE (1985) Adrenergic receptor characteristics of cardiac myocytes cultured in serum-free medium in comparison with serum supplemented medium. Biochem Biophys Res Commun 128: 376-382
4. Simpson P (1983) Norepinephrine-stimulated hypertrophy of cultured neonatal rat myocardial cells is an alpha$_1$-adrenergic response. J Clin Invest 72: 732-738
5. Simpson P (1985) Stimulation of hypertrophy of cultured neonatal rat heart cells through an α_1-adrenergic receptor and induction of beating through an α_1- and β_1-adrenergic receptor interaction. Circ Res 56: 884-894

6 Karliner JS, Barnes P, Hamilton CA, Dollery CT (1979) Alpha-adrenergic receptors in guinea pig myocardium: identification by binding of a new radioligand (^{3}H)-prazosin. Biochem Biophys Res Commun 90: 142–148
7. Brown JH, Buxton IL, Brunton LL (1985) α_1-adrenergic and muscarinic cholinergic stimulation of phosphoinositide hydrolysis in adult rat cardiomyocytes. Circ Res 57: 532–537
8. Woodcock EA, White LBS, Smith I, McLeod JK (1987) Stimulation of phosphatidylinositol metabolism in the isolated, perfused rat heart. Circ Res 61: 625–631
9. Schmitz W, Scholz H, Scholz J, Steinfath M, Lohse M, Purrunen J, Schwabe U (1987) Pertussis toxin does not inhibit the α_1-adrenoceptor-mediated effect on inositol phosphate production in the heart. Eur J Pharmacol 134: 377–378
10. Otani H, Otani H, Das DK (1988) α_1-adrenoceptor-mediated phosphoinositide breakdown and inotropic response in rat left ventricular papillary muscles. Circ Res 62: 8–17
11. Berridge MJ, Irvine RF (1984) Inositol trisphosphate, a novel second messenger in cellular signal transduction. Nature 312: 315–321
12. Simpson P, Bishopric N, Coughlin S, Karliner JS, Ordal C, Starksen N, White N, Williams L (1986) Dual trophic affects of the alpha$_1$-adrenergic receptor in cultured neonatal rat heart muscle cells. J Mol Cell Cardiol 18 (Suppl 5): 45–58
13. Karliner JS, Simpson P, Braun L, Honbo N, Woloszyn W (1985) Alpha$_1$-adrenoceptor regulation of phosphoinositide turnover in hypertrophied myocardial cells. Circulation 72 (Suppl III): 182
14. Moss J (1987) Signal transduction by receptor-response guanyl nucleotide-binding proteins: modulation by bacterial toxin-catalyzed ADP-ribosylation. Clin Res 35: 451–458
15. Karliner JS, Kagiya T, Simpson P (in press) Effects of pertussis toxin on α_1-agonist-mediated phosphatidylinositide turnover and myocardial cell hypertrophy in neonatal rat ventricular myocytes. Experientia
16. Henrich CJ, Simpson PC (1988) Differential acute and chronic response of protein kinase *C* in cultured neonatal rat heart myocytes to α_1-adrenergic and phorbol ester stimulation. J Mol Cell Cardiol 20: 1081–1085.
17. Starksen N, Simpson P, Bishopric N, Coughlin S, Lee WMF, Escobedo JA, Williams L (1986) Cardiac myocyte hypertrophy is associated with *c-myc* protooncogene expression. Proc Natl Acad Sci USA 83: 8348–8350
18. Bishopric NH, Simpson PC, Ordahl CP (1987) Induction of the skeletal α-actin gene in α_1-adrenoceptor-mediated hyertrophy of rat cardiac myocytes J Clin Invest 80: 1194–1199
19. Simpson PC (1988) Role of proto-oncogenes in myocardial hypertrophy. Am J Cardiol 62: 13G–19G
20. Lee HR, Henderson SA, Reynolds R, Dunnmon P, Yuan D, Chien KR (1988) α_1-adrenergic stimulation of cardiac gene transcription in neonatal rat myocardial cells. J Biol Chem 263:7352–7358
21. Scholz J, Schaefer B, Schmitz W, Scholz H, Steinfath M, Lohse M, Schwabe U, Purrunen J (1988) Alpha-1 adrenoceptor-mediated positive inotropic effect and inositol trisphosphate increase in mammalian heart. J Pharmacol Ther 245: 327–335
22 Culling W, Penny WJ, Cunliffe G, Flores NA, Sheridan DJ (1987) Arrhythmogenic and electrophysiological effects of alpha adrenoceptor stimulation during myocardial ischaemia and reperfusion. J Mol Cell Cardiol 19: 251–258
23. Bristow MR, Ginsburg R, Gilbert EM, Hersberger RE (1987) Heterogeneous regulatory changes in cell surface membrane receptors coupled to a positive inotropic response in the failing human heart. Basic Res Cardiol 82 (Suppl 2): 369–376
24. Bristow MR, Ginsburg R, Minobe W, Cubicciotti RS, Sageman WS, Lurie K, Billingham ME, Harrison DC, Stinson EB (1982) Decreased catecholamine sensitivity and β-adrenergic-receptor density in failing human hearts. N Eng J Med 307: 205–211
25. Vatner DE, Vatner SF, Fujii AM, Homcy CJ (1985) Loss of high-affinity cardiac beta-adrenergic receptors in dogs with heart failure. J Clin Invest 76: 2259–2264
26. Mukherjee A, Wong TM, Buja LM, Lefkowitz RJ, Willerson JT (1979) Beta adrenergic and muscarinic cholinergic receptors in canine myocardium. J Clin Invest 64: 1423–1428

27. Mukherjee A, Bush LR, McCoy KE, Duke KE, Hagler H, Buja LM, Willerson JT (1982) Relationship between β-adrenergic receptor numbers and physiological responses in canine myocardial ischemia. Circ Res 50: 735–741
28. Maisel AS, Motulsky HJ, Insel PA (1985) Externalization of β-adrenergic receptors promoted by myocardial ischemia. Science 230: 183–186
29. Thandroyen F, Muntz K, Rosenbaum T, Willerson JT, Ziman B, Buja LM (1989) β-Receptor-adenylate cyclase coupling in hypoxic neonatal rat ventricular myocytes. Am J Physiol 256: H1209–1217
30. Rhee HM, Tyler L (1985) Myocardial ischemic injury and β-adrenergic receptors in perfused working rabbit hearts. Adv Exp Med Biol 191: 281–288
31. Wolff A, Karliner JS (1988) H_2-histaminergic and β_1-adrenergic adenyl cyclase activation is maintained despite receptor changes and Gs dysfunction during acute myocardial ischemia. Clin Res 36: 328A
32. Marsh JD, Sweeney KA (1989) β-Adrenergic receptor regulation during hypoxia in intact cultured heart cells. Am J Physiol 256: H275–281
33. Rocha-Singh K, Karliner JS (1988) Effects of hypoxia and glucose deprivation on β-adrenergic receptors and cAMP generation in cultured neonatal rat myocardial cells. Clin Res 36: 312A
34. Vatner DE, Knight DR, Shen YT, Thomas JX Jr, Homcy CJ, Vatner SF (1988) One hour of myocardial ischemia in conscious dogs increases β-adrenergic receptors but decreases adenylate cyclase activity. J Mol Cell Cardiol 20: 75–82
35. Thandroyen FT, Muntz K, Ziman B, Rosenbaum T, Willerson JT, Buja LM (1987) Temporal changes in beta adrenergic receptor-adenylate cyclase activity during acute myocardial ischemia and reperfusion in canine myocardium. Clin Res 35: 576A
36. Freissmuth M, Schutz W, Weindlmayer-Gottel M, Zimpfer M, Spiss CK. (1987) Effects of ischemia on the canine myocardial β-adrenoceptor-linked adenylate cyclase system. J Cardiovasc Pharmacol 10: 568–574
37. Devos C, Robberecht P, Nokin P, Waelbroeck M, Clinet M, Camus JC, Beaufort P, Schoenfeld, Christophe J (1985) Uncoupling between beta-adrenoceptors and adenylate cyclase in dog ischemic myocardium. Naunyn Schmiedebergs Arch Pharmacol 331: 71–75
38. Longabaugh JP, Vatner DE, Vatner SF, Homcy CJ (1988) Decreased stimulatory guanosine triphosphate binding protein in dogs with pressure-overload left ventricular failure. J Clin Invest 81: 420–424
39. Horn EM, Barr ML, Morrow BS, Reemtsma K, Cannon PJ, Bilezikian JP (1987) N proteins from human myocardium and lymphocytes are coordinately reduced in congestive heart failure. Circulation 76 (Suppl IV): 309
40. Karliner JS, Scheinman M (1988) Adenylate cyclase activity coupled to the stimulatory guanine nucleotide binding protein in patients having electrophysiologic studies and either structurally normal hearts or idiopathic myocardial disease. Am J Cardiol 62: 1129–1130
41 Denniss AR, Colucci WS, Allen P, Marsh JD (1987) Distribution and function of beta adrenergic receptors in congestive heart failure. Circulation 76 (Suppl IV): 432
42. Bristow M, Ginsburg R, Strosberg, Montgomery W, Minobe W (1984) Pharmacology and inotropic potential of forskolin in the human heart. J Clin Invest 74: 212–223
43. Vatner DE, Lee DL, Schwarz KR, Longabaugh JP, Fujuii AM, Vatner SF, Homcy CJ (1988) Impaired cardiac muscarinic receptor function in dogs with heart failure. J Clin Invest 81: 1836–1842
44. Feldman AM, Cates AE, Veazey WB, Hershberger RE, Bristow MR, Baughman KL, Baumgartner WA, Van Dop C (1988) Increase of the 40,000-mol wt pertussis toxin substrate (G protein) in the failing human heart. J Clin Invest 82: 189–197
45. Glaubiger G, Lefkowitz JR (1977) Elevated β_1-receptor number after chronic propranolol treatment. Biochem Biophys Res Commun 78: 720–725
46. Cramb G, Griffiths NM, Aiton JF, Simmons NL (1984) Biochemical and physiological adaptation to chronic propranolol treatment in the rat. Biochem Pharmacol 33: 1969–1976

47. Aarons RD, Molinoff PB (1982) Changes in the density of beta adrenergic receptors in rat lymphocytes, heart and lung after chronic treatment with propranolol. J Pharmacol Exp Ther 221: 439–443
48. Heilbrunn SM, Shah P, Bristow MR, Valentine HA, Mullin AV, Ginsburg R, Fowler MB (1989) Increased 5-adrenergic receptor density and improved hemodynamic response to catecholamine stimulation during long-term metoprolol therapy in heart failure from dilated cardiomyopathy. Circulation 79: 483–490
49. Aarons RD, Nies AS, Gal J, Hegstrand LR, Molinoff PB (1980) Elevation of β_1-adrenergic receptor density in human lymphocytes after propranolol administration. J Clin Invest 65: 949–957
50. Hedberg A, Kempf F Jr, Josephson ME, Molinoff PB (1985) Coexistence of beta-1 and beta-2 adrenergic receptors in the human heart: effects of treatment with receptor antagonists or calcium entry blockers. J Pharmacol Ex Ther 234: 561–568
51. Baker SP, Potter LT (1980) Effect of propranolol on β-adrenoceptors in rat hearts. Br J Pharmacol 68: 8–10
52. Kennedy RH, Donnelly TE Jr (1982) Cardiac responsiveness after acute withdrawal of chronic propranolol treatment in rats. Gen Pharmacol 13: 231–239
53. Mugge A, Reupcke C, Scholz H (1985) Increased myocardial α-adrenoceptor density in rats chronically treated with propranolol. Eur J Pharmacol 112: 249–252
54. Chess-Williams RG, Broadley KJ (1984) Ex vivo examination of β_1-adrenoceptor characteristics after propranolol withdrawal J Cardiovasc Pharmacol 6: 701–706
55. Cooper G IV, Kent RL, McGonigle P, Watanabe AM (1986) Beta adrenergic receptor blockade of feline myocardium. J Clin Invest 77: 441–455
56. Golf S, Hansson H (1986) Effects of beta blocking agents on the density of beta adrenoceptors and adenylate cyclase response in human myocardium: intrinsic sympathomimetic activity favours receptor upregulation. Cardiovasc Res 20: 637–644
57. Manning AS, Yellon DM, Coltart DJ, Hearse DJ (1981) Abrupt withdrawal of chronic beta-blockade: adaptive changes in cyclic AMP and contractility. J Mol Cell Cardiol 13: 999–1009
58. Baumann G, Riess G, Erhardt WD, Felix SB, Ludwig L, Blumel G, Bloomer H (1981) Impaired beta-adrenergic stimulation in the uninvolved ventricle post-acute myocardial infarction: reversible defect due to excessive circulating catecholamine-induced decline in number and affinity of beta-receptors. Am Heart J 101: 569–581
59. Karliner JS, Stevens M, Honbo N, Hoffman JIE (1989) Effects of acute ischemia in the dog on myocardial blood flow, beta-receptors and adenylate cyclase activity with and without chronic beta-blockade, J Clin Invest 83: 474–481
60. Bristow MR, Kantrowitz NE, Ginsburg R, Fowler MB (1985) β-adrenergic function in heart muscle disease and heart failure. J Mol Cell Cardiol 17: 41–52
61. Myers JH, Horwitz LD (1978) Hemodynamic and metabolic response after abrupt withdrawal of long-term propranolol. Circulation 58: 196–201
62. Lindenfeld J, Crawford MH, O'Rourke RA, Levine SP, Montiel MM, Horwitz LD (1980) Adrenergic responsiveness after abrupt propranolol withdrawal in normal subjects and in patients with angina pectoris. Circulation 70: 704–711
63. Myers MG, Freeman MR, Juma ZA, Wisenberg G (1979) Propranolol withdrawal in angina pectoris: a prospective study. Am Heart J 97: 298–302
64. Croft CG, Rude RE, Gustafson N, Stone PH, Poole WK, Roberts R, Strauss HW, Raabe DS Jr, Thomas LJ, Jaffe AS, Muller J, Hoagland P, Sobel BE, Passamani ER, Braunwald E, Willerson JT, and the Milis Study Group (1986) Abrupt withdrawal of β-blockade therapy in patients with myocardial infarction: effects on infarct size, left ventricular function, and hospital course. Circulation 73: 1281–1290
65. Friedman LM, Byington RP, Capone RJ, Furberg CD, Goldstein S, Lichstein E, and Beta-blocker Heart Attack Trial Research Group (1986) Effect of propranolol in patients with myocardial infarction and ventricular arrhythmia. J Am Coll Cardiol 7: 1–8

Adrenergic Regulation During Exercise in Patients with Heart Failure

HIDEYUKI SATO[1], MASATSUGU HORI[1], AKIRA KITABATAKE[1], and MICHITOSHI INOUE[2]

Summary. Sympathetic nervous activity in chronic heart failure is augmented when compared with normal subjects at identical exercise workloads. There are two possible mechanisms for this alteration, relating to the somatic reflex evoked by chemosensitive muscle afferents; one is an exaggerated somatic exercise reflex resulting from decreased inhibitory influences of arterial and cardiopulmonary baroreceptors, and the other is an increase in hypoxic or related metabolic stimuli to muscle afferents caused by an inadequate perfusion of exercising muscle.

To clarify which mechanism is more important, we analyzed the somatic reflex from the response of plasma norepinephrine (NE) levels to graded muscle hypoxia induced by 3-stage bicycle exercise in 5 patients without organic heart disease and 27 patients with heart failure. Muscle hypoxia was assessed by mixed venous oxygen tension (PVO_2). Plasma NE levels increased exponentially with a decrease in PVO_2. The relationship between plasma NE levels and PVO_2 was not affected by the presence or the severity of heart failure, indicating that the somatic reflex is not exaggerated in heart failure. To simulate a low perfusion of exercising muscle, restudy was performed after an administration of 40 mg metoprolol in 6 of the patients. After the administration, PVO_2 decreased and plasma NE levels increased at each exercise level, and their relationship was exactly similar to that before administration. Thus, we conclude that the augmented sympathetic response to exercise in heart failure is attributed to an increase in hypoxic stimulus to muscle afferents caused by an inadequate perfusion of active muscle, but not to an exaggerated somatic exercise reflex.

Introduction

Sympathetic nervous activity is augmented both at rest and during exercise in chronic heart failure, as reflected by high plasma norepinephrine (NE) levels [1-5]. The increased sympathetic drive serves as an important compensatory mechanism in heart failure, increasing cardiac contractility and heart rate, redistributing blood flow to vital organs, and maintaining perfusion pressure in the face of a limited cardiac output. However, it may also exert harmful influences on the failing heart, by increasing cardiac work, precipitating arrhythmias, and presumably affecting a toxic effect on the failing myocardium [6]. In addition, the increase in renal sympathetic discharge activates the renin-angiotensin system, producing vasoconstriction and sodium retention; this causes increases in afterload and preload which

The First Department of Medicine[1], and Department of Medical Information Science[2]
Osaka University School of Medicine, 1-1-50 Fukushima, Fukushima-ku, Osaka, 553 Japan

cycle through the activation of the sympathetic nervous system.

The augmented sympathetic activity at resting state has been attributed to the decreased inhibitory influences on the vasomotor center from aortic and cardiopulmonary baroreceptors, the tonic activities of which are shown to decrease in heart failure [8, 9]. However, the mechanism(s) of the augmentation during exercise is not completely understood. In this article, we briefly review the literature and describe our data concerning this issue.

Normal Response of Sympathetic Nervous System to Exercise

During exercise there is a local metabolic vasodilatation in the exercising muscles, a vasoconstriction of the splanchnic and renal beds, and an increase in cardiac output, each proportional to the exercise intensity. These events permit an elevation in mean arterial pressure as the exercise becomes more severe, and this, together with the local vasodilatation, provides for adequate blood flow to the active muscles so that their metabolic requirements are met. In these adjustments the sympathetic nervous activation plays an important role, constricting vascular beds in nonexercising parts of the circulation and increasing heart rate and cardiac contractility. The normal response of the sympathetic nervous system to exercise has been recently reviewed in detail [10].

Two main components comprise the sympathetic response to rhythmic and static exercise [11]. One is related to the central cortical activity for the recruitment of motor units (i.e., central command) [12], and the other to a somatic reflex originating in the exercising muscles (i.e., exercise reflex) [13, 14]. A central command mechanism, providing information concerning the muscle mass involved in exercise to the vasomotor center, may initiate the sympathetic activation and be responsible for determining the beginning level of sympathetic activity. This mechanism is, however, unlikely to change markedly in heart failure. The somatic reflex serves as a feedback system from the exercising muscle by monitoring a flow error that is important in the fine moment-to-moment regulation of the sympathetic activity during exercise. Receptors with unmyelinated group IV afferent fibers (C fibers) in the contracting muscles are believed to respond to hypoxic or related metabolic stimuli, e.g., bradykinin, prostaglandins, and/or potassium, and to trigger reflex sympathetic discharge. For example, the inhibition of prostaglandin synthesis by indomethacin, a blocker of cyclooxygenase, attenuates the increase in mean arterial pressure and left ventricular dp/dt in response to electrical stimulation-induced contraction of hind limb in cats [15]. These substances also induce local vasodilation in exercising muscles. Thus, the same chemical stimuli could initiate both reflex vasoconstriction and direct local vasodilation.

Modulation of Somatic Exercise Reflex by Aortic and Cardiopulmonary Baroreceptors

Both arterial and cardiopulmonary baroreceptors exert a tonic inhibitory influence on the vasomotor centers, and restrain the sympathetic discharge under the basal resting state. There is also an interaction between the excitatory somatic exercise reflex and the inhibitory afferents from arterial and cardiopulmonary baroreceptors, although the set point at which baroreceptors act may be centrally reset to a higher range of blood pressure during exercise. The somatic reflex is markedly augmented in dogs when the input from arterial or cardiopulmonary baroreceptors is decreased [16]. In humans the vasoconstriction in the nonexercising limbs produced by isometric handgrip exercise is also augmented in the presence of lower body negative pressure, which decreases the sensory input from arterial and cardiopulmonary baroreceptor afferents [17]. However, a recent study using a microneurographic technique demonstrated that decreased inhibitory input from cardiopulmonary baroreceptors does not augment sympathetic outflow during static handgrip exercise in normal subjects [18]. This result challenges the concept of an important interaction between baroreceptor afferents and somatic exercise reflex.

Augmented Sympathetic Activation in Response to Exercise in Heart Failure and Its Possible Mechanism(s)

It has been shown that the sympathetic drive during exercise is markedly augmented in chronic heart failure. Plasma NE levels during a given submaximal exercise, as well as at rest, are greater in patients with heart failure than in normal subjects [1, 4]. Also the arterial and cardiopulmonary baroreceptors are desensitized, presumably because of the decreased distensibility of the cardiovascular wall or damage to these endings in heart failure [8, 9]. Therefore the augmented sympathetic response to exercise in heart failure has been attributed to the exaggerated somatic exercise reflex caused by decreased inhibitory input from arterial and cardiopulmonary baroreceptors [9]. However, direct evidence supporting this postulate is lacking.

There is an alternative explanation for the augmented sympathetic response to exercise in heart failure, which has received relatively little attention: a negative feedback loop mediated by chemosensitive muscle afferents attempting to overcome an inadequate perfusion to active muscle. The nutritional blood flow into the exercising muscle is limited in heart failure because of the decreased cardiac functional reserve. The sensory afferents in the active muscle in heart failure may be more strongly stimulated by more severe muscle hypoxia than in normal subjects, during the same exercise workload. Consequently, the sympathetic activity in response to a comparable exercise would be greater in heart failure. Also, this possibility has not been tested. Thus, we studied the relative importance of these two possible mechanisms in the regulation of sympathetic activity during rhythmic exercise in patients with heart failure. In this study, we analyzed a stimulus-response curve of the somatic exercise reflex from the relationship between mixed venous oxygen tension and plasma NE levels.

Analysis of Somatic Exercise Reflex Mediated by Sensory Afferents in Active Muscle in Patients with Heart Failure

Since oxygen extraction by metabolizing tissue is a passive diffusion process according to the capillary-tissue gradient of partial pressure of oxygen (PO_2), the end

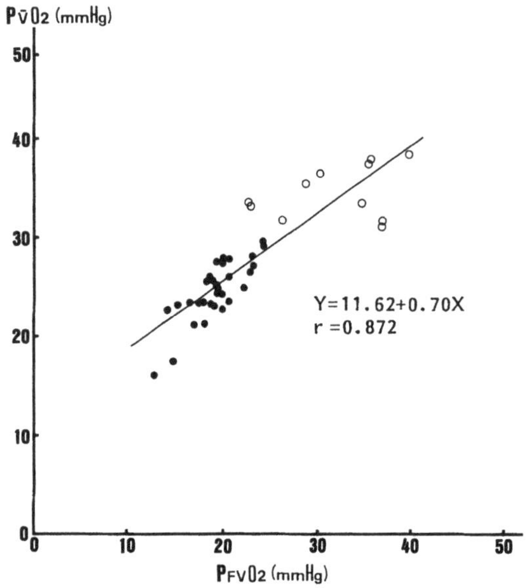

Fig. 1. The relationship between partial oxygen pressure in femoral venous blood (PFV_{O2}) and that in mixed venous blood (PV_{O2}) at rest (○) and during 3-stage graded supine bicycle exercise test (●) n = 10 patients with chronic heart failure. A close linear correlation was observed between PFV_{O_2} and PV_{O_2}

capillary PO_2 is likely to be equivalent to the tissue PO_2. Thus, femoral venous PO_2 during leg bicycle exercise can reflect the tissue hypoxia in exercising muscle. In addition, we observed in ten heart failure patients that the mixed venous PO_2 (Pv_{O2}) correlates well with femoral venous PO_2 ($P_{FV}{}_{O2}$) during bicycle exercise (Fig. 1). This result indicates that mixed venous PO_2 is also an index of the intensity of tissue hypoxia in exercising legs (i.e., a stimulus to the sensory afferents in active muscle).

Plasma NE levels have been used as an index of the sympathetic activity in the clinical setting, but plasma NE in circulating blood does not necessarily reflect the amount of NE released from nerve endings [19]. The plasma NE level represents only the net result of spillover into circulation minus the disappearance. Changes in the rate of disappearance may influence plasma NE levels without a change in spillover rate. Recently Hasking et al. demonstrated that NE clearance decreased in patients with heart failure [5]. However, their data still showed that even in such patients, plasma NE levels were largely determined by spillover rate. Plasma NE levels can be, therefore, employed as an index of sympathetic activity even in the presence of heart failure. Thus, the relation curve of plasma NE levels to mixed venous PO_2 during leg bicycle exercise is considered to reflect the stimulus-response curve of the somatic exercise reflex.

We obtained this response curve in five patients without organic heart disease as a control group, and in 27 patients with chronic heart failure. The etiology of the heart failure was isolated aortic regurgitation in 9 patients, old myocardial infarction in 7, dilated cardiomyopathy in 6, and hypertrophic cardiomyopathy in 6. Eight of 27

Table 1. Patient characteristics

	n	Age (years)	EF (%)	Peak VO_2 (ml/kg per minute)
Control	5	43 ± 13	64 ± 5	25 ± 7
NYHA I	8	49 ± 14	62 ± 7	19 ± 5
NYHA II	11	52 ± 10	47 ± 19[a,b]	19 ± 5
NYHA III	8	60 ± 10[a]	42 ± 17[a,b]	15 ± 3[a]

EF, left ventricular ejection fraction; Peak VO_2, peak oxygen consumption during supine bicycle exercise. Data are mean ± SD.
[a]$P < 0.05$ vs control.
[b]$P < 0.05$ vs NYHA I.

patients were classified in NYHA functional class I, 11 were in class II, and 8 were in class III. The patients' characteristics are shown in Table 1; the resting left ventricular ejection fraction and the peak oxygen consumption achieved during a symptom-limited bicycle exercise test decreased with worsening NYHA functional classification.

A 3-stage graded supine bicycle exercise test was conducted at one-third of the previously determined peak workload for 4 min (Ex-1), followed by a load of two-thirds peak workload for another 4 min (Ex-2), and then a final 4 min at peak workload (Ex-3), under the right heart catheterization. Hemodynamic measurements and mixed venous blood sampling were made at the 4 time points; at rest and at the end of each exercise level. In a preliminary study it was observed that the changes in plasma NE levels during exercise at a given workload did not reach a steady state until the 4th minute. Plasma NE levels in mixed venous blood were determined by the high performance liquid chromatography (HPLC) with electrochemical detection (ECD), and blood gas analysis was performed. In this HPLC-ECD system the sensitivity of the assay is 20 pg/ml plasma and intra-assay coefficient of variation is 5.6%.

Plasma NE levels increased and mixed venous PO_2 decreased with the intensity of exercise. As previously reported by several investigators, the plasma NE responses to an absolute exercise workload were augmented in patient groups according to the severity of heart failure when compared with the control group (Fig. 2). On the other hand, it appeared that the plasma NE levels of individuals increased along the same exponential curve, irrespective of the presence and/or the severity of heart failure, when they were plotted against corresponding mixed venous PO_2 (Fig. 3). This was more clearly illustrated when both plasma NE levels and mixed venous PO_2 were depicted on a logarithmic scale (Fig. 4). A linear inverse correlation between the logarithm of mixed venous PO_2 and that of plasma NE was observed on both pooled data ($r = -0.83$) and each subject ($r = -0.75$ to -0.99, mean $= -0.92$). The slope of the regression line reflecting the increase in plasma NE per unit decrease in PO_2 (i.e., the sensitivity of somatic exercise reflex) was calculated in each subject. Their mean showed no significant difference among groups (control, -2.3 ± 1.2; class I, -2.3 ± 0.9; class II, -2.9 ± 0.9; class III, -2.9 ± 1.1).

These results indicate that the somatic reflex evoked by hypoxic stimulus to sensory afferents in active muscle plays an important role in the regulation of sympa-

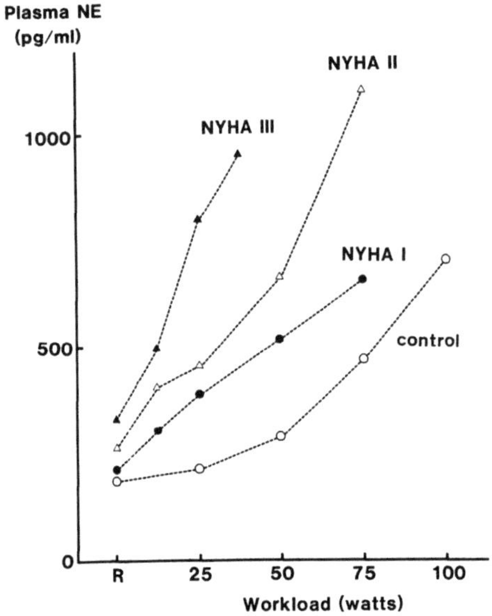

Fig. 2. The increase in mean values of plasma norepinephrine (*NE*) levels from rest (*R*) to the increased exercise workload in 5 patients without organic heart disease (control group) and 27 patients with heart failure subdivided into 3 groups according to NYHA functional classification: 8 were in NYHA class I, 11 were in class II, and 8 were in class III. Plasma NE levels at a given workload were higher in heart failure groups than in the control group along with worsening functional classification

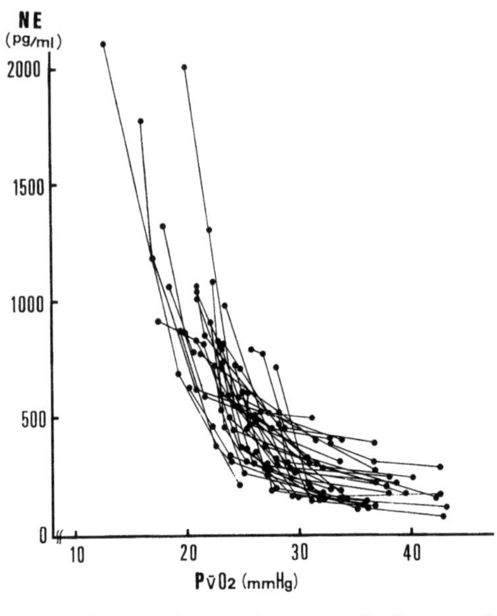

Fig. 3. The relationship between mixed venous PO_2 (PV_{O_2}) and plasma norepinephrine (*NE*) levels at rest and during leg bicycle exercise in all 32 patients, with or without heart failure. Plasma NE levels at the 4 time points (at rest and 3 levels of exercise) were plotted against corresponding values of PV_{O_2} in each patient. Plasma NE increased in an exponential manner as PV_{O_2} decreased from rest to exercise with increasing intensity. This relationship was apparently identical among patients

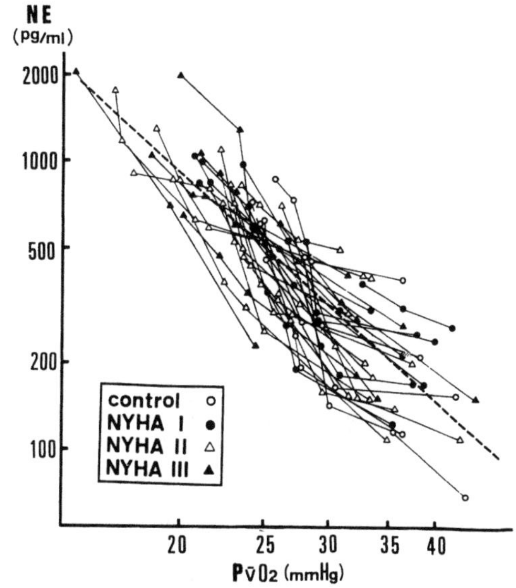

Fig. 4. The relationship between logarithm of mixed venous PO_2 (PV_{O_2}) and logarithm of plasma norepinephrine (*NE*) levels (data as in Fig. 3). A linear inverse correlation was observed on both pooled data (r = −0.835) and 4 time points of individuals (r = −0.75 to −0.99, mean −0.92). The relationship was not affected by the presence and/or the severity of heart failure. $P < 0.001$; $n = 123$; Y = 14.41 − 2.55 X

thetic activity during rhythmic exercise, and suggest that this regulatory mechanism is not altered even in patients with heart failure, challenging the concept that the somatic exercise reflex is exaggerated in heart failure. Thus, the increased sympathetic activity in response to a comparable exercise workload, observed in heart failure, could be attributed to the low tissue PO_2 in active muscle caused by an inadequate perfusion, rather than to the exaggerated somatic reflex caused by baroreceptor desensitization.

To confirm this postulate, we examined the effect of a reduction in mixed venous PO_2, induced by decreasing cardiac reserve, on plasma NE levels during exercise. The exercise test was repeated with the same workload and time intervals after an oral administration of 40 mg metoprolol, a cardioselective β_1-adrenoceptor antagonist, in 6 of 8 patients in NYHA class I. Metoprolol decreased heart rate, cardiac output, and mixed venous PO_2, but increased plasma NE levels at each exercise level (Fig. 5). The decreased mixed venous PO_2 was probably due to a reduction in blood flow into the active muscle. The plasma NE level during exercise was unlikely to be increased by a direct effect of metoprolol on the central nervous system or on the release of NE from nerve endings, because plasma NE levels did not increase in the resting state as shown in Fig. 5. It appeared that the increased plasma NE levels resulted from a reduction in mixed venous PO_2, since the quantitative relationship between mixed venous PO_2 and plasma NE levels after administration was quite similar to that before administration (Fig. 6). This finding indicates that the intensity of tissue hypoxia in active muscle primarily determines sympathetic activity

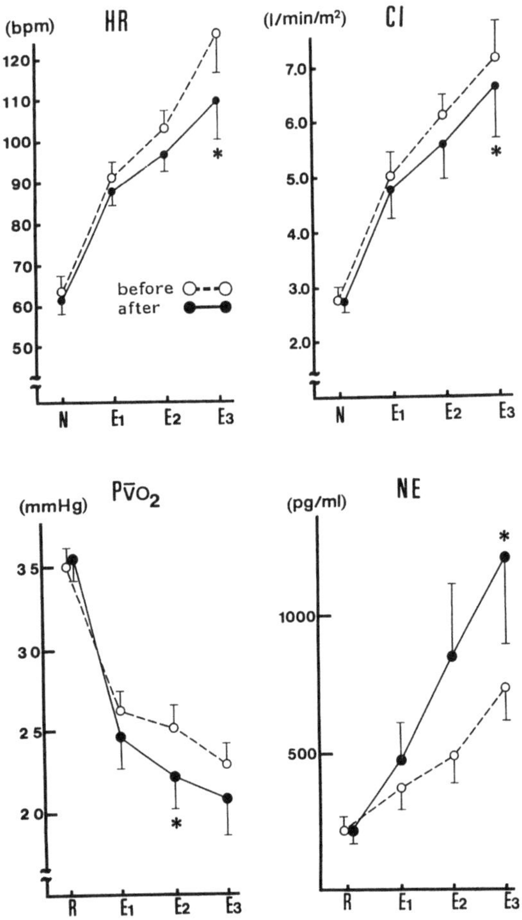

Fig. 5. The effect of metoprolol on heart rate (*HR*), cardiac index (*CI*), mixed venous PO_2, ($P\bar{V}o_2$), and plasma norepinephrine (*NE*) level at rest (*R*) and during a 3-stage graded bicycle exercise (E_1, E_2, E_3) in 6 of the heart failure patients. Values are mean, and SE is shown with *bars*. $^*P<0.05$ vs the values before administration

during rhythmic exercise, at least, except for the initial phase in which the "central command" may be responsible for the activation of the sympathetic nervous system.

Thus, we conclude that in patients with chronic heart failure, an inadequate perfusion of active muscle, which increases the hypoxic or related metabolic stimuli to chemosensitive muscle afferents, plays a critical role in the increased sympathetic nervous activity in response to exercise. The role of an exaggerated somatic exercise reflex resulting from arterial and cardiopulmonary baroreceptor desensitization is apparently less important.

Acknowledgment. Our study is supported in part by Grant-in-aid for Scientific Research 63480229 from the Ministry of Education, Science, and Culture, Japan.

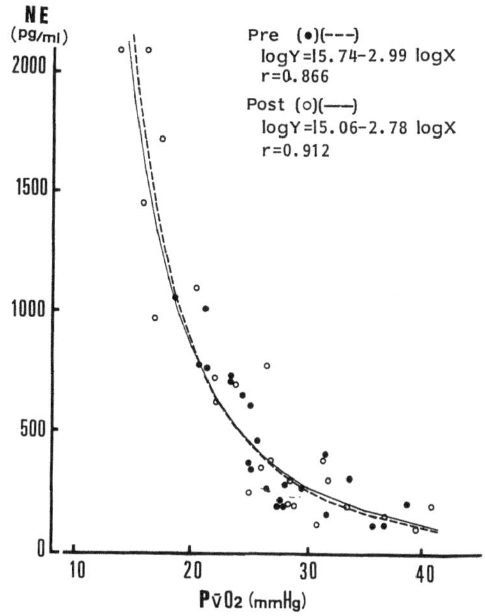

Fig. 6. The relationship between mixed venous PO_2 (PV_{O_2}) and plasma norepinephrine (*NE*) levels before (●) and after (○) metoprolol administration, which was obtained from 6 patients with heart failure. The regression curve after administration (*solid curve*) was exactly similar to that before administration (*dashed curve*)

References

1. Chidsey CA, Harrison DC, Braunwald E (1962) Augmentation of plasma norepinephrine response to exercise in patients with congestive heart failure. N Engl J Med 267: 650–654
2. Thomas JA, Marks BH (1978) Plasma norepinephrine in congestive heart failure. Am J Cardiol 41: 233–243
3. Minami M, Yasuda H, Yamazaki N, Kojima S, Nishijima H, Matsumura N, Togashi H, Koike Y, Saito H (1983) Plasma norepinephrine concentration and plasma dopamine-beta-hydroxylase activity in patients with congestive heart failure. Circulation 67: 1324–1329
4. Francis GS, Goldsmith SR, Ziesche S, Cohn JN (1982) Response of plasma norepinephrine and epinephrine to dynamic exercise in patients with congestive heart failure. Am J Cardiol 49: 1152–1156
5. Hasking GJ, Esler MD, Jennings GL, Burton D, Korner PI (1986) Norepinephrine spillover to plasma in patients with congestive heart failure: evidence of increased overall and cardiorenal sympathetic nervous activity. Circulation 73: 615–621
6. Bristow MR (1984) The adrenergic nervous system in heart failure. N Engl J Med 311: 850–851
7. Francis GS, Goldsmith SR, Levine TB, Olivary MT, Cohn Jn (1984) The neurohumoral axis in congestive heart failure. Ann Intern Med 101: 370–377
8. Hirsch AT, Dzau VJ, Creager MA (1987) Baroreceptor function in congestive heart failure: effect on neurohumoral activation and regional vascular resistance. Circulation 75 (SupplIV): IV-36 - IV-48
9. Abboud FM, Thames MD, Mark AL (1981) Role of cardiac afferent nerves in the regulation of the circulation during coronary occlusion and heart failure. In: Abboud FM,

Fozzard HA, Gilmore JP, Reis DJ (eds) Disturbances in neurogenic control of the circulation. American Physiological society, Bethesda, Maryland, USA, pp 65-86
10. Christensen NJ, Galbo H (1983) Sympathetic nervous activity during exercise. Annu Rev Physiol 45: 139-153
11 Mitchell JH (1985) Cardiovascular control during exercise: central and reflex neural mechanisms. Am J Cardiol 55:34D-41D
12 Kaufman MP, Rybicki K (1987) Discharge properties of group III and IV muscle afferents: their responses to mechanical and metabolic stimuli. Circ Res 61 (Suppl): I-60 - I-65
13 Millhorn DE, Eldridge FL, Waldrop TG, Kiley JP (1987) Diencephalic regulation of respiration and arterial pressure during actual and fictive locomotion in cat. Circ Res 61 (Suppl I): I-53 - I-59
14 Freud PR, Rowell LB, Murphy TM, Hobbs SF, Butler SH (1979) Blockade of the pressor response to muscle ischemia by sensory nerve block in man. Am J Physiol 237: H433-H439
15 Stebbins CL, Maruoka Y, Longhurst JC (1986) Prostaglandins contribute to cardiovascular reflexes evoked by static muscle contraction. Circ Res 59: 645-654
16 Thames MD, Abboud FM (1979) Interaction of somatic and cardiopulmonary receptors in control of renal circulation. Am J Physiol 237: H560-H565
17 Walker JI, Abboud FM, Mark AL, Thames MD (1980) Interaction of cardiopulmonary receptors with somatic receptors in man. J Clin Invest 65: 1491-1497
18 Scherrer U, Vissing SF, Victor RG (1988) Effects of lower-body negative pressure on sympathetic nerve responses to static exercise in humans: microneurographic evidence against cardiac baroreflex modulation of the exercise pressor reflex. Circulation 78: 49-59
19 Goldstein DS, McCarty R, Polinsky RJ, Kopin IJ (1983) Relationship between plasma norepinephrine and sympathetic neural activity. Hypertension 5: 552-559

Exercise Hemodynamics in Patients with Heart Failure

SHIGETAKE SASAYAMA and HIDETSUGU ASANOI[1]

Summary. The concept of ventriculo-arterial coupling provides a useful framework to understand the exercise response in normal and failing hearts. Ventricular and arterial properties are matched so as to optimize ventricular work efficiency in normal hearts and to maximize stroke work in moderately depressed hearts. Exercise capacity is generally reduced in failing hearts. Thereby, decreases in anaerobic threshold and peak oxygen consumption are inversely related to the resting plasma catecholamine levels.

During anaerobic exercise, stroke volume is augmented by an increase in contractility and an enhanced mechanical efficiency in normal subjects, whereas in patients with cardiac dysfunction, stroke volume is augmented through an increase in ventricular volume without any change in both ventricular and arterial volume elastance. Thus, the regulation of stroke volume during exercise shifts from a catecholamine-mediated reduction in end-systolic volume to a greater reliance on the Frank-Starling mechanism.

Introduction

Patients with chronic congestive heart failure generally develop the earliest manifestations of symptoms during physical activity; these substantially limit their capacity to exercise. Overall circulatory impairment is usually apparent in the resting state, and becomes increasingly prominent during graded exercise testing. The elevated metabolic requirements of the exercising muscle are generally met by an increase in venous return and in cardiac output.

This incremental blood flow during exercise provides the capability for performing exercise. However, there still remains considerable dispute concerning the major mechanism by which the increase in cardiac output is achieved during exercise: whether due to enhanced contractility or to the influence of the Frank-Starling mechanism [1]. It has also been shown that hemodynamic measurements at rest do not correlate with exercise capacity largely because systemic hemodynamics are affected by the complex interaction of many other factors which can modify left ventricular performance [2]. In this regard, we analyzed circulatory adjustments necessary to provide adequate flow to peripheral tissues. Normal and failing hearts, both at rest and under stressed conditions, were studied using a newly developed conceptual framework for quantifying the coupling of the ventricle and the arterial load [3, 4].

[1] The Second Department of Internal Medicine, Toyama Medical and Pharmaceutical University, 2630 Sugitani, Toyama, 930-01 Japan

Exercise and Catecholamine

The neural mechanism is known to be of great importance in mediating the initial response to exercise. The activity of the sympathetic nervous system is mediated by the release of the neural transmitter substance norepinephrine (NE) at the sympathetic nerve endings.

We observed significant inverse correlations between plasma norepinephrine level and exercise capacities determined by anaerobic threshold (ATge) or peak body oxygen consumption ($\dot{V}_{O_2}$) using a 1-min incremental sitting ergometer, in 26 patients with variable degree of heart failure (NE vs ATge, r = −0.59, $P < 0.01$; NE vs peak $\dot{V}_{O_2}$, r = −0.56, $P < 0.01$) (Fig. 1) [5].

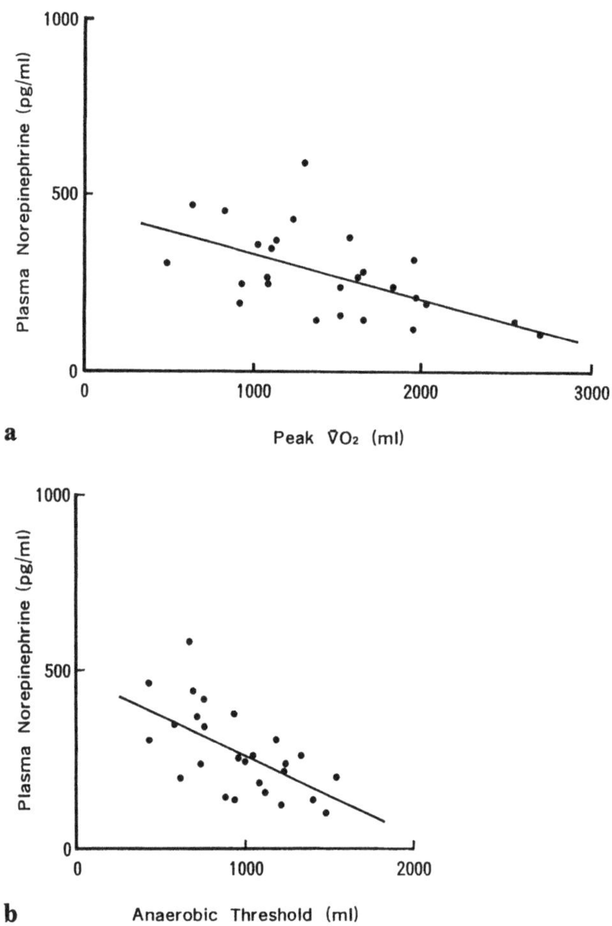

Fig. 1a, b. The plasma norepinephrine concentration progressively increased along with the increased severity of cardiac failure. Significant inverse correlation was observed between plasma norepinephrine and peak $\dot{V}_{O_2}$ (**a**) and between norepinephrine and anaerobic threshold (**b**). **a** y = − 0.12 × + 460; r = 0.56; $P<0.01$; **b** y = 0.23 × + 499, r = − 0.59, $P<0.01$. From Asanoi and Sasayama [5], with permission

$\dot{V}_{O_2}$ usually rises at maximal exercise associated with an increase in cardiac output in normal subjects, while it may fail to increase during exercise in patients with severe congestive heart failure. Some vasodilators or inotropic agents have been shown to increase cardiac output acutely during exercise but not to be associated with an immediate change in peak $\dot{V}_{O_2}$. Hence, a fixed vasodilating capacity of the peripheral vasculature, associated with a mismatch between the limited ability of the peripheral circulation to perfuse active muscles and impaired nutritive flow to skeletal muscle, is assumed to be the factor most responsible for the reduced maximal exercise capacity in these patients [6]. Thus, the presence of elevated plasma norepinephrine levels in these patients reflects a reduced peripheral perfusion due to impaired rise in cardiac output during exercise.

Left Ventricular Coupling with Arterial Load in Normal and Failing Heart

The ventricle transfers the mechanical energy of contraction to the intracavitary blood to provide adequate flow to the arterial system [7, 8]. In this way, the energy source and its load are matched when a maximal amount of energy is transferred from the source to the load. The ventricle is generally considered matched to a given load if it allows a maximal amount of external work against that load [3, 9, 10], while from the energetic point of view the matching between the ventricle and arterial system can be regarded as optimal when they interact to provide maximal cardiac efficiency [4, 11, 12]. In the case of the heart, the mechanical efficiency is defined as the ratio of stroke work to myocardial oxygen consumption per beat.

Sunagawa et al. [13] characterized the arterial system as an elastic chamber with a volume elastance, similar to the ventricle, which has the end-systolic volume elastance. When both ventricular and arterial end-systolic pressure-volume relationships are superimposed in the same pressure-volume plane, the stroke volume that the ventricle can eject from a given end-diastolic volume is represented as the intersection of these two lines. Within this framework, the external work of the ventricle becomes maximal, providing a maximal transfer of mechanical energy of contraction to the arterial system, when the slopes of the ventricular and arterial end-systolic pressure-volume relationships are identical. On the other hand, from the standpoint of economical fuel consumption, mechanical efficiency, from a given end-diastolic volume, has been shown to be maximal when arterial impedance is nearly one-half the cardiac output impedance [4]. On the other hand, from the standpoint of economical fuel consumption, mechanical efficiency, from a given end-diastolic volume, has been shown to be maximal when arterial impedance is nearly one-half the cardiac output impedance [4].

Recently, we studied, in resting human's, the coupling between the ventricular pump and arterial afterload in terms of the slope of the end-systolic pressure-volume relationship (E_{es}, ventricular elastance), and the slope of the arterial end-systolic pressure-stroke volume relationship (effective arterial elastance, E_a). Both normal subjects and patients with variably depressed ventricular function were studied, to clarify the physiologic controller of the heart and arterial system at rest and in stressful conditions [14].

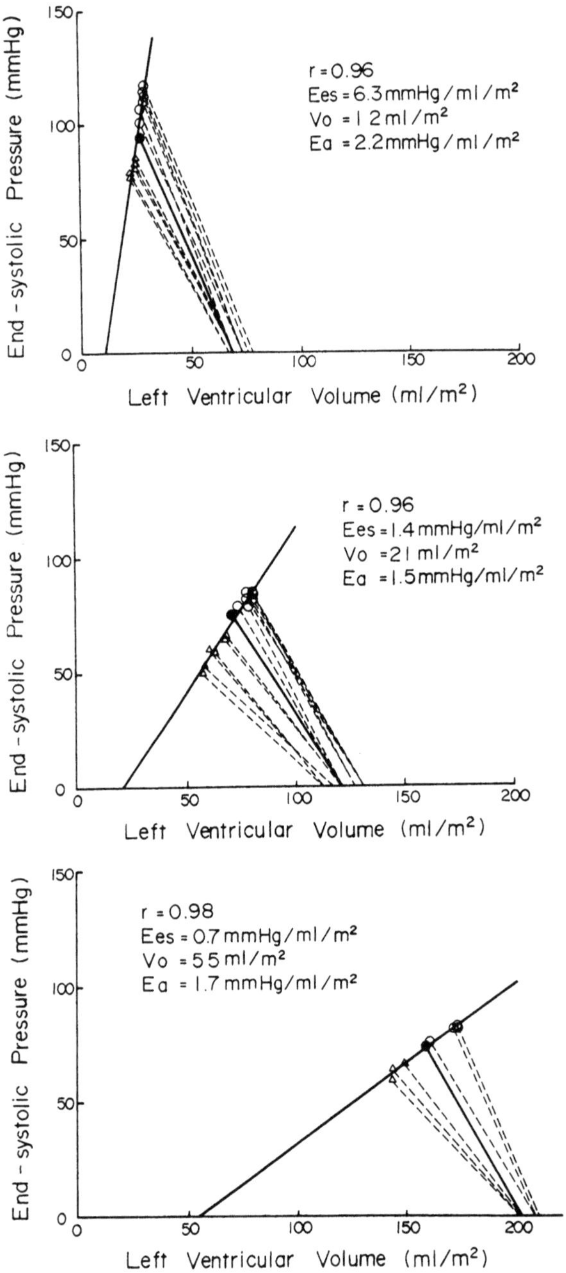

Fig. 2. Left ventricular end-systolic pressure-volume relationship and end-systolic pressure-stroke volume relationship in three representative cases. From *top* to *bottom*, normal, mildly depressed, and severely depressed ventricle. The end-systolic pressure volume relation was obtained by plotting more than 5 different dicrotic arterial pressures against corresponding left ventricular end-systolic volumes during pharmacological pressure change. End-systolic elastance (E_{es}) is greater than baseline arterial elastance (E_a) in the normal group, almost equal to baseline E_a in mild heart failure, and far less than baseline E_a in the severely depressed heart. V_o, volume axis intercept. From [14], with permission

According to the pressure-volume diagram proposed by Suga [15], we also determined the left ventricular stroke work (SW), end-systolic potential energy, and the ventricular work efficiency defined as stroke work/(stroke work + potential energy). In healthy men, the ventricle was in a good contractile state with E_{es} of 4.5 mmHg per ml/m^2 and ejection fraction (EF) of 60% or more. Their E_a was always less than E_{es}, thereby allowing a greater mechanical efficiency and a smaller SW for the given contractility than when E_{es} was equal to E_a. In those patients with mildly depressed ventricle with E_{es} of 2.5 mmHg per ml/m^2 and EF of 50%, E_a was nearly equal to E_{es}, permitting almost the maximal stroke work possible at a given preload. In those patients with severely depressed ventricle with E_{es} of 1.5 mmHg per ml/m^2 and EF of less than 40%, E_a was substantially larger than E_{es}, resulting in a higher E_a/E_{es} ratio (2.56) than in the normal group (0.46) or the mildly depressed heart group (0.90) (Fig. 2).

These results indicated that in the normal human ventricle, the cardiac performance was adjusted to arterial load to optimize the work efficiency rather than external work. In the ventricle with mild to moderate heart failure, ventricular and arterial properties were matched so that the ventricle could perform maximal external work at the expense of work efficiency; while in the ventricle with a severely depressed contractile state, the volume elastance of the ventricle at end-systole was far less than that of the arterial system, and these ventricles failed to maintain normal stroke work, the failure being associated with a further reduction in mechanical work efficiency. These results were surprisingly similar to the theoretical analysis of Burkhoff and Sagawa [4] of the arterial load that would produce maximal stroke work, and maximal mechanical efficiency of ventricular contraction. In that study, the coupling condition for maximal stroke work was shown to be $E_a = E_{es}$, whereas that for maximal efficiency was $E_a = E_{es}/2$.

Hemodynamic Parameters and Exercise Capacity

The previously mentioned conceptual framework to quantify ventricular-arterial interaction has great potential usefulness for gaining insight into the relevance of adaptational changes in congestive heart failure. We employed this coupling framework to understand the exercise response in the normal and failing heart [5].

In 10 symptomatic patients with congestive heart failure and in 10 normal subjects, left ventricular dimensions were analyzed by two-dimensional targeted echocardiograms with simultaneous measurement of direct arterial pressure. As stated in the previous section, the ventricular contractile properties were expressed by the slope and volume axis intercept of the end-systolic pressure-volume relationship (E_{es} and V_0 respectively) as follows:

$$ESP = E_{es}\,(ESV - V_0) \qquad (1)$$

where ESP and ESV are ventricular end-systolic pressure and volume respectively. The arterial input impedance properties were expressed by the slope of the arterial end-systolic pressure-stroke volume (SV) relationship (E_a).

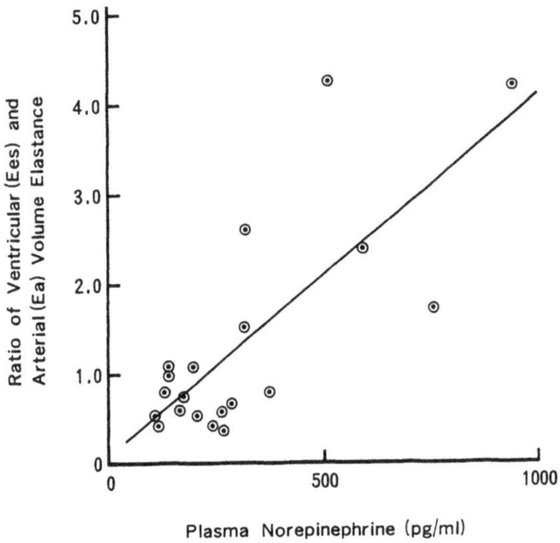

Fig. 3. Relation between E_a/E_{es} ratio and plasma norepinephrine. Increased severity in heart failure with increased plasma norepinephrine levels was characterized by a parallel increase in E_a/E_{es} ratio. This change is supposed to cause reduction in stroke volume. However, an augmented preload compensates for the reduction in stroke volume in these patients. $Y = 0.004 \times + 0.090$; $r = 0.76$; $P<0.01$. E_a, arterial elastance, E_{es}, end-systolic elastance

When ventricle and arterial systems are coupled, the resulting SV and ESP can be predicted as follows:

$$SV = (EDV - V_0)/(1 + E_a/E_{es}) \tag{2}$$

$$ESP = SV \cdot E_a$$

where EDV is the end-diastolic volume.

In patients with moderate heart failure, SV and ESP were maintained within the normal range over a wide range of resting plasma norepinephrine levels. On the other hand, E_a/E_{es} was linearly related to plasma norepinephrine levels (Fig. 3). It is apparent from Eq. 2 that an increase in the E_a/E_{es} ratio directly reduces stroke volume. However, early use of the Frank-Starling mechanism provides a fundamental compensation and permits the heart to maintain the normal stroke volume in these hearts.

Anaerobic Threshold and Ventricular-Arterial Coupling

The anaerobic threshold can be defined by the continuous determination of the respiratory exchange ratio during graded exercise [16]. Anaerobic metabolism provides only a small part of the energy at moderate workloads but becomes a greater energy source at heavy workloads. Above the anaerobic threshold, the rate of lactic acid production increases. Being buffered by sodium bicarbonate, it augments carbon dioxide production and the respiratory exchange ratio. The carbon dioxide formed is exhaled via the lungs, and ventilation increases still further in parallel with carbon dioxide production but ultimately exceeds it. In normal subjects and patients

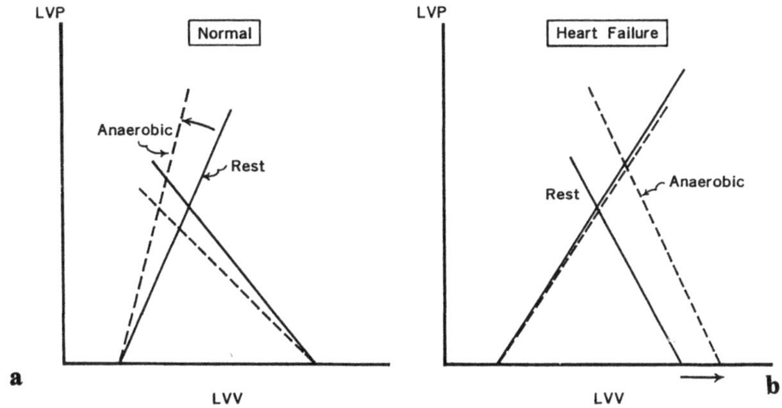

Fig. 4a, b. Schematic representation of ventriculoarterial coupling at rest (*solid line*) and during anaerobic exercise (*dashed line*) in the normal (**a**) and failing heart (**b**). In the normal heart, stroke volume is augmented from essentially unchanged end-diastolic volume by enhanced contractility, with a resultant increase in E_{es} and a decrease in E_a. In contrast, in the failing heart an increase in stroke volume during anaerobic exercise is mediated by an increase in preload without detectable changes in E_{es} and E_a. *LVP*, left ventricular pressure; *LVV*, left ventricular volume; E_{es}, end-systolic elastance; E_a, arterial elastance

with chronic congestive heart failure (EF averaged 36% ± 12%), the anaerobic threshold was determined during graded threshold exercise as the point at which the linear relationship between ventilation and oxygen uptake was lost. Hemodynamic and dimensional (by radionuclide angiography) analyses were also performed at workloads 30% above and 30% below ATge, and ventricular-load coupling was analyzed in the same manner as mentioned before [5].

In normal subjects, stroke volume plays a role in the increase in cardiac output during exercise. In dogs running spontaneously in the field at speeds exceeding 20 mph, stroke volume rose approximately 50% from the preexercise control level in the standing position [17]. Stroke volume can increase either due to an increase in preload, or an increase in the contractile state. The former would operate generally during low levels of exercise while the latter acts to reduce left ventricular end-systolic dimensions at high levels of exercise. This response is represented by an increase in E_{es} and a decrease in E_a, and therefore mechanical efficiency is markedly augmented (Fig. 4a).

In patients with severe cardiac dysfunction, changes in end-diastolic volume and stroke volume were variable during aerobic exercise, while stroke volume was generally augmented during anaerobic exercise. This response was mediated essentially by an increase in end-diastolic volume rather than a decrease in end-systolic volume. Consequently, E_{es} and E_a remained unchanged both during aerobic and anaerobic exercise (Fig. 4b). The failure of E_{es} to rise during anaerobic exercise despite the fact that the sympathetic nervous system is called into play is a result of reduced contractile reserve in these hearts, presumably due to decreased effectiveness of β-adrenergic modulation of myocardial contractility, increased ischemia, or loss of contractile elements. Thus, in advanced heart failure, the regulation of stroke volume during exercise shifts from a catecholamine-mediated reduction in end-systolic volume to a greater reliance on the Frank-Starling mechanism.

References

1. Vatner SF, Pagani M (1976) Cardiovascular adjustments to exercise: hemodynamics and mechanisms. Prog Cardiovasc Dis 19: 91-108
2. Meiler SEL, Ashton JJ, Moeschberger ML, Unverferth DV, Leier CV (1987) An analysis of the determinants of exercise performance in congestive heart failure. Am Heart J 5: 1207-1217
3. Sunagawa K, Maughan WL, Burkhoff D, Sagawa K (1983) Left ventricular interaction with arterial load studied in isolated canine heart. Am J Physiol 245: H773-H780
4. Burkhoff D, Sagawa K (1986) Ventricular efficiency predicted by an analytical model. Am J Physiol 250: R1021-R1027
5. Asanoi H, Sasayama S (1989) Relationship of plasma norepinephrine to ventricular-load coupling in patients with heart failure. Jpn Circ J 53: 131-140
6. Wilson JR, Martin JL, Schwartz D, Ferraro N (1984) Exercise intolerance in patients with chronic heart failure: role of impaired nutritive flow to skeletal muscle. Circulation 69: 1079-1087
7. Piene H, Sund T (1982) Does normal pulmonary impedance constitute the optimal load for the right ventricle? Am J Physiol 242: H154-H160
8. Yamakoshi K (1985) Interaction between heart as a pump and artery as a load. Jpn Cir J 49: 195-205
9. Piene H, Sund T (1979) Flow and power output of right ventricle facing load with variable input impedance. Am J Physiol 237: H125-H130
10. Van den Horn GJ, Westerhof N, Elzinga G (1985) Optimal power generation by the left ventricle: a study in the anesthetized open thorax cat. Circ Res 56: 252-261
11. Suga H, Igarashi Y, Yamada O, Goto Y (1985) Mechanical efficiency of the left ventricle as a function of preload, afterload, and contractility. Heart Vessels 1: 3-8
12. Elzinga G, Westerhof N (1980) Pump function of the feline left heart: changes with heart rate and its bearing on the energy balance. Cardiovasc Res 14: 81-92
13. Sunagawa K, Maughan WL, Sagawa K (1985) Optimal arterial resistance for the maximal stroke work studied in isolated canine left ventricle. Circ Res 56: 586-595
14. Asanoi H, Kameyama T, Sasayama S (to be published) Ventriculo-arterial coupling in normal and failing heart in humans. Circ Res
15. Suga H (1979) Total mechanical energy of a ventricle model and cardiac oxygen consumption. Am J Physiol 236: H498-H505
16. Le Jemetel TH, Mancini D, Gumbardo D, Chadwick B (1985) Pitfalls and limitations of "maximal" oxygen uptake as an index of cardiovascular functional capacity in patients with chronic heart failure. Heart Failure 1: 112-118
17. Vatner SF, Franklin D, Higgins CB, Patrick T, Braunwald E (1972) Left ventricular response to severe exertion in untethered dogs. J Clin Invest 51: 3052-3060

New Inotropic Agents for Treatment of Heart Failure

KEITARO HASHIMOTO[1]

Summary. New positive inotropic agents have been synthesized and they are under intensive research to examine whether they are different from older drugs like digitalis and catecholamines in their mechanisms of action and in their clinical effectiveness. Among the positive inotropic agents, those developed by Japanese pharmaceutical laboratories were reviewed mainly in terms of their possible mechanism of action in the cardiac cells. OPC-8212 is orally active and has no chronotropic effect, which may prove beneficial in the treatment of heart failure. It increases intracellular cAMP by PDE inhibition, increases Ca current, and prolongs the duration of action potentials. OPC-8490, with a similar mechanism of action, has vasodilator and positive chronotropic action. MCI-154 is an orally active positive inotropic agent with vasodilator action. The mechanism of action also involves an increase in cAMP, but in addition, sensitization of the contractile elements to Ca has been demonstrated to play a partial role in the inotropic effect. RS-1893, ZSY-27, ZSY-39 and FK664 are orally active positive inotropic agents for which cAMP mechanisms are postulated, but their effectiveness was demonstrated only in experimentally produced heart failure.

Introduction

Heart failure is the state of the heart where perfusion to the organs becomes insufficient under stressful or even resting conditions. The causes of the inadequate pumping ability of the cardiac muscles are multiple, but in many cases therapy for pathophysiologic causes is not always available, except for cases that can be corrected by cardiac or vascular surgery, or cases of secondary endocrinological disorders. Instead, most heart failure treatments use drugs to (1) strengthen cardiac contractility using inotropic agents, (2) decrease blood volume and preload by diuretics, and (3) reduce afterload by vasodilators. Recently the classical approach to increase contractility has again attracted attention after introduction of the so-called new cardiotonic drugs, such as amrinone. The positive inotropic agents used so far for the treatment of cardiac failure have been digitalis glycosides and catecholamines. Digitalis glycosides are effective positive inotropic agents having no positive chronotropic effects on the sinoatrial node, but inducing side effects of induction of tachyarrhythmias in doses slightly higher than for therapeutic use. Catecholamines

[1] Department of Pharmacology, Yamanashi Medical College, 1110 Tamaho, Yamanashi, 409-38 Japan

increase contractility, and also increase the sinoatrial rate which is already high in cardiac failure due to the compensatory reflex of the sympathetic nervous system from decreased myocardial contractility. The new cardiotonic agents, which are different from these drugs in their mechanisms of action, are under investigation and expected to act on cardiac muscle already treated by digitalis or become a substitute for long-term oral treatment with digitalis. The present review discusses some of the newer cardiotonic drugs developed by Japanese pharmaceutical laboratories which are under preclinical or early clinical trials.

Cardiac Contraction and Sites of Action of Positive Inotropic Agents

Cardiac muscle contraction occurs with the transient increase of intracellular calcium which binds itself to troponin and a result facilitates the interaction between actin and myosin (Fig. 1). The positive inotropic effect may well be the result of increase in the intracellular Ca transient, or increase in sensitivity of the contractile elements (site 1) to intracellular Ca. The intracellular Ca transient is regulated by membrane-bound and intracellular mechanisms. Membrane-bound mechanisms include Ca influx through Ca channels (site 2) and Na and Ca exchange (site 3). Ca channels are regulated by membrane voltage and cyclic AMP (cAMP), and can be modulated either through adrenergic β-receptors (site 4) or directly affected by binding to subunits of the channel proteins. Catecholamines bind to β-receptors and increase intracellular cAMP which (1) increases the probability of the Ca channel being open, resulting in an increase in the inward Ca influx, and (2) also

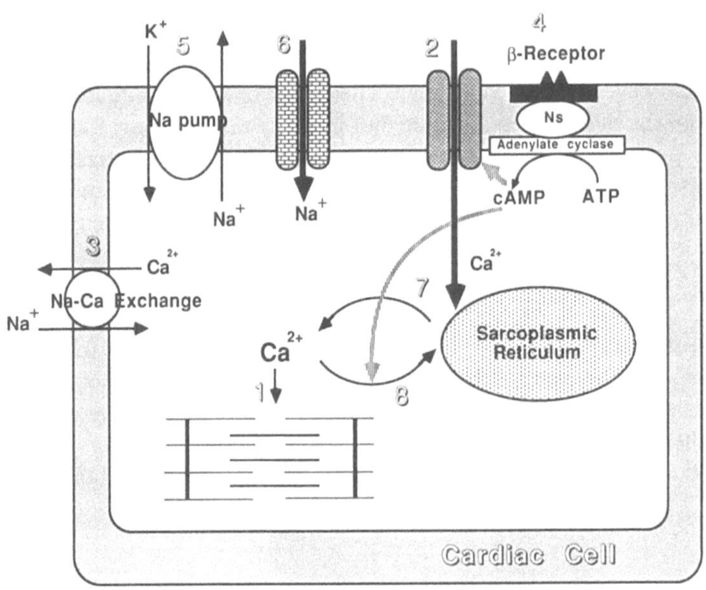

Fig. 1. Possible sites of action of positive inotropic agents

modulates the intracellular Ca transient. Agents called Ca agonists, such as BAY K 8644, can act directly on the Ca channel and increase the probability and duration of opening of the channel. Unfortunately Ca agonists also act on smooth muscle Ca channels and often increase vascular resistance, which is an undesirable property for positive inotropic agents. Na-Ca exchange (site 3) is normally an electrogenic Ca efflux system, which if inhibited increases intracellular Ca concentration. The driving force for the exchange is a transmembrane Na concentration gradient. Thus this exchange mechanism is controlled by Na-, K-adenosine triphosphatase (ATPase) (site 5), which works as a Na pump and brings Na out of cells and brings back K lost during action potentials. Digitalis inhibits this Na pump and decreases the Na gradient and subsequently decreases Ca efflux and produces a positive inotropic effect. Other membrane bound mechanisms include an increase in intracellular Na resulting from uptake through Na channels (site 6), either via an increase in the frequency of channel opening or via inhibition of rapid inactivation of Na channels. Na channel openers, like aconitine or veratrine, are known to produce arrhythmias besides their positive inotropic effects. Anthopleurin-A and other similar toxins are known to delay Na channel inactivation and increase action potential duration, and have positive inotropic effects. Other agents which increase action potential duration by acting on other current systems should also show positive inotropic effects. Another important membrane bound mechanism is activation of adrenergic β-receptor (site 4), which increases intracellular cAMP and phosphorylates Ca channels or phosphorylates phospholamban, a protein responsible for Ca uptake to the sarcoplasmic reticulum (SR). The resulting increase in the Ca influx may increase the rate of firing of the sinoatrial node. Activation of adenylate cyclase and the resulting increase in the cAMP level can be induced by directly acting agents like forskolin.

Intracellular regulation of the Ca transient is governed by the release of stored Ca from the SR, which is initiated by the increased Ca level attained by Ca influx through the Ca channels. It is believed that the greater the Ca influx, the greater the release of stored Ca (site 7), and this increase in the intracellular Ca triggers the contractile process. The SR takes up released Ca in the face of an upstream Ca concentration gradient via phospholamban (site 8), which is phosphorylated by cyclic AMP to increase Ca uptake. The increased Ca uptake may result in an increase in the stored Ca in the SR, and subsequently, in a greater amount of Ca release.

These membrane and intracellular mechanisms and the increase in the sensitivity to Ca of the contractile proteins are targets for positive inotropic agents discussed later [1].

Experimental Approaches to Elucidate Positive Inotropic Effect and the Mechanisms of Action

In Vitro Physiological Experiments

Isolated atrial or ventricular muscle strips from various animals are suited for isometric or isotonic contractile force recording under electrically stimulated conditions. Contraction curves give information with regard to the contraction phase and relaxation phase, as well as the resting tension. Catecholamines enhance the relaxation phase, and digitalis in higher doses increases the resting tension. These simple preparations are also suited for pharmacological approaches, such as the use of β-blockers and muscarinic agonists. The use of the latter compounds, especially carbachol, is now widely used to elucidate cAMP-dependent positive inotropic effects of drugs, because only a positive effect due to an increase in the intracellular cAMP is blocked by a muscarinic agonist [2]. In addition to the classical technique of superfusion of the saline solutions to isolated cardiac preparations, the blood perfusion technique through coronary arteries is quite useful because of well-preserved physiological contractions and the maintenance of stable conditions which enables the comparison of drugs or drug interactions in one experiment.

Multicellular or single myocardial cell preparations are often used for electrophysiological experiments, and provide important information on drug effects on ionic Na and Ca channels, or indirectly on the states of the Na pump or Na-Ca exchange mechanisms, because both are electrogenic and can be recorded as changes in the membrane potential. Isolated myocytes are quite useful for analysis of ionic channels in combination with the patch electrode voltage clamp technique, but it is still difficult to record contractions from this preparation.

Recent progress in this field include an introduction of Ca sensitive dyes [2] and skinned cardiac fibers. The Ca sensitive dyes, like aequorin, opened the possibility of recording the intracellular free Ca transient, which can be recorded simultaneously with contractile responses. The technique of chemically skinned fiber revealed possible mechanisms of action which are not related to membrane bound mechanisms [3].

In Vivo Physiological Preparation

Demonstration of the positive inotropic effect of drugs in whole animal experiments is important before any clinical trials are begun. Both the anesthetized or conscious animal experiments are used. Strain gauge recording of epicardial myocardium, pressure recordings of isovolumic intraventricular pressure and its speed of elevation (dP/dt) or endodiastolic pressure, or ultrasonic dimension recording are the routine recording techniques. Simultaneous recording of the heart rate and blood pressure or peripheral vascular resistance provides important information as to the possible usefulness of the positive inotropic agents [4]. Those drugs with vasodilatory action or without tachycardia may be regarded as desirable. It is still controversial whether these cardiohemodynamic experiments should be done in chronic experimental heart failure model animals. Stable animal models suitable for the analysis of positive inotropic drugs are limited, especially for long-term oral administration treatment.

Biochemical Experiments

Biochemical approaches are used to elucidate drug actions on Na-, K-ATPase activity, cAMP formation, and Ca pump activity of the sarcoplasmic reticulum. Receptors or channel proteins are now targets for binding studies of isotope-labelled ligands.

New Inotropic Agents

OPC-8212 (Otsuka)

OPC-8212, 3, 4-dihydro-6-[4-(3, 4-dimethoxybenzoyl)-1-piperazinyl]-2 (1H)-quinolinone, is a newly synthesized orally active positive inotropic agent used on isolated cardiac atrial and ventricular muscles from various species, except rats. The positive inotropic effect occurred in the concentration range of 3×10^{-6}–$3 \times 10^{-4} M$, almost equipotent to amrinone [5]. It is also evident in in vivo experiments as demonstrated by increases in left ventricular or right ventricular dP/dt and an increase in the velocity of circumferential fiber shortening, both in anesthetized and conscious animals. In anesthetized dogs, OPC-8212 (1–10 mg/kg i.v.) exerted positive inotropic effects on normal and failed hearts induced by drugs, such as propranolol and diltiazem, or coronary stenosis or acute constriction of the aorta [6, 7]. The effects are unique in that there were no changes in the heart rate and blood pressure. This pharmacological profile has been observed also in conscious dogs after either intravenous or oral administration with both normal and failed hearts, the latter of which was produced by tricuspid insufficiency and pulmonary stenosis [8]. OPC-8212 has been examined on canine ventricular arrhythmia models and it did not worsen digitalis-induced arrhythmias [9]. These preclinical data are reflected in preliminary clinical trials, which show beneficial effects.

The mechanism of the inotropic effect is not related to β-receptor stimulation or inhibition of the Na pump and seems to be mainly due to elevated intracellular cAMP [2]. An increase of the cyclic AMP level before and during the positive inotropic effect has been demonstrated, and also carbachol induced the suppression of the inotropic effect. However carbachol did not completely suppress the increase in the tension, and also aequorin signals did not show parallel changes with inotropic effects. These observations can be explained by the existence of additional mechanisms for the positive inotropic effect. OPC-8212 increased the Ca current, which may be due to increased cyclic AMP, but it also affected the K channel and Na channels. Delayed inactivation of the Na channel increases the duration of action potential in Purkinje fibers and possibly produces a positive inotropic effect. Decreases in the K current and i_h current might have offset the positive chronotropic effect resulting from the increased Ca current, which is so far a unique property among the new positive inotropic agents affecting phosphodiesterase activity [10]. One serious problem in elucidating the mechanisms of action of OPC-8212 is the effects of the solvent, sulfolane, which inhibits phosphodiesterase and sensitizes contractile systems to Ca [2].

OPC-8490 (Otsuka)

Another new quinolinone derivative OPC-8490, 3, 4-dihydro-6-(4-oxo-4-phenylbutyl)-1-piperazinylcarbonyl)-2(1H)-quinolinone, has positive inotropic effects on isolated myocardial tissue from various species except rats (10^{-6}–3×10^{-4} M) [11]. It is effective on conscious heart failure models produced by tricuspid insufficiency and pulmonary stenosis in a similar dose range to that of OPC-8212 (0.3–3 mg/kg i.v.), but it had a vasodilatory effect without increasing the heart rate.

The mechanism of action may involve phosphodiesterase (PDE) III inhibition, but carbachol did not completely suppress the positive inotropic effect, suggesting another mechanism of action. Electrophysiologically, prolongation of the duration of action potential has been observed.

MCI-154 (Mitsubishi Chemical)

MCI-154, 6-[4-(4'-pyridyl) aminophenyl]-4, 5-dihydro-3 (2H)-pyridazinone hydrochloride, is a newly synthesized orally active potent positive inotropic agent on isolated cardiac muscle of various mammalian species (3×10^{-7}–3×10^{-4} *M*) [12]. In vivo experiments revealed that MCI-154 in the dose range of 0.3–100 μg/kg in anesthetized and conscious dogs has positive inotropic effects which are simultaneous with vasodilatory and slight positive chronotropic effects. MCI-154 has been shown to be effective also in acute heart failure models induced by pentobarbital, propranolol, or verapamil, and in those cases the positive chronotropic effect became less prominent [12]. This cardiovascular profile of MCI-154 is similar to that of theophylline or sulmazole and also to amrinone except for lesser vasodilatory action, but the positive inotropic effect of MCI-154 is nearly 300 times more potent than amrinone. In the blood-perfused papillary muscle preparation, MCI-154 has been shown to have a long-lasting effect even after intraarterial administration [13].

The mechanism of the positive inotropic effect involves PDE inhibition and an increase in cAMP, but PDE inhibition was weak and carbachol did not decrease the positive inotropic effect of MCI-154 in isolated preparations [2, 14]. These results suggest the existence of other mechanisms. Electrophysiologically MCI-154 seems to have no effect on the Na channel or action potential duration, except for a small increase in Ca current, which may be related to increased cAMP. Recently, an increase in the Ca sensitivity has been proposed for the effect of MCI-154 [3]. In the chemically skinned guinea pig papillary muscle, MCI-154 shifted the pCa-tension curve to the left and also increased the peak tension. The effect occurred in a rather high 10^{-4} *M* concentration, but the effect was stronger than that of sulmazole. The study also demonstrated no effects on cardiac sarcoplasmic reticulum.

RS-1893 (Sankyo)

RS-1893, 2-[2-chloro-4-(2, 3, 4, 5-tetrahydro-3-oxo-6-pyridazinyl)]-phenoxy-*N*-(2-morpholinoethyl)-acetamide, is a newly synthesized orally active potent positive inotropic agent, three times more potent than milrinone in isolated canine ventricular muscle [15]. In anesthetized and conscious dogs, RS-1893 (1–30 μg/kg i.v. and 0.03–0.1 mg/kg p.o.) increased LVdP/dt simultaneously with decreases in the blood pressure and peripheral vascular resistance and also slightly increased the heart rate. The beneficial effect has also been demonstrated in conscious dogs with heart failure produced by adriamycin [16].

The mechanism of action seems to be inhibition of PDE-III and it was demonstrated to be 15 times more potent than milrinone. It is not known at the moment whether other mechanisms are also involved in the inotropic effect of RS-1893.

ZSY-27, ZSY-39 (Zenyaku Kogyo)

ZSY-27, 5-methyl-6-(4-pyridinyl)-2H-1, 4-thiazine-3 (4H)-one hydrochloride, is a newly synthesized positive inotropic agent [17] and has been shown to be almost equipotent on isolated guinea pig and isolated canine preparations to milrinone. The increased contractile force of cat papillary muscle caused by 10^{-6}–10^{-4} *M* ZSY-27 was markedly decreased by 5×10^{-6} *M* carbachol. In anesthetized dogs ZSY-27 (0.03–1 mg/kg i.v.) showed a positive inotropic effect, and it was accompanied by a decrease in the blood pressure and slight increase in the heart rate. Beneficial effects were also observed in acute pentobarbital-induced heart failure in dogs without much change in the heart rate [18]. Interestingly, the vasodilatory effect of ZSY-27 is observed in pulmonary circulation, and it is expected to be used for treatment of pulmonary hypertension [19].

The mechanism of action does not seem to be related to the β-receptor or the Na pump. Mostly through PDE inhibition, like other PDE inhibitors, such as theophylline and papaverine, ZSY-27 increased pancreatic exocrine secretion. A similar PDE inhibitor, ZSY-39, 4-methyl-5-(4-pyridinyl)-thiazole-2-carboxyamide, seems to have another mechanism of positive inotropic action, though its cardiovascular profile and potency seem to be almost equal to those of ZSY-27.

FK664 (Fujisawa)

FK664, 6-(3, 4-dimethoxyphenyl)-1-ethyl-4-mesitylimino-3-methyl-3, 4-dihydro-2 (1H)-pyrimidinone, is a newly synthesized potent positive inotropic agent used on isolated cardiac tissues (at higher than 10^{-7}*M*) and on anesthetized dogs (1–100 μg/kg/min i.v. infusion). Its effects can be observed in acute heart failure induced by dextran, propranolol, and coronary ligation, which was accompanied by vasodilatory action especially on the venous vasculature [20].

The mechanism of action does not involve β-receptor or Na pump inhibition, but may partly involve PDE inhibition. An electrophysiological study demonstrated blocking effects of nifedipine and ryanodine on the positive inotropic effect, suggesting a role of enhanced Ca entry in the mechanisms of action.

References

1. Wetzel B, Hauel N (1988) New cardiotonic agents—a promising approach for treatment of heart failure. Trends Pharmacol Sci 9: 166–170
2. Endoh M, Yanagisawa T, Taira N, Blinks JR (1986) Effects of new inotropic agents on cyclic nucleotide metabolism and calcium transients in canine ventricular muscle. Circulation 73 (Suppl III): 117–133
3. Kitada Y, Narimatsu A, Matsumura N, Endo M (1987) Increase in Ca^{++} sensitivity of the contractile system by MCI-154, a novel cardiotonic agent, in chemically skinned fibers from the guinea pig papillary muscles. J Pharmacol Exp Ther 243: 633–638
4. Taira N (1984) New positive inotropic agents: their modes and mechanisms of action. In: Abe H et al. (eds) Regulation of cardiac function. Japan Sci Soc Press, Tokyo, pp 293–303

5. Yamashita S, Hosokawa T, Kojima M, Mori T, Yabuuchi Y (1984) In vitro and in vivo studies of 3,4-dihydro-6-[4-(3, 4-dimethoxybenzoyl)-1-piperazinyl]-2 (1H)-quinolinone (OPC-8212), a novel positive inotropic drug, in various animals. Arzneimittel Forschung 34: 342–346
6. Hori M, Inoue M, Tamai J, Koretsune Y, Kitakaze M, Iwai K, Ito H, Kitabatake A, Kamada T (1986) Cardiotonic activity of a new inotropic agent, 3,4-dihydro-6-[4-(3, 4-dimethoxybenzoyl)-1-piperazinyl]-2 (1H)-quinolinone (OPC-8212), in the dog with and without β-blocker and Ca^{2++}-antagonist pretreatment. Jpn Circ J 50: 37–44
7. Maruyama Y, Nishioka O, Watanabe J, Keitoku M, Satoh S, Isoyama S, Ashikawa K, Ino-Oka E, Takishima T (1986) Effects of OPC-8212, a new positive inotropic agent, and dobutamine on left ventricular global and ischemic regional functions and coronary hemodynamics under coronary artery stenosis. J Cardiovasc Pharmacol 8: 161–169
8. Hashimoto K, Yabuuchi Y, Yamashita S, Mori T (1984) Positive inotropic effect of 3, 4-dihydro-6-[4-(3, 4- dimethoxybenoyl)-1-piperazinyl]-2(1H)-quinolinone (OPC-8212) in the dog with experimentally-induced right-sided heart failure. Arzneimittel Forschung 34: 390–393
9. Hashimoto K, Mitsuhashi H (1986) Effects of OPC-8212, a new positive inotropic agent, on canine ventricular arrhythmias. Br J Pharmacol 88: 915–921
10. Yanagisawa T, Ishii K, Taira N (1987) Antitachycardic effect of OPC-8212, a novel cardiotonic agent, on tachycardiac responses of guinea pig isolated right atria to isoproterenol and histamine. J Cardiovasc Pharmacol 10: 47–54
11. Mori T, Yamashita S, Hosokawa T, Yabuuchi Y (1988) Cardiovascular effects of OPC-8490, a new positive inotropic agent with vasodilator action. (abstract) Jpn J Pharmacol 46: 130P
12. Narimatsu A, Kitada Y, Satoh N, Suzuki R, Okushima (1987) Cardiovascular pharmacology of 6-[4-(4'-pyridyl) aminophenyl]-4,5-dihydro-3(2H)-pyridazinone hydrochloride, a novel and potent cardiotonic agent with vasodilator properties. Arzneimittel Forschung 37: 398–406
13. Hosono M, Taira N (1987) Cardiac and coronary vasodilator effects of the novel cardiotonic agent, MCI-154, assessed in isolated, blood-perfused dog heart preparations. J Cardiovasc Pharmacol 10: 692–698
14. Kitada Y, Narimatsu A, Suzuki R, Endoh M, Taira N (1987) Does the positive inotropic action of a novel cardiotonic agent, MCI-154, involve mechanisms other than cyclic AMP? J Pharmacol Exp Ther 243: 639–645
15. Miyake S, Shiga H, Koike H, Mizuno H, Yorikane R, Oda T, Iijima Y, Kumakura S, Morisawa Y (1988) Cardiovascular pharmacology of RS-1893, an orally active cardiotonic agent with arterial and venous vasodilator actions. Jpn J Pharmacol 47: 273–283
16. Shiga H, Miyake S, Koike H (1988) Echocardiographic study of cardiotonic effects of RS-1893 in conscious beagles with congestive heart failure. (abstract) Jpn J Pharmacol 47: 215P
17. Chiba S, Furukawa Y, Saegusa K, Ogiwara Y (1986) Cardiovascular effects of ZSY-27 [5-methyl-6-(4-pyridinyl)-2H-1, 4-thiazine-3(4H)-one-hydrochloridel] in the dog cross-circulated atrial preparation. Jpn Heart J 27: 749–757
18. Tanaka H, Tajimi K, Kasai T, Kobayashi K, Nakamura K, Okuaki A (1987) Cardiovascular effects of ZSY-27 in dogs with pentobarbital-induced heart failure. Circulation Control 8 (Suppl): 274–275
19. Tajimi K, Tanaka H, Kasai T, Kobayashi K, Okuaki A (1989) Selective pulmonary vasodilatory effect of ZSY-27 in dogs with pulmonary hypertension due to pulmonary embolism. Circ Care Med 17: 163–165
20. Esumi K, Sudo Y, Ozaki T, Maeda K, Ono T (1988) Effect of FK664, a new cardioprotective agent, on acutely induced congestive heart failure in anesthetized dogs (abstract) FASEB J 2: A1817

Subject Index

MIX
Papier aus verantwortungsvollen Quellen
Paper from responsible sources
FSC® C105338

If you have any concerns about our products,
you can contact us on
ProductSafety@springernature.com

In case Publisher is established outside the EU,
the EU authorized representative is:
Springer Nature Customer Service Center GmbH
Europaplatz 3, 69115 Heidelberg, Germany

Printed by Libri Plureos GmbH
in Hamburg, Germany